CULINARY

Herbs

&

Spices

OF THE WORLD

CULINARY
Herbs
&
Spices
OF THE WORLD

Ben-Erik van Wyk

The University of Chicago Press

Chicago and London

Kew Publishing

Royal Botanic Gardens, Kew

Conceptualized and developed by Briza Publications, South Africa www.briza.co.za

Ben-Erik van Wyk is a professor of botany at the University of Johannesburg in South Africa. He has a research interest in systematic botany and plant utilization and currently holds a National Research Chair in indigenous plant use. This latest book is part of a successful international series, including *Medicinal Plants of the World* (2004) and *Food Plants of the World* (2005) that have been translated into several languages.

Great care has been taken to maintain the accuracy of the information contained in this work. However, neither the publisher, the editors nor authors can be held responsible for any consequences arising from use of the information contained herein. The views expressed in this work are those of the individual authors and do not necessarily reflect those of the publisher or of the Board of Trustees of the Royal Botanic Gardens, Kew.

Joint publication with the Royal Botanic Gardens, Kew, Richmond, Surrey TW9 3AB, UK.
www.kew.org
and
The University of Chicago Press, Chicago 60637
The University of Chicago Press, Ltd., London

22 21 20 19 18 17 16 15 14 13   1 2 3 4 5
University of Chicago Press ISBN-13: 978-0-226-09166-2 (cloth)
University of Chicago Press ISBN-13: 978-0-226-09183-9 (e-book)
Kew Publishing/Royal Botanic Gardens ISBN-13: 978-1-84246501-1 (cloth)

Library of Congress Cataloging-in-Publication Data

Van Wyk, Ben-Erik. author.
    Culinary herbs and spices of the world / Ben-Erik van Wyk.
        pages : color illustrations ; cm
    Includes index.
        ISBN 978-0-226-09166-2 (cloth : alkaline paper)—ISBN 978-0-226-09183-9 (e-book) 1. Herbs—
Guidebooks. 2. Spices—Guidebooks. 3. Cooking (Herbs)—Guidebooks. 4. Cooking (Spices)—Guidebooks.
5. Herbs—Utilization—Guidebooks. I. Title.
        SB351.H5.V36 2013
        641.6'57—dc23
                                                                                    2013020735

DOI: 10.7208/chicago/9780226091839.001.0001

Note that some of the terms used in this book may refer to registered trade names even if they are not indicated as such.

Project manager: Reneé Ferreira
Inside design and typesetting: Ronelle Oosthuizen – Folio Graphic Studio
Reproduction: Resolution Colour, Cape Town
Printed and bound by Tien Wah Press (Pte.) Ltd, Singapore

Contents

Preface

This book is aimed at providing a broad bird's eye view of all or most of the commercially relevant culinary herbs and spices of the world. It is not a recipe book but a quick reference guide to the physical appearance, correct names, botany, geographical origin, history, cultivation, harvesting, culinary uses and flavour chemistry of more than 120 different herbs and spices from all the well-known culinary traditions of the world. An attempt is made to paint a new picture of the fascinating complexity and great wonders that are associated with an everyday activity, namely the enjoyment of food.

It is a blessing to be amongst those who are able to experience food as much more than just a means of staying alive. Mealtimes provide the opportunity for family bonding and social interaction. It is a time for taking delight in the shapes and colours of the meal, carefully prepared and skilfully presented by the expert chef. We take time out to experience the complex interactions and synergisms (and yin yang effects) of sweet, bitter, salty, sour and savoury (umami) tastes. Finally, it is perhaps also a time to marvel at the wonderful aromas that emanate from the kitchen, and to reflect on the underlying principles of flavour perception and the complicated evolutionary and cultural interactions that have resulted in the endless diversity of aroma compounds and their sensory effects.

For many of us (including this botanist), the journey of culinary discovery is just beginning. Judged by the levels of culinary and cultural diversity I have been confronted with during many years of researching and several months of writing, the end of the journey is not yet in sight, perhaps not even for the most experienced of food gurus among us.

I hope that the book succeeds in conceptualising some of the complexities of food cultures and in providing new perspectives on the diversity of flavours and aromas. It is my wish to add to the sensual delights an intellectual and scientific stimulation that some readers will hopefully enjoy almost as much as their favourite dishes.

Ben-Erik van Wyk
Johannesburg, August 2013

Introduction

Spices have become freely available today and form the basis of a grand new adventure – a culinary exploration of the world, and a journey to discover new tastes and recipes from far corners of the earth. Some of us are fortunate enough to travel the world and experience the sights and sounds that go along with local food in various parts of the globe. But the Information Age allows even the most sedentary among us to enjoy the marvellous diversity of cuisines from around the world – via the internet and the numerous food channels on television. We can all enrich our everyday lives by exploring new tastes and flavours and the many exotic products and recipes that have become part of our wonderful modern world.

From my own culinary and botanical perspective, the world can be roughly divided into seven major "cuisine families": 1: Sub-Saharan Africa, where food is often considered rather bland by outsiders and where flavour was traditionally provided by the ingredients themselves rather than by herbs and spices. The traditional staple foods are yams (*fufu*) and cereals such as finger millet, pearl millet and sorghum, eaten in the form of porridges and gruels. 2: The Middle East, representing an early fusion of Eastern and Western cooking traditions, and the origin of many present-day European culinary practices. No wonder, this is after all the place of origin of bread wheat and bread. 3: South Asia (India and surrounding countries), where people eat food by hand (normally using only the right hand) and where spices (and spiciness) have been developed to a fine art. This is the traditional home of black pepper, cinnamon, ginger, cardamom and masala. 4: East Asia (mainly China, Japan and Korea), a region of extreme culinary diversity and customs, strongly characterized by chopsticks and fermented (savoury) sauces, and rice as the staple food. 5: Southeast Asia (Malaysia, Thailand, Vietnam and Indonesia), where a fork (in the left hand) and spoon (in the right hand) are often used, and where elements of both South and East Asian cooking have been incorporated into local traditions. It is the birthplace of many spices, including cloves, nutmeg and mace. 6: Europe and many of its former colonies, where the knife and fork are the main utensils for eating. Traditional spices are few and far between, hence the need for an overland and maritime spice trade and its dramatic human cultural upheavals, still with us today. 7: The New World, with marvels such as chilli pepper, vanilla and chocolate, not to mention corn (maize), beans, pumpkins and tomatoes. Today it is the ultimate mosaic of fusion foods, ranging from authentic Inca, Mexican and Canadian-aboriginal to Creole, Cajun, Tex-Mex, Chinese-American, African-American Soul Food and many more.

The adventure of exploring the culinary diversity of the world also involves the age-old tradition of growing your own vegetables and herbs, not only to have access to fresh food items but also to enjoy the pleasure of being self-sufficient or to experience the fulfilment that comes from creating a beautiful herb garden. Many city-dwellers today have limited space, but people often grow fresh herbs in containers on the balcony or even on the window sill. The modern herb garden sometimes takes the shape of a large ceramic or terracotta pot with a lemon tree (or cherry tomato) as centre-piece, surrounded by parsley, basil, thyme, chives and other favourite herbs.

Another part of the new adventure is to learn more about the terminology and intricate relationships between the biology, botany, physiology and chemistry of tastes and flavours. New information is becoming available on an unprecedented scale but is not always reliable and rarely presents a complete synthesis and bird's eye view of a subject, such as culinary herbs and spices. The aim of this book is therefore to supply a broad new perspective of the scientific (botanical and chemical) principles of tastes and flavours and to make an interesting and colourful contribution to the culinary exploration of the world.

History

Herbs and spices are as intriguing today as they were 400 years ago when the exploration of the world was driven by the fabulous wealth created by a lucrative international trade. The early silk and spice routes from east to west were monopolized by Arab traders and various countries and city states in the western Mediterranean region, who all shared in the rich bounty.

Spices first reached Constantinople and Alexandria, from where they were taken to Venice, Naples and Genoa in Italy, and from there to the rest of Europe. Spices were in high demand because of the poor quality of the food and the lack of a means to preserve meat, other than salt. During the medieval spice craze, shortages in supply resulted in a diversity of local spices that were used as substitutes or adulterants. The role of local, indigenous herbs and spices such as alecost and Melegueta pepper is now just a distant memory, as they were almost completely replaced by hops and black pepper respectively. Also long forgotten is the fact that the enormous spice wealth in northern Italy and the rich people who patronized architects, artists, authors and philosophers actually created the Renaissance. The overland monopolies were broken by the development of alternative sea routes, and colonial powers such as Portugal, Spain, the Netherlands, France and Britain all controlled and defended their access to spices and associated source areas at one time or another. Key moments were 12 October 1492, when the Spanish explorer Columbus reached the Bahamas and "discovered" *aji* (chilli peppers) and 20 May 1498, when the Portuguese sailor Vasco da Gama stepped ashore at Calicut in India, thus establishing a sea route to the east around the Cape of Good Hope. This was the Age of Discovery, an exciting but brutal time during which the full

extent of the earth and its rich bounty of culinary delights became much better known. The profound effects that the spice trade has had on determining where and how we live and eat today is no longer much thought about but it is true to say that spices were the driving force that changed and opened up the world.

The famous East Indies (including India, Sri Lanka, Malaysia and Indonesia) supplied important tropical spices such as cardamom, cinnamon, cloves, ginger, mace, nutmeg and especially pepper to satisfy the huge demand in Europe, where the lack of a tropical climate prevented local production. The same is true for chocolate, vanilla and allspice from the New World, including Mexico and Central America. Maize, potatoes and chilli peppers were distributed by Portuguese sailors from the New World to the Old World (Europe, Africa and Asia), where they quickly became important crop plants and consequently a non-indigenous part of many local food cultures. The East Indian spice monopoly was eventually broken by horticultural innovations, led by the French, who established plantations of pepper, nutmeg, cinnamon and other valuable spices on tropical islands such as French Guiana, French Polynesia, Réunion, Seychelles and Madagascar. The use of refrigeration and other means of food preservation further reduced the central role of spices in international trade.

Regions of origin and culinary traditions

A global view of the culinary traditions of the world, an almost impossible task, is presented below. In general, the same or similar types of dishes and ingredients are found in unrelated cultures, despite the incredible diversity of recipes and ingredients. Pungent spices for example, each with its own unique chemistry, are found everywhere, from horseradish and mustard in Europe to Melegueta pepper in Africa, black pepper in India and chilli pepper in the New World. One can trace the ancient origins of many "modern" food items such as meatballs, pizza, shawarma, salad, stew and tapas, as well as condiments (fermented sauces and vegetables, pickles and relishes) and drinks (cider, mead, wine and spirits). The discussion focuses mainly on the relatively poorly known Asian cuisines, the original home of many well-known (and poorly known) spices, with only brief mention of the main features of the more familiar Western traditions.

Africa

Indigenous herbs and spices such as *ajowan*, *buchu*, coffee, *geisho*, Guinea pepper, Indian borage, *karkade*, *korarima*, *koseret*, Melegueta pepper and tamarind are widely used in the many cooking traditions in Africa but, as in many other parts of the world, chilli peppers and other exotic spices have become popular or even dominant. Moroccan, Ethiopian and South African (Cape) cuisines are becoming more widely known and it is likely that others will follow. In North Africa, Roman, Turkish, French and Italian conquerors all contributed to the local cooking traditions. Harissa is a famous chilli-based purée (*tabal*) used for spicing mutton and semolina dishes, as well as highly aromatic soups (*harira* in Morocco, *brudu* in Tunisia and *chorba* in Algeria). Specialities include couscous with *tajine* or *kefta* followed by pastries and sweetmeats made from puff pastry, almonds, dates and honey. The extreme diversity in West African cooking ranges from bland-tasting *fufu* to fiery meat dishes (ragout and *canari*), spiced with chilli peppers (*pili-pili*) or Melegueta pepper, *atokiko* (mango stones), Guinea pepper, tamarind, *soumbala* (dried fruit rind), *tô* (millet paste) and *lalo* (powdered baobab leaves). Palm oil, peanuts and coconut are typical ingredients. West African traditions contributed to the Creole cookery of the West Indies. Ethiopian cuisine is the best known in East Africa. It is characterized by *injera* (teff bread) and various spicy dishes (e.g. *shiro*) in which *berbere*, a hot spice mixture, plays an important role. Chilli pepper has replaced black pepper but indigenous spices such as *ajowan* and *korarima* (Ethiopian cardamom) are essential ingredients. Also typical is *koseret*, a herb used with *kifto* (beef tartare). The national dish of Eritrea is *zegeni* (mutton with chilli pepper paste

and vegetables). In southern Africa, the cooking tradition of the original San people was largely replaced by ancient and more recent immigrants, mainly Khoi, Bantu, Indian and European. Cape cookery, based on Dutch and Malay (Indonesian) traditions, is especially well known. Coriander, black pepper, nutmeg and *blatjang* (*atjar*) are essential components, but dishes are rarely very hot (except for chilli-based Indian curries). Specialities include *braaivleis* (barbeque), *bobotie* (spicy mince meat), *sosaties* (skewered meat) and *potjiekos* (slow-cooked stew in a three-legged cast iron pot), served with samp and beans, yellow rice (coloured with turmeric) or *pap* (white maize porridge).

North and Central Europe

There are surprisingly few spices indigenous to this region and exotic products now dominate. Notable exceptions are horseradish, the traditional pungent ingredient accompanying meat dishes, the widely used caraway fruits and juniper berries and the less well-known blue fenugreek of Georgian cookery. Culinary herbs are more diverse, and include absinth, angelica, caraway leaves, chives, garden mint, peppermint and celery, as well as a large number of species from southern Europe and the western Mediterranean that were introduced by the Romans and distributed through monastery gardens. Scandinavians use horseradish and dill with their dishes, which are based mainly on fish, pork, mutton, smoked reindeer, potatoes, beetroot, cucumber, fruit and berries, cream, cheese and butter. Spices such as cumin and cardamom feature prominently. The Russian cuisine reflects a diversity of cultures, with Ukrainian cereals, Armenian wine, Caucasian fruits and Georgian tea, enriched with dill, galangal and other herbs and spices. The British Empire left its mark on many parts of the world, not only its language but also culinary traditions such as the famous English breakfast, curry powders inspired by India and commercial sauces such as English mustard, Worcestershire sauce, brown and mint sauces as well as piccalilli (vegetable pickle) and relishes.

The Netherlands is the country of cheese, herrings, mussels, pancakes, *hutspot* (vegetable stew) and *snert* (pea soup) but the impact of the spice trade can be seen in the popularity of spices and dishes of Indonesian origin. Confectionery, sweets (e.g. *drop*) and liqueurs flavoured with anise, juniper, mint and other spices are particularly popular. German culinary delights are many and varied, but the most delicious are basic items such as bread, pork sausages, potato dishes, cabbage dishes (e.g. sauerkraut) and white asparagus, enjoyed with generous amounts of excellent beer, cider, wine, herbal tea (e.g. German chamomile) or bitter liqueur. Beer-brewing calls for enormous quantities of hops. In Austria, pepper, sage and juniper berries are commonly used to prepare game. Switzerland is best known for delicious items such as cheese fondue, rösti and chocolate. In Polish cuisine, mushrooms and traditional sausages are prominent, often spiced with pepper, fresh garlic and caraway seeds. Hungary is best known for its sweet (mild) paprika, used in Hungarian stew (goulash).

South Europe

The Mediterranean region is arguably the most diverse and exciting hotspot of culinary diversity, with herbs and spices that date back to Greek and Roman times. These include indigenous spices (e.g. aniseed, capers, carob, coriander, cumin, fennel, liquorice, mustard, saffron, sumac) and herbs (e.g. celery, chervil, fennel, hyssop, lavender, marjoram, myrtle, oregano, parsley, rocket, rosemary, sage, savory and thyme). Portugal has a distinctive cuisine characterized by the liberal use of fresh herbs and spices together with cabbage, rice, potato, sardines and salt cod (*bacalhau*). Spanish cookery has become better known in recent years through the popularity of tapas – savoury or spicy dishes presented as small open sandwiches, often with dry ham (e.g. *serrano*), sweet red peppers, black olives, stuffed peppers and various seafood items. Famous Spanish dishes include paella, liberally spiced with saffron. The use of chocolate and chilli peppers in cooking dates back to colonial times. Typical Spanish drinks include sangria (red wine punch) and *horchata de chufa*, an almond-flavoured cold drink made from crushed tiger nuts (*chufa*). French cuisine is so overwhelmingly and fascinatingly complicated that it is impossible to summarize it in a few words. Sauces have been developed to a fine art. Larousse (who proclaims Paris as the Mecca of gastronomy) quotes Curnonsky: "Sauces comprise the honour and glory of French cookery." Herbs and spices are skilfully blended and often used in fixed combinations (e.g. *fines herbes* and *quatre épices*). Pepper and cloves are typical ingredients in marinades, cinnamon and nutmeg in wine sauces, saffron in bouillabaisse, aniseed and cumin in confectionery, and coriander and juniper in meat dishes. Desserts often include chocolate, vanilla, cream and processed fruits. Provençal cooking (the culinary tradition of Provence) is characterized by a predominance of olive oil, tomato and garlic, with *herbes de Provence* (thyme, rosemary, bay, basil, savory, and nowadays also lavender). Italian food has become part of the global village, as seen in the popularity of pizza, pasta and risotto, together with salami and Parma ham, Parmesan and Gorgonzola, not to mention Italian ice cream. Pasta and rice dishes are particularly interesting and varied. Garlic, tomatoes, mozzarella cheese, truffles and saffron often feature prominently. Popular herbs include basil, thyme, sage, oregano, rosemary and parsley. Pesto is a famous Genoese sauce made with pounded fresh basil, parsley and marjoram. Greek cuisine shows many regional variations but common features include the use of fish and mutton with Mediterranean vegetables and the liberal use of herbs, lemon juice and olive oil. An Eastern influence is noticeable in the eating of meze or small appetizers, enjoyed with ouzo. Fish is often prepared with herbs such as fennel, anise or coriander (cilantro). Feta is the best-known of the many Greek cheeses which, together with lettuce, onions, herbs and ripe olives, make a Greek salad. Turkish cuisine was inspired by both European and Eastern traditions and has influenced the cooking traditions of many countries in the Middle East and North Africa, as well as Greece and Russia. Turkish specialities include pilaf rice, lamb kebab, meze, stuffed aubergines, dried figs, coffee and patisserie (e.g. halva, baklava and Turkish delight), sometimes flavoured with rose water or orange-flower water.

Ethiopian *injera*, accompanied by traditional spicy condiments

Italian pesto and its ingredients

Middle East

Many of the best-known spices and herbs had their origins in the Middle East, which is often referred to as the Cradle of Agriculture and the Cradle of Civilization. There is indeed a golden thread of cultural continuity that links modern dishes such as Italian pizza, French ragout and Spanish tapas to ancient Mesopotamian dishes such as Babylonian flatbread, Persian *khoresh* and meze, respectively. Middle Eastern cuisines are based on local ingredients such as olives, chickpeas, dates, honey, dried limes, pitas, pomegranate, saffron, sesame seeds, sumac, parsley, mint, thyme and *za'atar*, However, it has been influenced by African okra, Mongolian dumplings, New World tomatoes, as well as Indian and East Indian spices (black pepper, cloves, cumin, garlic, turmeric).

Mesopotamian and Iraqi cuisine has a long history dating back 10 000 years with influences from neighbouring countries such as Iran, Syria and Turkey. Meals are started with appetizers and salads, known as *mezza*. It is the region of origin of kebabs (grilled meat, marinated in garlic, lemon and spices), falafel (fried chickpea patties served in a pita with salad and spicy sauces or pickles) and stuffed vegetable dishes such as *dolma* and *mahshi*. Other typical dishes include *gauss* (a wrap of grilled meat similar to *döner kebab*), and various lamb and rice dishes. Tahini is a popular paste made from hulled and roasted sesame seeds. The culinary traditions of Persia are equally diverse and ancient. Iranian stew (*khoresh* or *khoresht*) comes in many different variations but is often made with lamb and liberal amounts of saffron and other spices. Vegetarian *koreshts* are not uncommon and

may comprise almost entirely of herbs, such as *khoresht-e gormeh sabzi*. A plate of fresh herbs, called *sabzi khordan* (made with ingredients such as basil, cilantro, cress, fennel, fenugreek, peppermint, radishes, *za'atar*, savory, tarragon and Welsh onion) is an essential accompaniment to lunch and dinner. Levantine cuisine represents the culinary traditions of the Levant, which includes modern-day Jordan, Israel, Lebanon, Palestine, Syria, parts of southern Turkey and northern Iraq. Aleppo was the main cultural and commercial centre during the days of the spice trade. Essential components of Syrian cooking include *baharat* or Aleppo seven-spice mixture, roasted red pepper paste and pomegranate molasses. *Mezzas* often include *baba ghanoush* (a dip or paste made from aubergines and spices with a smoky flavour) and hummus (chickpea and sesame paste). Syrian *muhammara* is a spicy red bell pepper and walnut dip made with pomegranate molasses. Tabbouleh usually accompanies the meze. It is a type of salad with finely chopped parsley, mint, baby lettuce, tomatoes, cucumbers, onion and garlic, mixed with bulgur (wheat groats) or couscous and seasoned with olive oil, lemon juice and salt. The shawarma, now an international fast food, is another Levantine invention, as are *kibbeh*, meatballs made from mince meat, chopped onions and bulgur. Israeli dishes today often include chilli-based hot sauces introduced by immigrant Jews, such as *skhug* from the Yemen, harissa from Tunisia and *filfel chuma* from Libya. *Amba* is a mango pickle from Iraqi and Indian origin. These sauces may be blended with hummus and tahini and served with appetizers, stews, grilled meats, egg dishes, sandwiches and falafels.

East Asia

East Asian cuisine includes mainly the Chinese, Japanese and Korean traditions, which are all ultimately derived from or inspired by China. Chinese cooking has also greatly influenced the cooking of Taiwan, Malaysia, Singapore, Indonesia and even America. East Asian cuisine and its influence can most easily be identified by the use of chopsticks as eating utensils and the preparation of food into bite-sized pieces. Other give-aways are the prominence of rice as the staple food, the liberal use of fermented soy and fish sauces during cooking as well as eating, and the popularity of tea (in many variations, but mainly green or black) as an everyday drink.

The interactions of Chinese traditions with regional and local indigenous customs in Asia over thousands of years have created a marvellously complicated culinary system in which the principles of yin and yang (opposites that balance each other) feature prominently. This does not only apply to the dishes themselves, but also to the different courses of a meal. There is often a combination (or succession) of hot and cold, spicy and mild, crispy and soft, pickled and fresh, sour and sweet, dark and light-coloured and so on. (This I learnt from a memorable evening of fine dining with the late Lannice Snyman in a Chinese restaurant in Singapore in 1998, when the first edition of her best-selling book, *Rainbow Cuisine*, was being printed). Peking duck for example, is typically served as two courses, first the dark, crispy skin and then the pale, soft meat. Food is judged by four criteria, matching the four main senses used in eating: colour (visual), aroma (olfactory), texture (tactile) and taste (gustatory). Chinese cuisine is also characterized by an extreme diversity of ingredients and dishes (almost anything is eaten), techniques of preservation (salting, drying, fermenting and pickling), as well as cooking techniques (boiling, braising, smoking, stewing, baking, steaming and stir-frying) and regional eating styles.

The most famous Chinese spice is "five-spice powder" or "five-spices powder" (*wu xiang fen*), made from star anise, fennel fruits, cassia bark, cloves and Sichuan pepper. There are many variations but star anise is always a key ingredient. Less well-known spice mixtures commonly used in China (see SPICE MIXTURES on page 28) include "spice liquid" (*lu shui*), "major spices" (*da liao*), "fish spices" (*yu xiang*) and "hot peanut sauce" (*sha cha jiang*).

There are eight major culinary traditions in China, namely Anhui, Cantonese, Fujian, Hunan, Jiangsu, Shandong, Sichuan and Zhejiang. Cantonese cooking is known for dim sum (small hearty dishes) such as dumpling, soups, lotus leaf rice and stir-fried vegetables. Cantonese cuisine has spread to many parts of the world, where countless versions of chow mein ("fried noodles") and chop suey ("assorted pieces", stir-fried meat and vegetables) have been popularized. Also well known is Cantonese "sweet and sour pork", originally a Jiangsu dish called "pork in a sugar and vinegar sauce". In the same way, numerous versions of Mongolian hotpot have become popular within China and other parts of Asia. The Sichuan style (from southwestern China) is known for hot and spicy dishes, resulting from the liberal use of strong flavours, including Sichuan pepper or *hua jiao* (*Zanthoxylum piperitum*), ginger, garlic and "facing heaven pepper" (*chao tian jiao*), which accurately describes the erect fruits of *Capsicum frutescens* (as opposed to the down-facing fruits of all other species). Examples of the more than 4 000 dishes of Hunan cuisine (many of them hot and spicy) include fried chicken with Sichuan pepper sauce and smoked pork with dried long green beans. A few more examples of regional delicacies include bamboo shoots and mushrooms (Anhui, Fujian and Zhejiang), sweet and sour carp and braised abalone (Shandong) and Nanjing's Jinling salted dried duck (one of many dishes of Jiangsu cuisine).

Other features of Chinese cuisine include Chinese noodles, made from rice flour or wheat flour, tofu ("soy cheese") and fermented tofu (often smelly, not unlike blue cheeses), numerous indigenous vegetables (bok choy, *gai lan*, mustard greens – too many to list), rice vinegar, oyster sauce, bean-based sauces (hoisin sauce, peanut sauce, yellow bean sauce) and prawn crackers as a popular snack. Finally, there is an astounding diversity of tea types, as an integral part of the culinary tradition: green, oolong, black, scented, white or compressed, each with numerous different variations.

Japanese cuisine is characterized by an appreciation and respect for the individual tastes of food items and the seasonality (*shun*) of the ingredients. Despite its simplicity, it has developed into a complex system of culinary traditions and

Exotic spices on display (Abu Dhabi)

Chinese noodles

techniques based on many different and sometimes unique ingredients. In the Western world, Japanese food is strongly associated with sushi – bite-sized food items in countless variations, created from vinegared sticky rice (*sushi-meshi*) and some savoury ingredient or filling (*neta*), which may or may not include raw fish, and usually wrapped in a sheet of *nori* (edible seaweed). Sushi is often confused with sashimi (raw fish), which is eaten with various condiments and dips, such as soy sauce and wasabi. The traditional Japanese meal comprises steamed white rice and miso soup, with side dishes of fish, pickles (*tsukemono*) and vegetables.

The taste of individual food items is appreciated and great care is taken to keep them apart. Spices and condiments (*yakumi*) are therefore used rather infrequently, and for very specific purposes (e.g. to remove an unwanted smell during cooking). Spices and herbs are most often added to the dish as a garnish (or served on the side). These may include grated ginger or daikon radish, wasabi, mitsuba (*Cryptotaenia japonica*) or shiso leaves (*Perilla frutescens*) and *nori* (edible seaweed). Food is generally flavoured only with dashi, soy sauce, mirin, vinegar, sugar and salt. Mirin is a strongly flavoured type of sweet rice wine, not unlike sherry. It is one of the most popular condiments in Japanese cuisine. The best-known spice is *shichimi togarashi* ("seven flavour chilli pepper") or simply *shichimi*, which is a mixture of ground chilli pepper with smaller amounts of

ground sansho (*Zanthoxylum piperitum*), roasted orange peel, black sesame seed, white sesame seed, hemp seed, ground ginger and *nori* seaweed. It may also include poppy seed, rape seed, shiso or yuzu peel. *Yuzu* is a hybrid between mandarin (*Citrus reticulata*) and the Ichang papeda (*C. ichangensis*). On its own, chilli powder is called *ichimi togarashi* ("one flavour chilli pepper") or simply *ichimi*.

Miso soup is a Japanese culinary staple made from dashi and miso. Dashi is a fish stock (broth) made from kombu (a type of edible kelp) and *kezurikatsuo* (preserved and fermented shavings of skipjack tuna, sometimes substituted by young bonito). The fish is rich in sodium inosinate and contributes umami flavour. Miso is a traditional Japanese seasoning paste made from fermented soybeans (sometimes with rice or barley added). Fish is often served raw as sashimi or as sushi. Seafood and vegetables may be in the form of tempura – deep-fried in a light batter. Soba noodles (made from buckwheat) and udon noodles (made from wheat flour) are usually eaten alone and not as part of a meal. Another common dish (especially in winter) is *oden*, made from various ingredients stewed in dashi broth. The ingredients may include fish cakes, hard-boiled eggs, daikon radish, tofu and yam cake (*konnyaku*), made from the root of the konjac plant (*Amorphophallus konjac*). Beef sukiyaki has become a favourite, named after the sukiyaki method of soaking beef in a marinade (sukiyaki sauce) and then regularly basting the meat with the sauce while it is grilled. In recent years, Chinese ramen (noodle soup) and fried dumplings have become popular, as well as Western-style hamburgers (but often adapted to local taste preferences).

Korean cuisine resembles those of China and Japan but with important differences. Food is eaten with stainless steel chopsticks, oval in transverse section. Rice and soup are eaten with a long-handled, shallow spoon (the rice or soup bowl is never lifted and brought to the mouth as in China and Japan). The staple food comprises rice, vegetables and especially meats, for which Korea has become well known. Beef, chicken and pork are highly prized food items, as are kimchi or kimchee (fermented napa cabbage), which is served at almost every meal. Seafood, soups, stews and noodles also feature prominently amongst the hundreds of traditional Korean dishes. Rice (historically, millet, barley and other cereals were the main staple foods) is not only eaten as a bowl of steamed rice but is ground to flour to make countless variants of rice cakes (*tteok*), as well as numerous other dishes (not to mention rice wines and rice vinegars). Sesame oil plays an important role. Soybeans are used for a wide variety of purposes, including the production of soybean pastes used as condiments and collectively known as *jang*. *Doenjang* (fermented soybean paste) and its more pungent version called *cheonggukjang* are widely used. The local version of soy sauce is called *ganjang*. *Gochujang* is a special spicy paste, made from red chilli powder, glutinous rice powder, fermented soybean powder and salt. Mung beans and adzuki beans are used in many different ways – as sprouts, porridge, pancakes, noodles, additives for soups and stews, as well as the jelly-like food item known as *muk*. Spices include garlic, ginger, chilli pepper, black pepper, *sansho* pepper and mustard. Onions, leeks, scallions and perilla leaves are also popular. Seasonings include *saeujeot*, made from salted baby shrimps. Korea has numerous regional cuisines and a vibrant, dynamic street food culture.

South Asia

The cuisines of India and those of its close neighbours such as Nepal, Sri Lanka, Pakistan and Bangladesh represent an astounding mosaic of regional traditions, based not only on many interesting local food ingredients and spices, but also on religious and cultural principles.

These include vegetarianism, an ancient but now increasingly popular trend (not only in India but also in many other parts of the world) and the Hindu taboo on eating beef. Food is eaten by hand. The staple food is often roti, the typical Indian bread made from *atta* (stoneground wholewheat flour), used alongside rice. This age-old cuisine has been strongly influenced by colonialism and cultural exchanges with other societies, especially the introduction of New World chilli peppers and potatoes by the Portuguese and the modern-day concept of curry by the British. India's contribution to the spice trade and the wealth it created in Europe can easily be underestimated but its far-reaching impact on the use of spices in all modern-day Western food cultures can hardly be overlooked. Especially prominent are the ancient influence seen in Southeast Asian cuisines and the more recent influence on the British and Caribbean food cultures. In a speech on British identity in 2001, the British foreign secretary proclaimed that chicken tikka masala is a "true British" national dish! Chilli pepper (*mirch*) has become a feature of the stereotype concept of Indian cuisine, but spices such as asafoetida (*hing*), coriander (*dhania*), cumin (*jeera*), black pepper (*kali mirch*), fenugreek (*methi*), ginger (*adrak*), turmeric (*haldi*), cardamom (*elaichi*), curry leaves (*karipata*), cinnamon (*dalchini*) and mustard seeds (*sarso*) are equally typical and highly characteristic. Garam masala, *goda masala* and *sambar podi* are among the best known and most widely used spice mixtures (see SPICE MIXTURES on page 28). Curry leaf is especially important in Gujarati, South Indian and Sri Lankan cuisines. Rose water, cardamom, saffron and nutmeg are widely used to flavour sweet dishes and desserts. Pickles (*achar*), often made from green mangoes, and various forms of chutney (made from different fruits and herbs) are widely used as condiments. The astounding regional diversity includes spicy and searing hot Bengali dishes, vegetarian Gujarati food (with roti, rice and vegetables), intensely flavoured seafood with a frequent use of coconut and kokum in Goa, vegetables and dhals with *kalonji* (nigella) in Odisha, tandoori food from the Punjab (especially tandoori chicken, naan bread and masalas rich in onion, garlic and ginger) and finally South Indian and Sri Lankan Tamil food, with rice and lentils, curry leaves, tamarind, chutneys and sambar. The word "curry" is derived from the Tamil *kari* (meaning "sauce"). Pakistani

Lunchbox with sushi and sashimi
(Yokohama, Japan)

Spices at the Anjuna flea market, Goa, India

dishes reflect the cultural diversity of the country and the influence of Muslin culinary traditions, especially halal. Commonly used spices include coriander fruit powder, cumin, chilli powder, cardamom (both green and brown), cinnamon, cloves, bay leaves, nutmeg, mace and black pepper. Garam masala is used in many dishes. Meat (especially goat, lamb and chicken) is much more popular in Pakistan than in India and is used to make everyday food items such as korma (a type of spicy curry), *aloo gosht* (spiced meat and potato stew) and especially kebabs, in countless variations. Food in Bangladesh is typically very hot and spicy and closely resembles Indian food. A popular spice mixture, called *panch phoron*, is usually composed of *radhuni* (*Trachyspermum roxburghianum*), *jeera* (cumin), *kalo jira* (nigella), *methi* (fenugreek) and *mouri* (anise).

Southeast Asia

The cuisines of Southeast Asia (Thailand, Vietnam, Laos, Cambodia, Malaysia, Singapore, Indonesia and the Philippines) represent a blend of culinary influences from India and China, integrated into the already diverse local food customs. Distinctive features of this region include the abundance of exotic tropical fruits, the diversity of culinary herbs and spices that are used, the importance of rice as a staple food and the common use of a fork and spoon as eating utensils. The fork, held in the left hand, is used to push food onto the spoon held in the right hand.

Thai food is aromatic, spicy and complex, prepared with a balance between sweet, sour, bitter and salty. The staple food is jasmine rice or sticky rice, with many shared side dishes, all served at the same time. Herbs and spices are typically used fresh (rather than dried). Herbs include lime leaves (*bai makrut*), pandan leaves (*bai toei*), cilantro (*phak chi*), lemongrass (*takhrai*) and Thai basils, as well as spearmint, shallots and garlic. Typical fresh spices are galangal (*kha*), turmeric (*khamin*), fingerroot (*krachai*) and peppercorns (*phrik thai*), while popular dried spice mixtures include five-spice powder (*phong phalo*) and curry powder (*phong kari*). Thai dishes and soups often contain coconut milk, fresh turmeric and lime leaves. The famous hot and sour *tom yam* soup contains lime leaves (with galangal, lemongrass, lime juice and, of course, chillies). The Chinese influence is seen in the use of a wok for deep-frying or stir-frying, the importance of soy sauce and dishes such as rice pudding (*chok*), fried rice noodles (*kuai-tiao rat na*) and stewed pork with rice (*khao kha mu*). Indian *kuai* influence is seen in yellow curry (*kaeng kari*); Muslim (Persian) influence in massaman curry (*kaeng matsaman*). Sour tastes may come from tamarind or coconut vinegar, while sweetness often originates from palm sugar and coconut sugar. Typical of Thai cuisine are fermented fish sauce (*nam pla*), oyster sauce (*namman hoi*), chilli pastes (*nam phrik*), curry pastes (*phrik kaeng*), light soy sauce (*si-io khao*), dark soy sauce (*si-io dam*) and soy paste (*taochiao*). A variety of sauces and condiments in small containers are served with the meal, especially various forms of the chilli-based *nam phrik* (the equivalent of Malaysian and Indonesian sambals) used for dipping. Popular examples are *nam pla phrik* (fish sauce, chopped chillies, garlic and lime juice) and *nam phrik num* (green chillies, coriander leaves, garlic and shallots). Dishes popularized by Thai restaurants include *tom yam kung* (soup with prawns), *tom yam kai* (soup with chicken), *pad thai* (a stir-fried rice noodle dish, commonly served as street food), massaman curry (made from beef, chicken, duck or tofu and potatoes, with coconut milk, fish sauce, chilli sauce, tamarind, peanuts, bay leaves, cardamom, cinnamon, star anise and palm sugar), *som tam* (shredded unripe papaya, prepared as a spicy salad) and *khao phat* (a stir-fried jasmine rice dish).

Vietnamese cuisine is characterized by an abundance of fresh herbs and vegetables, many different types of soup and a colourful presentation of food. It relies on rice, fish sauce, soy sauce, shrimp paste and fresh herbs and spices for the distinctive tastes and flavours. Commonly used herbs include lemongrass, spearmint, peppermint, Vietnamese mint, culantro (long coriander), basil and lime leaves, while ginger, Saigon cinnamon (*Cinnamomum loureiroi*) and black cardamom (*Amomum costatum*) are common spices. It is considered important to include all five traditional taste elements (spicy, sour, bitter, salty and sweet) and all five senses (sight, sound, taste, smell and touch). Northern Vietnamese food typically contains black pepper rather than chilli pepper and is therefore less spicy. Signature dishes include *bún riêu* (a soup made with rice vermicelli, meat, shrimp paste and tomatoes), *bánh cuôn* (a rice noodle roll made from rice batter, pork, mushrooms and shallots), *bún bò* (a lemongrass flavoured soup made from thick rice vermicelli and beef) and *bánh xèo* (a type of savoury pancake made from rice flour and stuffed with meat and vegetables). Most dishes are served with generous helpings of fresh herbs, which include unusual ingredients such as fish mint (*Houttuynia cordata*), rice paddy herb (*Limnophila aromatica*), common knotgrass (*Polygonum aviculare*), Vietnamese balm (*Elsholtzia ciliata*) and shredded banana flower, together with more familiar items such as raw bean sprouts, green onion leaves, basil, cilantro, culantro, water spinach, Vietnamese mint and perilla leaf. Chinese and French influences are visible not only in some of the ingredients of Vietnamese food but also in many of the dishes.

Malaysian cuisine is a mixture of Malay, Chinese, Indian, Thai and other culinary traditions. The national dish is considered to be *nasi lemak* (steamed rice with coconut milk and served with fried anchovies, hard-boiled eggs, sliced cucumber and the typical sambal – spicy chilli paste). Another favourite dish is *nasi goreng* (fried rice). Curry and meat stew (*rending*) are also popular, as are many different types of seafood. Various noodles, made from rice, wheat or mung beans are common, as well as Indian-style bread. A diversity of fruits form an important part of the cuisine, including durian (the "King of fruits"), mangosteen (the "Queen of fruits"), lychee, rambutan, longan and mango. Many dishes start of as a *rempah* (a mixture of fresh herbs and spices that are pounded and then briefly fried in a little oil to bring out the flavours). A very popular dish in Malaysia is satay – marinated and skewered meat, grilled over an open fire and dipped in *bumbu kacang*, a special spicy peanut sauce. *Asam laksa* is an example of a dish of Chinese origin – rice noodles served in soup made from fish, onion, tamarind, torch ginger flower, basil, pineapple and cucumber. Desserts often have coconut milk and palm sugar as main ingredients.

Singapore, where eating is considered a national passion, is the place to experience street food and the ultimate in fusion food, elaborated from Malay, Chinese, Indian, Indonesian, Philippine, British and many other culinary traditions. An example of hybrid food is the local *laksa* or *katong laksa*, a noodle dish with seafood, chicken or eggs in a rich coconut curry sauce. Eating chilli crab in Singapore is an unforgettable experience. It is made from a hard-shelled mud crab cooked in tomato and chilli sauce and served on a bed of steamed rice.

Indonesia is the world's largest archipelago. Amongst the 18 000 islands are the famous Spice Islands (Maluku or the Moluccas), the home of cloves and nutmeg. Since ancient times there have been many influences on the local cuisine, especially from India, China and the Middle East but more recently also from Europe (Spain, Portugal and the Netherlands). Meals are usually served at room temperature, with rice as staple food, accompanied by soup, salad, sambals (spicy relishes, such as *sambal ulek*), *krupuk* (traditional crunchy crackers), *emping melinjo* (crackers made from *Gnetum gnemon*) and *kripik* (crisps made from banana or cassava). A typical Indonesian spice (*rempah*) and spice mixture (*bumbu*) may comprise the indigenous *pala* (nutmeg and mace), *cengkeh* (cloves) and *laos* (galangal), or several other popular spices introduced in ancient times:

Satay with peanut sauce

Ingredients of Thai *tom yam* soup

black pepper, candlenut, cinnamon, coriander, lemongrass, shallot, tamarind and turmeric from India; and garlic, ginger and scallions from China. *Kecap asin*, the common (salty) soy sauce and *kecap manis*, the uniquely Indonesian soy sauce sweetened with palm sugar, are also important flavourings and components of marinades used for grilled meat or fish, satay and *semur* (Indonesian stew). Coconut and coconut milk are important ingredients of many dishes. Peanuts, introduced by the Portuguese and Spanish during the 16th century, have become integrated into Indonesian dishes and are used in *sambal kacang* (roasted and ground chilli pepper and peanuts) and the well-known peanut sauce (*bumbu kacang*), used as dipping sauce with satay and as garnish with *gado-gado*. The sauce is made from fried peanuts, chilli pepper, black pepper, sweet soy sauce, coconut sugar, garlic, shallots, ginger, tamarind, lemon juice, lemongrass, salt and water. Signature dishes include *nasi goreng* (fried rice, the origin of the Dutch *Rijsttafel* or "rice table"), *gado-gado* (a salad of boiled vegetables dressed with peanut sauce), satay (grilled skewered meat) and *soto* (traditional soups made from broth, meat and vegetables). There are many other typical dishes such as *pisang goreng* (fried bananas), *lumpia goreng* (fried spring rolls) and *bami* (fried noodles). Street food is very common and includes

sweet and savoury dishes, desserts, fruit juices, fruit crisps and various snacks, cakes and pastries, sold from bicycles and carts at the roadside.

The Philippines comprise a group of over 7 000 islands named after Philip II of Spain. The cuisine is based on steamed rice as staple food but bread is also quite common. It reflects various cultural and ethnic influences, including Malay, Spanish, Chinese, Indonesian and many others. Dishes typically have salty and sweet or salty and sour elements. Dipping sauces are common, including vinegar, soy sauce, calamansi juice (*Citrus madurensis*) or combinations of the three. Fish sauce, fish paste, shrimp paste and crushed ginger are used as condiments. Coconut and coconut milk are important ingredients of many savoury and sweet dishes. Fruits and vegetables are widely used, including bananas, guavas, mangoes, papayas, pineapples and calamansi, as well as water spinach, napa cabbage, eggplants, tomato, onions and garlic. There are hundreds of traditional dishes, often made with meat (chicken, pork, beef) or seafood stewed with vegetables in tomato, soy or pineapple sauce. Well-known examples include *lechón* (roasted whole pig), *adobo* (chicken or pork fried in oil, soy sauce, vinegar and garlic), *kare-kare* (oxtail and vegetables stewed in peanut sauce), *sinigang* (fish with vegetables in sour sauce), *pancit* (noodles) and *lumpia* (spring rolls).

North America

The cuisine reflects the dynamic and multicultural origins of the people that make up North American society today. Almost all food traditions of the world are represented. Every conceivable combination of ancient and modern trends in cooking has been adopted somewhere by someone to come up with new culinary concepts, from Canadian pizza ("pizza québécoise") to California rolls ("USA sushi") and jambalaya ("Creole or Cajun paella").

Canadian maple syrup, with aboriginal roots, is perhaps the best known but other contenders for Canadian national dishes include *poutine* (potato fries topped with cheese curds and gravy), butter tarts (a Canadian speciality of flaky pastry with a semi-solid filling of butter, sugar, syrup and egg), smoked and dried salmon, Kraft Dinner (a packed mixture of dry macaroni and cheese) and Montreal-style bagels (handmade, wood-fired, and topped with poppy seeds or sesame seeds). Many typically American food items are derived or adapted from the culinary traditions of Germany (hamburgers, hotdogs, chicken-fried steak, apple pie), Italy (pizza, pasta), Mexico (Tex-Mex tacos, burritos, enchiladas) and China (chow mein, General Tso's chicken and fortune cookies). In the state of Louisiana, Cajun and Creole cooking have become popular, reflecting French, Acadian and Haitian influences but resulting in uniquely American dishes. These include *boudin* (white sausage made from pork or seafood, sometimes rolled into a ball), crawfish étouffée (a shellfish stew made with a roux and served on rice), gumbo (a stew of seafood or chicken with celery, bell peppers and onions, thickened with African okra, Choctaw filé powder or a dark French roux and served on rice) and jambalaya (similar to gumbo but made with long-grain rice in the same way as paella or risotto). Other American culinary concepts that have become popular throughout the world include peanut butter, popcorn, Coca-Cola, muffins, brownies, doughnuts and cupcakes.

Latin America

The cuisines of Latin American countries are distinctive in having chilli peppers as common flavour ingredient, often eaten with various bread-like containers made from the indigenous Mexican corn (maize) and stuffed with a diversity of meat and vegetable creations. *Chicha*, a fermented or non-fermented beverage made from grains or fruits, often accompanies a meal. There are notable exceptions, such as Caribbean cuisine (based on rice, not maize) and Cuban food (also based on rice but traditionally without chilli).

Mexican cuisine represents a rich cultural heritage derived from the indigenous Aztec and Mayan traditions and Spanish influences. The cuisine is based on indigenous crops that have become international commodities: maize (corn), beans, pumpkins and tomatoes, with a rich diversity of herbs and spices, the most famous of which are chilli peppers, chocolate and vanilla. Mexican food in various regional forms (e.g. Tex-Mex, Cal-Mex) has become well known, not only throughout North and South America, but also in the rest of the world. Mexican restaurants have become one of the most popular amongst those serving ethnic cuisines. Famous traditional dishes are tacos (maize tortillas folded around food), quesadillas (tortillas filled with cheese and other ingredients), enchiladas (maize tortillas rolled around various fillings and topped with chilli sauce), burritos (wheat tortillas folded into a cylindrical shape to completely enclose a filling of meat, beans and/or other ingredients), tamales (starchy maize dough – *masa* – steamed or boiled in a leaf wrapper and filled with a seasoned filling of meat, cheese, vegetables, fruits and often chilli peppers) and moles (the generic name for various sauces or dishes derived from them). In Mexico, moles are associated with festivals and celebrations. The most famous of the moles is *mole poblano* (some say it is the national dish of Mexico). It is made from about 20 ingredients but has dried mulato pepper and chocolate as the main ingredients. The chocolate serves to counteract the heat of the chillies and contributes to the dark colour of the sauce. Mulato peppers have a distinctive taste – they are medium spicy and dark brown to blackish (usually sold dry). An example of a regional speciality from Puebla is *chiles en nogada*, which consists of a green poblano chilli filled with *picadillo* (a spicy meat and fruit mixture), topped with *nogada* (a white, walnut-based cream sauce) and sprinkled with pomegranate seeds. The green chilli, white sauce and red pomegranate give the three colours of the Mexican flag.

Peru has a varied and interesting cuisine that has remained relatively unknown to the outside world. Lima is sometimes considered to be the culinary capital of the Americas, where African and Chinese

Spice market, São Paulo, Brazil

Traditional Mexican food

immigrants have added Creole cuisine and other cooking styles. The ancient Inca traditions and food ingredients, blended with Spanish influences have resulted in many unique features. Most notable are tubers (potatoes, oca, ulluco), fruits (avocado, cherimoya, lucuma, pineapple) and indigenous meats (taruca, llama, guinea pig). The main spice is *aji* (*Capsicum pubescens*) and there are many unique dishes such as ceviche (seafood in citrus juice) and Andean *pachamanca* (meat, potato and broad bean stew), often accompanied by *chicha*.

South Americans love meat and various forms of *asado* (barbecue) are popular in many countries. Regional specialities include Colombian *bandeja paisa*, Costa Rican and Nicaraguan *gallo pinto*, Ecuadorian *hornado*, El Salvadorian *pupusa*, Honduran *baleadas*, Uruguayan *dulce de leche* and Paraguayan *sopa paraguaya*, to name only a few.

Oceania

Polynesia is made up of more than 1 000 islands in the triangle between Hawaii, New Zealand and Easter Island. The Polynesian staple food is *poi*, a type of porridge or dough made from the corms of the taro plant (*Colocasia esculenta*). Traditional dishes include *ota ika* (raw fish marinated in citrus juice and coconut milk) and *rewana paraoa* (sourdough bread made from potatoes). Polynesian cuisine is less well known than Tiki cuisine, a 20th century theme used in Polynesian-style restaurants and clubs, mainly in the USA. It was inspired by

Tiki carvings and mythology but is an American form of kitsch rather than authentic Polynesian.

The cuisines of New Zealand and Australia are based on British traditions, but with some Polynesian (Maori) and Asian influences that have become more evident in recent years. Fish and chips, meat pies, custard squares and pavlova are considered typical dishes. Australian food is based on Irish and British traditions but there has been a revival of interest in so-called bush tucker (the food of the indigenous Australians). Typical are roast dinners, barbeques, fish and chips and meat pies, but modern Australia features many exotic influences such as organic and biodynamic food, haute cuisine and nouvelle cuisine. The traditional billy tea (of Waltzing Matilda fame), flavoured with a gum leaf, has been enriched by excellent wine, beer and coffee (e.g. flat white, an Australian invention – cafe latte with no foam). Also typical are kangaroo meat, macadamia nuts, the damper (wheat flour bread baked in the coals of a campfire) and Anzac biscuits, made from rolled oats and coconut (associated with the Australian and New Zealand Army Corps of World War I, hence ANZAC). Popular condiments include tomato sauce, barbecue sauce and Vegemite. Indigenous culinary herbs and spices that have become more widely known in recent years include lemon myrtle (*Backhousia citriodora*), mountain pepper (*Drimys lanceolata*), lemon-scented tea (*Leptospermum citratum*) and mint bush (*Prostanthera rotundifolia*).

Cultivation, harvesting and processing

Culinary herbs are amongst the easiest and most rewarding plants to grow. All that is required are sufficient sunlight and fertile, well-drained soil. Most herbs need warm temperatures to flourish, but a hot window sill or small greenhouse will allow success even in very cold regions.

The traditional herb garden has a formal layout, often bordered by carefully trimmed hedges of box (*Buxus sempervirens*), but the modern trend is to incorporate culinary herbs as ornamental plants in regular garden beds. Many herbs are attractive garden plants, with interesting texture and colourful leaves or flowers. Herbs are ideally suited for very small spaces and can be grown in decorative pots and containers, strategically placed in a sunny spot or against a warm wall. When selecting a container, bigger is definitely better, because small pots dry out rapidly and require regular watering. When larger spaces are available, a formal herb garden may be considered, or specialized gardens for kitchen herbs, salad herbs or aromatic plants (for potpourri). A good source of inspiration and knowledge may be the local herb society and occasional visits to some of the famous herb gardens around the world.

The general principles of gardening also apply to herb gardening and there are many excellent gardening books giving details on the local requirements for propagation and cultivation, which differ slightly for different horticultural zones. In general, most herbs are easily grown in a warm temperate climate, while humid tropical conditions are needed for most spices. Although many herbs can tolerate drought and poor soil conditions, it is much better to feed and water them regularly. Use generous amounts of compost to improve clay or sandy soil. The acidity of the soil should ideally be around neutral to slightly alkaline (pH 6 to 7.5). A slightly acidic pH is also fine, but adding a little lime may rapidly improve the vigour of herbs that occur naturally on limestone soils (such as thyme, oregano and sage). A flourishing plant is not only more attractive but much less prone to pests and diseases.

Almost all culinary herbs (annuals and perennials) can easily be grown from seeds, sown at the start of the growing season (although timing is often not critical, as long as the temperature is sufficiently high for the seeds to germinate). Annual and biennial herbs such as basil, dill and parsley are grown from seeds, while perennials such as thyme,

sage and rosemary are more often propagated by cuttings or by division. Softwood cuttings are generally made in spring, semi-hardwood cuttings in summer and hardwood cuttings in autumn. The majority of perennial herbs are surprisingly easy to grow from cuttings. Commercial rooting powder may be used but is often not even necessary. An easy way to multiply perennial herbs is to plant them fairly deep so that the bases of the side shoots are well below the soil surface. These can be separated from the main stem once they have formed roots. Clump-forming species such as chives and lemongrass are readily multiplied by division – at any time during the growing season. Despite the ease with which herbs can be propagated, many people do not have the time and patience to create their own plants. The solution is simple and becoming very popular – an instant herb garden, created by buying the required plants in containers from a local nursery or garden centre.

Leafy herbs can be harvested and dried throughout the growing season but the best flavour often develops just before or during flowering. At this time the highest yields of essential oil can be expected. Loosely tie the cut stems in bundles and hang them for a few days in a cool, well-ventilated area until dry. It may be necessary to spread paper sheets below the bundles in case the leaves fall off during drying. Alternatively, simply place the cut stems upside down in large paper bags and leave them open to dry. Once dry, the stems can be shaken or broken to reduce the volume, and the dried herb transferred to airtight, labelled containers for storage. Make sure that the material is completely dry otherwise it may go mouldy. Most herbs are commercially available as dry products (spice herbs) but many of us share the sentiments of celebrity chefs and famous restaurants by using fresh herbs whenever practically possible. Fresh herbs can also be preserved in oil or vinegar and may be processed into sauces and pastes that can be kept for up to one year in the refrigerator once the container is opened. An excellent way of preserving tender culinary herbs is to place them (alone or in required combinations) in small

The herb garden at the Royal Botanic Gardens, Kew

plastic bags in the freezer. When taken out for use they can easily be crushed by hand before they thaw. Another method is to place finely chopped herbs into the ice-cube tray of the freezer and to top up the cubes with water. This is also a good way of adding decorative mints to a cocktail or fruit salad. Herbs can be used in the form of herbal tea, herbed butter, herbed vinegar, herbed jellies, candied flowers and potpourris.

Spices, on the other hand, are mostly not practical to grow on a small scale because they are mass-produced in tropical countries at relatively low cost. They are generally used as dry products and are readily available at modest prices from supermarkets or specialized spice shops. Modern drying techniques such as freeze-drying are used to ensure that little or no flavour is lost. Since flavour compounds are often volatile, spices (and especially powdered spices) should be kept in air-tight containers and stored in a cool, dark place. Whole spices retain their flavour for much longer periods and can be grated or powdered in the kitchen as required. They may also be added whole during the cooking process and removed from the dish before it is served. Muslin sachets or stainless steel mesh balls (infusers) are commonly used. It is important to check expiry dates and to buy products from reputable dealers who do not stockpile their supplies for several years but buy fresh raw materials on a regular basis.

A greenhouse is ideal for cold climates

Salad herbs and herb mixtures

Salad herbs are edible green leaves that comprise all or most of a prepared salad. These leaves are used for food and not only in small quantities for flavour. Salad herbs are typically eaten raw and do not require cooking or stir-frying, as in the case of **potherbs** (spinaches). Potherbs are here considered to be vegetables and are excluded from the following list, but note that some are used for both purposes (e.g. Swiss chard, celery cabbage, spinach and turnip). **Microgreens** are the seedlings of salad herbs and other edible greens that are harvested within a few days after germination. They are carefully cut at ground level when less than 25 mm (1 in.) long and comprise a short stem, two cotyledons and the first two true leaves. Herbs commonly used for microgreen production include red beetroot and white or variously coloured Swiss chard, lettuce and cress.

List of salad herbs

angelica (*Angelica archangelica*)
anise hyssop (*Agastache foeniculum*)
basil (*Ocimum basilicum*)
bear's garlic (*Allium ursinum*)
beefsteak plant (*Perilla frutescens*)
borage (*Borago officinalis*)
caraway (*Carum carvi*)
celery (*Apium graveolens*)
celery cabbage (*Brassica napa* var. *pekinensis*)
chervil (*Anthriscus cerefolium*)
chicory (*Cichorium intybus*)
Chinese chives – *see* garlic chives
Chinese parsley – *see* cilantro
chives (*Allium schoenoprasum*)
chop suey greens (*Chrysanthemum coronarium*)
cilantro (*Coriandrum sativum*)
coriander leaves – *see* cilantro
corn mint – *see* field mint
corn salad (*Valerianella locusta*)
cress (*Lepidium sativum*)
crown daisy – *see* chop suey greens
dandelion (*Taraxacum officinale*)
dhania (*Coriandrum sativum*)
dill (*Anethum graveolens*)
endive (*Cichorium endivia*)
fennel (*Foeniculum vulgare*)
fenugreek (*Trigonella foenum-graecum*)
field mint (*Mentha arvensis*)
French tarragon – *see* tarragon
garden mint (*Mentha spicata*)
garden nasturtium (*Tropaeolum majus*)
garlic chives (*Allium tuberosum*)
green pepper, sweet pepper (*Capsicum annuum*)
horse mint – *see under* spearmint
Indian borage (*Plectranthus amboinicus*)
Indian cress (*Tropaeolum majus*)
Italian parsley (*Petroselinum crispum* var. neapolitanum)
Japanese bunching onion (*Allium fistulosum*)
kangkong – *see* water spinach
laksa leaf (*Persicaria odorata*)

land cress (*Barbarea verna*)
leeks (*Allium ameloprasum*)
lemon balm (*Melissa officinalis*)
lemon thyme (*Thymus ×citriodora*)
lovage (*Levisticum officinale*)
Malabar spinach (*Basella alba*)
mint (*Mentha* species)
mitsuba (*Cryptotaenia japonica*)
mustard greens (*Brassica juncea*)
nasturtium (*Tropaeolum majus*)
onion (*Allium cepa*)
orache (*Atriplex hortensis*)
parsley (*Petroselinum crispum*)
peppermint (*Mentha piperita*)
pineapple mint (*Mentha suaveolens* 'Variegata')
purslane (*Portulaca oleracea*)
radish (*Raphanus sativus*)
rhubarb (*Rheum rhabarbarum*)
rocket (*Eruca sativa*)
salad burnet (*Sanguisorba minor*)
shallot (*Allium cepa* var. *ascalonicum*)
shiso (*Perilla frutescens*)
sorrel, French sorrel (*Rumex scutatus*)
sorrel, garden sorrel (*Rumex acetosa*)
sow thistle (*Sonchus oleraceus*)
spearmint (*Mentha spicata*)
spilanthes, pará cress (*Spilanthes acmella*)
spinach (*Spinacea oleracea*)
sweet balm – *see* lemon balm
sweet basil (*Ocimum basilicum*)
sweet cicely (*Myrrhis odorata*)
sweet pepper (*Capsicum annuum*)
Swiss chard (*Beta vulgaris*)
tarragon, estragon (*Artemisia dracunculus*)
turnip, Japanese greens (*Brassica rapa*)
Vietnamese coriander / mint (*Persicaria odorata*)
water spinach (*Ipomoea aquatica*)
watercress (*Nasturtium officinale*)
Welsh onion (*Allium fistulosum*)
wild rocket (*Diplotaxus tenuifolia*)
winter cress – *see* land cress
winter purslane (*Montia perfoliata*)

Bundles of mixed herbs for pickling cucumbers (St Petersburg, Russia)   Mitsuba (*Cryptotaenia japonica*)

Classical herb combinations have been developed in many parts of the world to flavour dishes and sauces or to enjoy as salads. The best-known examples are from Europe but there are undoubtedly many more examples from Asian culinary traditions that are not yet well known outside their regions of origin.

Bouquet garni refers to a bunch of aromatic herbs that is traditionally used (especially in French cooking) to flavour a sauce or stock. The herbs (typically parsley, thyme, bay leaves, rosemary, sage and even cloves) are tied together in a bunch or wrapped in a piece of bacon or cheesecloth. This allows for the easy removal of the herbs and prevents them from dispersing into the dish.

Fines herbes are a mixture of chopped, aromatic but mild-tasting "fine herbs" that are popular in French and Mediterranean cuisines. The classical mixture comprises fresh leaves of chives, chervil, parsley and tarragon but nowadays other herbs such as basil, bay leaf, cress, cicely, fennel, lemon balm and marjoram, rosemary and thyme are sometimes added. It is traditionally used to flavour sauces, cream cheeses, omelettes, sautéed vegetables and some meat dishes.

Herbes à soupe are traditionally used in France to flavour soups and stews and include the tops (leaves) of various vegetables such as carrots, celery, parsley and radishes.

Herbes de Provence are a Provençal mixture of fresh or dried thyme, rosemary, bay leaf, basil and savory that was traditionally used to flavour grilled meat. The concept originally simply referred to the herbs typical of Provence but nowadays commercial mixtures under this name are widely sold in supermarkets. These often include lavender and

fennel to create mixtures that are not necessarily part of traditional French cuisine.

Herbes vénitiennes are a mixture of aromatic herbs (tarragon, chervil, parsley and sorrel) traditionally used in France to flavour butter.

Persillade is a classic French seasoning made of chopped parsley (*persil* in French) and garlic and traditionally added at the end of the cooking time.

Pickle herbs are a mixture of fresh leaves that is used to flavour pickled cucumbers. Fresh dill herb is an essential ingredient, often accompanied by raspberry leaves, green onions and various other herbs depending on local preferences.

Potherbs in the French tradition included chard, lettuce, orache, purslane, sorrel and spinach. It was used not only to flavour soups and stews, but also as garnishes, salad ingredients and vegetables. Nowadays the term is more widely used for almost any green leafy vegetable than is eaten by itself as cooked spinach or incorporated into other dishes.

Sabzi khordan is an Iranian (Persian) mixture of fresh herbs (served with lunch and dinner) that typically includes basil, cilantro, cress, fennel, fenugreek, peppermint, radishes, *za'atar*, savory, tarragon and Welsh onion.

Tabbouleh is a Syrian salad made from chopped herbs (parsley, mint and lettuce), bulgur (cracked wheat) and other ingredients.

Culinary herbs

Culinary herbs are the leaves or leafy twigs of aromatic plants that are used to add flavour to a dish. Included here are also leaf bases that form bulbs, in the case of garlic and onion. Herbs that are commonly used or sold as dry products (spice herbs) are marked with an asterisk (*). The aroma compounds of herbs are typically volatile, so that much of the flavour may be lost when herbs are dried. A good supply of fresh herbs is therefore just as important as other fresh ingredients for making an excellent dish.

List of culinary herbs

absinth (*Artemisia absinthium*)*
Aleppo rue (*Ruta chalepensis*)
alexanders (*Smyrnium olusatrum*)
allspice leaves (*Pimenta dioica*)*
angelica (*Angelica archangelica*)*
anise hyssop (*Agastache foeniculum*)
apple mint (*Mentha suaveolens*)
baldmoney (*Meum athamanticum*)
balm (*Melissa officinalis*)*
basil, sweet basil (*Ocimum basilicum*)*
bay leaf, Indian, tejpat (*Cinnamomum tamala*)
bay, bay leaf, sweet bay (*Laurus nobilis*)*
beefsteak plant (*Perilla frutescens*)
bergamot, bee balm (*Monarda didyma*)
black lovage – *see* alexanders
black thyme – *see* spiked thyme
boldo (*Peumus boldus*)
Bolivian coriander (*Porophyllum ruderale*)
borage (*Borago officinalis*)
Bowles's mint (*Mentha villosa* var. *alopecuroides*)
buchu (*Agathosma betulina*)*
calamint (*Calamintha nepeta*)
caraway (*Carum carvi*)
catnip (*Nepeta cataria*)
celery (*Apium graveolens*)*
chervil (*Anthriscus cerefolium*)*
Chinese chive – *see* garlic chive
Chinese onion (*Allium chinense*)
Chinese parsley – *see* cilantro
chives (*Allium schoenoprasum*)
chop suey greens (*Chrysanthemum coronarium*)
cicely – *see* sweet cicely
cilantro (*Coriandrum sativum*)
clary sage (*Salvia sclarea*)
coriander leaf – *see* cilantro
corn mint – *see* field mint
Corsican mint (*Mentha requinii*)
cress, garden cress (*Lepidium sativum*)
Cretan dittany (*Origanum dictamnus*)
crown daisy – *see* chop suey greens
culantro, long coriander, eryngo (*Eryngium foetidum*)
curry leaf (*Murraya koenigii*)*
curry plant (*Helichrysum italicum*)

dandelion (*Taraxacum officinale*)
dhania (*Coriandrum sativum*)
dill, dill weed (*Anethum graveolens*)*
dittany – *see* Cretan dittany
eau de cologne mint (*Mentha ×piperita* var. *citrata*)
epazote (*Chenopodium ambrosioides*)
eryngo, culantro, long coriander (*Eryngium foetidum*)
estragon, tarragon, French tarragon (*Artemisia dracunculus*)*
Ethiopian thyme (*Thymus schimperi*)
fennel (*Foeniculum vulgare*)*
fenugreek (*Trigonella foenum-graecum*)*
field mint (*Mentha arvensis*)*
fishwort, fish mint (*Houttuynia cordata*)
French tarragon, estragon (*Artemisia dracunculus*)*
garden cress, cress (*Lepidium sativum*)
garden mint (*Mentha spicata*)*
garden nasturtium (*Tropaeolum majus*)*
garlic (*Allium sativum*)
garlic chive (*Allium tuberosum*)
geisho (*Rhamnus prinoides*)*
ginger mint, Scottish mint (*Mentha gracilis*)
Greek oregano (*Origanum vulgare* subsp. *hirtum*)*
herb bennet, wood avens (*Geum urbanum*)
hoary basil, perennial basil (*Ocimum americanum*)
holy basil, tulsi (*Ocimum tenuiflorum*)
horse mint (*Mentha longifolia*)
hyssop (*Hyssopus officinalis*)*
Indian bay leaf, tejpat (*Cinnamomum tamala*)
Indian borage (*Plectranthus amboinicus*)
Indian cress (*Tropaeolum majus*)
Indonesian bay leaf, salam leaf (*Syzygium polyanthum*)*
Italian parsley (*Petroselinum crispum* var. *neapolitanum*)
Japanese bunching onion (*Allium fistulosum*)
Korean mint (*Agastache rugosa*)
koseret (*Lippia adoensis*)*
laksa leaf (*Persicaria odorata*)
laurel – *see* bay, bay leaf
lavender (*Lavandula angustifolia*)
lemon balm (*Melissa officinalis*)*
lemon myrtle (*Backhousia citriodora*)
lemon thyme (*Thymus ×citriodora*)

A selection of common culinary herbs. Top row: chives, Vietnamese mint, lemon balm, bay leaf, basil, purple basil, lime leaf, dill and cilantro. Bottom row: sage, lemon verbena, French lavender, spearmint, peppermint and pennyroyal

lemon verbena (*Aloysia triphylla*)*
lemongrass (*Cymbopogon citratus*)*
lime leaf (*Citrus hystrix*)*
lotus leaf (*Nelumbo nucifera*)
lovage (*Levisticum officinale*)*
makrut lime (*Citrus hystrix*)*
marjoram (*Origanum majorana*)*
melinjo (*Gnetum gnemon*)
Mexican pepperleaf (*Piper auritum*)
mint (*Mentha* species)*
myrtle (*Myrtus communis*)*
nasturtium (*Tropaeolum majus*)
onion (*Allium cepa*)
oregano (*Origanum vulgare*)*
pandan, pandan leaf (*Pandanus amaryllifolius*)*
papeda – *see* lime leaf
parsley (*Petroselinum crispum*)*
pennyroyal (*Mentha pulegium*)
peppermint (*Mentha piperita*)*
perilla (*Perilla frutescens*)
pimento leaves (*Pimenta dioica*)*
pineapple mint (*Mentha suaveolens* 'Variegata')
pineapple sage (*Salvia elegans*)
pot marjoram (*Ocimum onites*)
rhubarb (*Rheum rhabarbarum*)
rice paddy herb (*Limnophila aromatica*)
rocket (*Eruca sativa*)
rosemary (*Rosmarinus officinalis*)*
round-leaf mint (*Mentha ×rotundifolia*)
rue (*Ruta graveolens*)*
Russian tarragon – *see* tarragon
sage (*Salvia officinalis*)*
salad burnet (*Sanguisorba minor*)
salam leaf, Indonesian bay leaf (*Syzygium polyanthum*)*
sansho leaf (*Zanthoxylum piperitum*)*
savory – *see* summer / winter savory

Scottish mint, ginger mint (*Mentha gracilis*)
shallot (*Allium cepa* var. *ascalonicum*)
shiso (*Perilla frutescens*)
skirret (*Sium sisarum*)
sorrel, French sorrel (*Rumex scutatus*)
sorrel, garden sorrel (*Rumex acetosa*)
sorrel, red-veined sorrel (*Rumex sanguineus*)
sorrel, sheep sorrel (*Rumex acetosella*)
spearmint (*Mentha spicata*)*
spiked thyme, black thyme (*Thymbra spicata*)
spilanthes, pará cress (*Spilanthes acmella*)
sugar-leaf, stevia (*Stevia rebaudiana*)
summer savory (*Satureja hortensis*)
sweet balm (*Melissa officinalis*)*
sweet basil (*Ocimum basilicum*)*
sweet bay – *see* bay, bay leaf
sweet cicely (*Myrrhis odorata*)
tansy (*Tanacetum vulgare*)*
tarragon, French tarragon, estragon (*Artemisia dracunculus*)*
tarragon, Russian tarragon (*Artemisia dracunculoides*)
Thai basil – *see* basil
thyme (*Thymus vulgaris*)*
tulsi (*Ocimum tenuifolium*)
Vietnamese balm, kinh gioi (*Elsholtzia ciliata*)
Vietnamese coriander (*Persicaria odorata*)
Vietnamese mint, rau ram (*Persicaria odorata*)
water mint (*Mentha aquatica*)
Welsh onion (*Allium fistulosum*)
wild rocket (*Diplotaxus tenuifolia*)
wild thyme (*Thymus serpyllum*)
winter savory (*Satureja montana*)
wood avens, herb bennet (*Geum urbanum*)
woodruff (*Galium odoratum*)
wormwood – *see* absinth
za'atar, zatar (*Origanum syriacum*)*

Spices

Spices are various plant parts that are added to food in small amounts with the aim of improving the flavour. Some may also add gusto, "kick" or zest, because spices not only act on our taste buds but also have subtle and not so subtle stimulant effects. In the following list, the part or parts used are indicated by superscripts, respectively **r** for roots, **rh** for rhizomes and stems, **b** for bark, **fl** for flower (**flb** for flower buds, **fls** for styles), **fr** for mature fruits (including fleshy cones), **fri** for immature fruits, **s** for seeds, **sa** for seed arils and **e** for exudates and resins (usually obtained from stems). Note that spices derived from leaves are listed under CULINARY HERBS on page 24. Many of them are used fresh as culinary herbs but are also commercially available in the form of dried spice herbs.

ajowan, ajwain, bishop's weed (*Trachyspermum ammi*)[fr]
almond (*Prunus dulcis*)[s]
allspice (*Pimenta dioica*)[fr]
amchur, mango powder (*Mangifera indica*)[fri]
anardana (*Punica granatum*)[s]
angelica fruits (*Angelica archangelica*)[fr]
anise (*Pimpinella anisum*)[fr]
annatto (*Bixa orellana*)[s]
asafoetida (*Ferula assa-foetida*)[e]
bird chilli (*Capsicum frutescens*)[fr]
black cumin (*Bunium persicum*)[fr]
black cumin, incorrect use (*Nigella sativa*)[s]
black lime (*Citrus limon*)[fr]
black mustard (*Brassica nigra*)[s]
black pepper (*Piper nigrum*)[fr, fri]
black seeds (*Nigella sativa*)[s]
Brazilian pepper (*Schinus terebinthifolius*)[fr]
brown mustard (*Brassica juncea* var. *juncea*)[s]
cacao, chocolate (*Theobroma cacao*)[s]
camphor (*Cinnamomum camphora*)[e]
canarium, Chinese black- (*Canarium pimela*)[fr]
candlenut (*Aleurites moluccana*)[s]
capers (*Capparis spinosa*)[fli, fri]
caraway (*Carum carvi*)[fr]
cardamom (*Elettaria cardamomum*)[fr, s]
cardamom, Chinese black- (*Aframomum costatum*)[fr, s]
cardamom, Ethiopian (*Aframomum corrorima*)[s]
cardamom, Indian black (*Amomum subulatum*)[fr, s]
carob (*Ceratonia siliqua*)[fr]
cassia (*Cinnamomum aromaticum*)[b]
cayenne pepper (*Capsicum annuum, C. frutescens*)[fr]
celery (*Apium graveolens*)[fr]
Ceylon cinnamon – *see* cinnamon
chamomile (*Chamaemelum nobile*)[fl]
charoli (*Buchanaria lanzan*)[s]
chilli (*Capsicum frutescens, C. annuum*)[fr]
Chinese anise (*Illicium verum*)[fr]
Chinese keys (*Boesenbergia rotunda*)[rh]
Chinese olive – *see* canarium
Chinese pepper – *see* Sichuan pepper
cicely – *see* sweet cicely
cinnamon (*Cinnamomum verum*)[b]

clove(s) (*Syzygium aromaticum*)[flb]
coconut (*Cocos nucifera*)[fr]
coffee (*Coffea arabica*)[s]
cola (*Cola acuminata*)[s]
coriander (*Coriandrum sativum*)[fr]
cumin (*Cuminum cyminum*)[fr]
cumin, black (*Bunium persicum*)[fr]
devil's dung – *see* asafoetida
Dijon mustard – *see* brown mustard
dill (*Anethum graveolens*)[fr]
elder flowers (*Sambucus nigra*)[fl]
elderberries (*Sambucus nigra*)[fr]
Ethiopian cardamom (*Aframomum corrorima*)[s]
fennel (*Foeniculum vulgare*)[fr]
fenugreek (*Trigonella foenum-graecum*)[s]
galangal, greater (*Alpinia galanga*)[rh]
galangal, lesser (*Alpinia officinarum*)[rh]
galangal, small (*Kaempferia galanga*)[rh]
garden nasturtium (*Tropaeolum majus*)[fl, fr]
ginger (*Zingiber officinale*)[rh]
golpar (*Heracleum persicum*)[fr]
grains of paradise (*Aframomum melegueta*)[s]
greater galangal – *see* galangal
Guinea pepper (*Xylopia aethiopica*)[fr]
hazelnut (*Corylus avellana*)[s]
hibiscus (*Hibiscus sabdariffa*)[fl]
hop(s) (*Humulus lupulus*)[fri]
horseradish (*Armoracia rusticana*)[r]
Indian mustard – *see* brown mustard
Indonesian mango ginger – *see* mango ginger
Japanese pepper (*Zanthoxylum piperitum*)[fr]
juniper berries (*Juniperus communis*)[fr, fri]
kalonji (*Nigella sativa*)[s]
kokum, kokam (*Garcinia indica*)[fr]
korarima (*Aframomum corrorima*)[s]
lemon peel (*Citrus limon*)[fr]
lesser galangal – *see* galangal
licorice (*Glycyrrhiza glabra*)[rh]
liquorice (*Glycyrrhiza glabra*)[rh]
long pepper (*Piper longum*)[fri]
lovage (*Levisticum officinale*)[fr]
mace (*Myristica fragrans*)[sa]
makrut lime (*Citrus hystrix*)[fr]

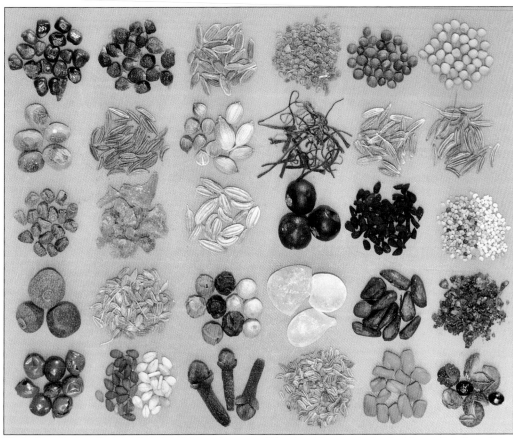

A selection of spices. Top row: korarima, Melegueta pepper, dill, celery, black mustard and white mustard. Second row: charoli, caraway, coriander, saffron, cumin and black cumin. Third row: cardamom, asafoetida, fennel, juniper, nigella and poppy seeds (blue and white). Fourth row: allspice, anise, pepper (green, black, white), mastic, anardana and sumac. Fifth row: pink pepper, sesame (brown and white), cloves, ajowan, fenugreek and Sichuan pepper. See also page 31.

mandarin peel (*Citrus reticulata*)^fr
mango ginger (*Curcuma amada*)^rh
mango ginger, Indonesian (*Curcuma mangga*)^rh
mastic (*Pistacia lentiscus*)^e
Melegueta pepper (*Aframomum melegueta*)^s
mustard – *see* black, brown or white mustard
myrtle (*Myrtus communis*)^fr
nutmeg (*Myristica fragrans*)^s
nigella (*Nigella sativa*)
paprika (*Capsicum annuum*)^fr
pepper (*Piper nigrum*)^fr,fri
pimento (*Pimenta dioica*)^fr
pine nuts (*Pinus* species)^s
pink pepper (*Schinus molle, S. terebinthifolius*)^fr
pomegranate seeds – *see* anardana
poppy (*Papaver somniferum*)^s
Roman chamomile (*Chamaemelum nobile*)^fl
rose petals (*Rosa* species)^fl
rue (*Ruta chalepensis*)^fr
safflower (*Carthamus tinctorius*)^fl

sansho (*Zanthoxylum piperitum*)^fr
saffron (*Crocus sativus*)^fls
sesame (*Sesamum indicum*)^s
Sichuan pepper (*Zanthoxylum piperitum, Z. simulans*)^fr
small galangal – *see* galangal
spilanthes (*Spilanthes acmella*)^fl
star anise (*Illicium verum*)^fr
sumac (*Rhus coriaria*)^fr
sweet cicely (*Myrrhis odorata*)^fri
Tabasco pepper (*Capsicum frutescens*)^fr
tailed pepper, cubebs (*Piper cubeba*)
tamarind (*Tamarindus indica*)^fr
torch ginger (*Etlingera elatior*)^fli
turmeric (*Curcuma longa*)^rh
vanilla (*Vanilla planifolia*)^fr
wasabi (*Wasabia japonica*)^rh
white ginger (*Mondia whitei*)^r
white mustard (*Sinapis alba*)^s
white pepper – *see* pepper
zedoary (*Curcuma zedoaria*)^rh

Spice mixtures

Spice mixtures (also called mixes or blends) are combinations of dried spices and herbs prepared according to traditional recipes and often associated with particular types of dishes. Many traditional spice mixtures can be found in countries of the Middle and Near East, perhaps because they are at the confluence of overland spice trade routes and have been exposed to both Eastern and Western cooking traditions over centuries. Unique spice mixtures are sometimes confined to a single country, region, town, restaurant or even a household. When specific mixtures of spices and herbs are regularly used (e.g. in a popular dish), it is convenient to buy such pre-made bslends from spice vendors or grocery stores (often as packed and branded products). Well-known examples include curry powder, garam masala, Chinese five spices, Thai seven spices, garlic salt and pumpkin pie spice.

Spice mixtures are sometimes roasted or toasted to bring out the flavour. Ingredients are added in a specific sequence, starting with woody spices such as cinnamon and cloves and ending with delicate herbs such as parsley and chervil. In Indian cuisine, spices are often fried in oil or ghee, a process known "tempering" (English), *chaunk* (Hindi), *baghaar* (Oriya), *bagar* (Bengali) and *phoṛon* (Maithili). The heat causes the spices to start popping and the flavours are released into the oily matrix. The food ingredients are then added to the fried spices and become coated in the spicy oil.

Spice mixtures may be added to the food before, during or after preparation. They are used whole or powdered, but also as pastes (e.g. *bumbus*), sauces, marinades, spice rubs or tisanes (infusions or decoctions of herbs, spices, or other plant material in water). Whole spices (or bunched herbs, e.g. *bouquet garni*) ares often removed from the dish just before it is served.

A brief overview of only the most famous spice mixtures of the world is given below. Note that sauces and pastes are not included here.

Spice mixtures and seasonings

Advieh or *adwiya* is a spice mixture used in traditional Persian (Iranian) cuisine and in the surrounding regions. The main ingredients are usually cardamom, cinnamon, cloves, cumin, rose petals, turmeric and ginger but regional variations may include ground black pepper, coriander, *golpar*, mace, nutmeg, saffron and sesame.

Aleppo seven-spice or **Lebanese seven-spice** is a traditional Middle Eastern (Syrian or Lebanese) spice mixture comprising allspice, cinnamon, coriander, cardamom, black pepper, nutmeg and cloves.

Apple pie spice usually comprises cinnamon, nutmeg and allspice (sometimes also cardamom) and is used to flavour apple pie.

Baharat (*Bahārāt* is the Arabic plural for *bahār* or "spice") is a mixture of finely ground spices used in Arabian (Persian) and Turkish cuisines. It is a hot and spicy mixture of dried red chilli peppers or paprika with allspice, black pepper, cardamom, cinnamon or cassia, cloves, coriander, cumin and nutmeg. The powder is rubbed into fish or meat or it can be used as a marinade (mixed with olive oil and lime juice). It is often added to soups and stews and can also be used as a condiment. Turkish *baharat* has mint as a main ingredient, while Tunisian *baharat* is a mixture of dried rosebuds with cinnamon and pepper. *Kebsa* spice (Gulf-style *baharat*, found along the Persian Gulf) also includes black lime (*loomi*) and saffron.

Bengali five-spice mixture – *see **panch phoron***

Berbere is the well-known spice powder and key ingredient of Ethiopian cuisine. There is no such thing as a "standard" *berbere* – various other herbs or spices may be added. However, according to a traditional recipe, the basic ingredients are 1½ cups korarima seeds, 2½ cups dried red chilli pepper, ½ cup each of salt, red onion (chopped) and garlic, 4 teaspoons each of black cumin, ajowan and finely chopped ginger and lastly 1 teaspoon each of cloves, black pepper (or tailed pepper/cubeb), cinnamon, coriander, savory, cumin, fenugreek seeds, sesame seeds and sacred basil (and/or rue).

Buknu or *buknu masala* is a North Indian mixture of oil, salt, black salt, ginger powder, turmeric, cumin, pepper, asafoetida, various dried fruits and other ingredients.

Examples of spice mixtures

Bumbu is the term used for Indonesian spice mixtures, often used in the form of seasoning pastes (wet stir-fry mixtures). They are variously composed depending on the intended dish. *Bumbus* usually contain onions, garlic and chilli with herbs such as lemongrass, *salam* leaf (Indonesian bay leaf) and spices such as galangal and ginger. *Bumbu kacang* or "peanut sauce" is a famous example, commonly used with signature dishes such as *gado-gado* and satays. The Balinese version of *bumbu* is called *jangkap* ("the flavour of root") and is made from ingredients such as galangal (*laos*), lesser galangal (*kencur*), ginger, turmeric, lemongrass, coriander and cumin.

Chaat masala or **chat masala** is a spice mixture used in India and Pakistan to flavour fast food, usually sold at carts. The masala is added by the street vendor according to the taste of the customer. It is a sweet-sour spice and condiment that has the following typical ingredients: *amchoor* (dried mango powder), black pepper, chilli powder, coriander, cumin, dried ginger, *hing* (asafoetida), *kala namak* (black salt) and salt. It may be served with food (and drinks) on a small plate or bowl and is sprinkled on salads and egg dishes. A version with less coriander, cumin and ginger but more *amchoor*, asafoetida, black salt and chilli pepper is known as *fruit chaat masala* and is used on chopped fruits or fresh vegetables.

Chaunk is an oily mixture of fried spices added to various dishes before, during or after preparation. The terms *chaunk* or *baghaar* ("tempering" in English) is also used for the cooking technique used in Bangladesh, Pakistan and India to produce these spice mixtures. The spices are fried in oil in a particular sequence, starting with more woody materials such as cinnamon and cloves and ending with more delicate ones such as coriander leaves and curry leaves.

Chilli powder (also known as chilli spice or chilli mix) is not just powdered chilli peppers but a spicy blend that includes various other spices such as coriander, cumin, cloves, allspice, oregano and garlic, as well as seasoned salt and powdered soup bouillon. It is often sold in the United States in small sachets for convenient instant use in "Tex-Mex" cooking, especially the famous Mexican *chili con carne*. Homemade chilli powder is prepared by roasting the required amount of dried chilli peppers (mild or hot, according to taste) and other main ingredients on low heat for a few minutes before the mixture is powdered.

Chinese five-spice powder (*wu xiang fen*) is the most famous spice mixture of China, made from star anise (*ba jiao*), fennel seeds (*xiao hui xiang*), cassia bark (*rou gui*), cloves (*ding xiang*) and Sichuan pepper (*hua jiao*) in an approximate weight ratio of 5:5:4:3:1. In southern China, *Cinnamomum aromaticum* is sometimes replaced with *C. loureiroi* and cloves with mandarin peel. The powdered spice mixture is rubbed into meat or often mixed with syrup or honey and spread over the meat (especially duck, chicken and pork) before cooking. The use of five-spice powder has spread from China to various parts of Asia and is also important in Vietnamese cuisine.

Chinese major spices (*da liao*) is a combination of spices with star anise as the main ingredient, often used with meat. The mixture varies from place to place in China but star anise is always included.

Crab boil is a term for commercial or home-made spice mixtures used to flavour the water in which crabs or other shellfish are boiled. Typical ingredients are salt, hot peppers, lemon, bay leaf, garlic, mustard seeds, coriander and allspice. These are sold in mesh bags or as liquid concentrate.

Curry powder and the modern use of the word "curry" are Western inventions for a spice mixture of Indian inspiration that was popularized throughout the British Empire during the 19th and 20th centuries. It is similar to some spice mixtures used in India, such as garam masala and especially the Tamil *sambar* powder. The main ingredients are chilli peppers, coriander, cumin, fenugreek seeds and turmeric in various combinations – some hot but mostly quite mild. Subtle and unique variations in taste may be created by adding other ingredients such as asafoetida, caraway, cardamom, cinnamon, cloves, fennel seeds, garlic, ginger, mustard seeds, nutmeg, long pepper and black pepper. Before the introduction of South American chilli peppers to South Asia, the hot taste of spice mixtures came from black pepper. There are very many regional and local variants, from fiery hot to very mild, but the yellow colour contributed by turmeric powder is perhaps the most characteristic feature. Curry powder is added to a dish at the beginning of cooking (unlike masala, which is added as an additional seasoning). Thai curries differ from Indian curries in that they are quick-cooking (not stewed) and that they start not with curry powder but with a green paste (a mixture of green herbs

with chilli and other spices, pounded in a pestle and mortar).

Dhansak masala is an Indian (Parsi, Gujarati) spice mixture used to season *dhansak*, a special type of curry stew made of meat (usually goat or mutton) and lentils, vegetables (including pumpkin), garlic, ginger and spices. This masala is a dry-roasted and powdered mixture of chilli peppers with bay leaves, black pepper, green or black cardamom, cinnamon, cloves, coriander, cumin, fenugreek seeds, mace and turmeric.

Duqqa, dukka or ***dukkah*** is an Egyptian spice mixture made from crushed (pounded) nuts, herbs and spices. Typical ingredients include salt, hazelnuts, pepper, sesame, coriander, cumin, caraway, chickpeas, marjoram, mint and *za'atar* (*Origanum syriacum*). The spice is used as a seasoning for stews and other dishes and can also be mixed with olive oil and eaten with bread (as a dip or a spread).

Five-spice powder – *see* **Chinese five-spice powder**

French spice mixture. The classical spice mixture widely used in French cookery comprises 20 g cloves, 20 g nutmeg, 12 g white pepper, 10 g bay leaf, 10 g coriander, 10 g thyme, 6 g cayenne pepper, 5 g marjoram and 5 g rosemary. The recipe given by Larousse for Provençal cooking is a mixture of 25 g each of basil, bay leaf, nutmeg, rosemary and thyme; 20 g each of cloves, summer savory and white pepper; 10 g coriander and 3 g lavender.

Gâlat dagga – *see* **qâlat daqqa**

Garam masala (literally "hot spices" in Hindi) is a popular blend of ground spices commonly used in Indian and other South Asian cooking traditions. There are many regional variations but the basic ingredients are dried red chilli peppers, dried garlic, ginger powder, sesame, mustard seed, turmeric, coriander, Indian bay leaves (tejpat), star anise and fennel. Some versions have cloves, black and white pepper, cinnamon, cumin and cardamom and various herbs added. Garam masala is usually relatively mild but has an intense spicy taste. The spices are usually toasted to bring out the flavour and aroma and the ground spices may be mixed with water, vinegar or coconut milk to form a paste. In recent years, garam masala has become readily available in supermarkets as commercially prepared ground mixtures.

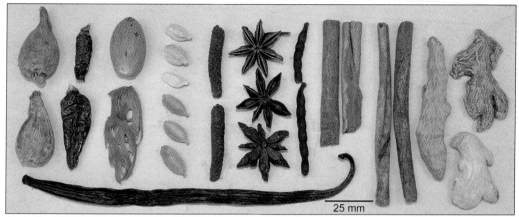

A selection of spices: Melegueta pepper, chilli pepper, nutmeg and mace, green and bleached cardamom, long pepper, star anise, Guinea pepper, cassia, cinnamon, turmeric and ginger. Bottom: vanilla

Goda masala or *kala masala* is a popular Indian (Marathi) sweet-tasting spice blend made of red chilli peppers, green chillies, garlic, ginger, mustard and onion.

Hawaij is a Yemenite ground spice mixture mainly used in coffee and in soups, stews, rice and vegetable dishes and also as meat rub. Typical ingredients are cardamom, cumin, black pepper and turmeric but some may also include caraway, cloves, coriander, nutmeg, onion powder and saffron. *Hawaij* used in coffee, cake and desserts typically has aniseed, cardamom, fennel seeds and ginger.

Italian seasoning, a classic blend of dried herbs used for Italian cuisine, includes dried basil, oregano, sage and rosemary and sometimes marjoram and thyme.

Jerk spice is a very hot Jamaican spice mixture made mainly from allspice ("pimento") and Caribbean red peppers, also known as Scotch bonnet peppers (*Capsicum chinense*). The mixture includes cinnamon, cloves, garlic, nutmeg, scallions, thyme and salt. Jerk is a uniquely Jamaican cooking style in which meat (typically pork or chicken) is dry rubbed or soaked in a marinade and then grilled over hardwood charcoal in a steel drum known as a jerk pan.

Kaala masala is a strong and spicy Maharashtrian (western Indian) mixture of cinnamon, cloves, co-conut, coriander, cumin, cinnamon sticks, kalpasi (*Didymocarpus pedicellatus*), sesame and chilli peppers. *Kaala* means "black" in reference to both the dark spices used (e.g. cloves and cinnamon) and the dark colour obtained after roasting.

Kebsa spice – *see baharat*

Khmeli suneli is a blend of dried herbs and spices used in Georgia and the Caucasus region. These include basil, bay leaf, black pepper, blue fenugreek or dried fenugreek leaves, celery, coriander, dill, fenugreek seeds, hyssop, marigold petals (florets), marjoram, mint, parsley, thyme and hot peppers. It is an ingredient of traditional mutton dishes and sauces.

Lemon pepper or **lemon pepper seasoning** is made from lemon zest crushed with black peppercorns which is then baked and dried. Jars of commercial lemon pepper also include small amounts of salt, sugar, onion, garlic, citric acid, additional lemon flavour, cayenne pepper and other spices. It is used mainly on fish, chicken and pasta.

Masala is the Hindi word for "spices" and refers to various mixtures of spices (sometimes in combination with aromatic herbs) that are typical of South Asian cookery. *Garam masala* is perhaps the most popular and best known type of masala but there are many other types, including *chaat masala*, *dhansak masala*, *kashmiri masala*, *sambar masala*, *tandoori masala* and *tikka masala*.

Mitmita is a hot powdered spice traditionally used in Ethiopia to season *kitfo* (chopped raw beef, a type of beef carpaccio). It is similar to *berbere* in appearance and contains large amounts of chilli pepper, with Ethiopian cardamom seeds, cloves, salt, and sometimes also cinnamon, cumin and ginger. *Mitmita* can be sprinkled on food or used as a condiment or dip with *injera* and other dishes.

31

Mixed spice, also known as **pudding spice**, is a mixture of allspice, cinnamon and nutmeg that is commonly used in desserts (especially baked fruits) and in baking. Small amounts of cloves, coriander, mace and ginger are sometimes added. It is popular in Britain and the former colonies of the British Empire. The Dutch *koekkruiden* or *speculaaskruiden* is similar but contains cardamom.

Montreal steak seasoning or **Montreal steak spice** is a Canadian spice mixture (of eastern European origin or inspiration) that is used to flavour steak and grilled meats. It was originally used as a dry rub mixture for smoked meat. The main ingredients are black pepper, cayenne pepper flakes, coriander, dill seeds, garlic and salt.

Mulling spices is a spice mixture used in Europe and North America to prepare mulled wine, hot apple cider and other hot drinks and juices. The drink is heated with the spices in a pot and then simply strained and served warm. Such "mulled" drinks are typically enjoyed during autumn and winter. There are regional and individual preferences as to the combination of spices to be used – often allspice, cinnamon, cloves and nutmeg, but sometimes also cardamom, star anise or peppercorns. Dried fruits or berries (e.g. raisins, apple rings, orange or tangerine rind) are often added to the mixture.

Old Bay Seasoning is a branded condiment developed in the United States in 1939, mainly to flavour crab, shrimp and other seafood (but nowadays also used on eggs, French fries, fried chicken, popcorn, potato chips, salads and other food items). The mixture includes allspice, bay leaf, black pepper, cardamom, celery salt, cloves, ginger, mace, mustard, nutmeg, paprika and crushed red pepper flakes.

Panch phoron or *panch phoran* (literally "five spices") is a Bengali five-spice blend of whole fenugreek, nigella, fennel, cumin and black mustard or *radhuni* seeds (last-mentioned can be substituted with *ajowan* or celery seeds) Unlike most spice mixtures, *panch phoran* is not powdered but always used whole. The spices are fried in oil or ghee to release the flavours (i.e., tempered). The food ingredients (typically vegetables, chicken, beef, fish or lentils) are then added and become coated with the oily spice mixture as part of the cooking method to prepare the required dish.

Powder-*douce* or *poudre-douce* ("sweet powder") is the term used for medieval spice mixtures that may have contained cinnamon, grains of paradise, ginger, nutmeg and sugar.

Pudding spice – *see* **mixed spice**

Pumpkin pie spice is an American and Canadian blend of cinnamon, clove, nutmeg, ginger and allspice, used to flavour pumpkin pie, a traditional sweet dessert made from pumpkin-based custard baked in a single pie shell.

Qâlat daqqa or *gâlat dagga* is a North African (Tunisian) five-spice mixture comprising finely ground black pepper, grains of paradise, cloves, cinnamon and nutmeg. It is moderately hot and sweetish and is used to flavour meat dishes (as dry rub or marinade), stews and vegetable dishes.

Quatre épices (literally "four spices") is a French blend of ground pepper (black or white), powdered cloves, grated nutmeg and ground cinnamon. Some recipes suggest allspice instead of pepper or ginger instead of cinnamon. It is used in stews, terrines and game dishes.

Ras el hanout or *Rass el hanout* (literally "head of the shop" in Arabic, implying the best of the spices on offer) is a Moroccan and North African spice mixture used to flavour meat, rice and savoury dishes. There is no fixed recipe or combination but black pepper, cardamom, chilli peppers, cinnamon, cloves, coriander, cumin, nutmeg and turmeric are typical ingredients. In some areas the mixture may also include cubebs, dried rosebuds, grains of paradise or sumac. The ingredients are often toasted before they are finely ground and the powders then mixed.

Recado rojo or *achiote* **paste** is a Mexican spice mixture that includes annatto (= *achiote*, *Bixa orellana*), allspice, black pepper, cinnamon, cloves, cumin, garlic, Mexican oregano (*Lippia graveolens*) and salt. The red colour of meat or vegetable dishes seasoned with *recado rojo* comes from the annatto seeds.

Sambar masala or *sambar podi* is a dry spice powder mixture made from powdered pigeon pea dhal (and/or various other types of dhal) combined with dried red chilli peppers, tamarind, coriander, cumin, curry leaves, fenugreek seeds, turmeric, asafoetida and black pepper (optional). It is used to prepare *sambar*, a South Indian and Sri Lankan Tamil vegetable stew.

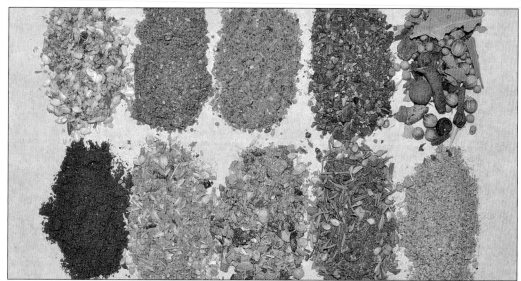

Examples of spice mixtures. Top row: Egyptian dukkah, Indian sambar powder, Indian garam masala, Ethiopian tea spice, and pickle spice. Bottom row: Ethiopian berbere, Cape Malay curry powder, Thai seven-spice, za-atar and barbeque spice

Shichimi, also called **Shichimi togarashi** ("seven flavour chilli pepper"), is a common Japanese spice mixture containing seven ingredients: coarsely ground red chilli peppers (the main ingredient), ground sansho (Sichuan pepper), roasted orange peel, black sesame seeds, white sesame seeds, hemp seeds, ground ginger and *nori* or *aonori* (edible seaweeds). *Ichimi togarashi* ("one flavour chilli pepper") is ground red chilli pepper with no other ingredients.

Tabil, typical of Arab cookery, is a mixture of three parts coriander and one part caraway, crushed with red peppers and garlic, which is dried in the sun and then powdered. It is used to spice mutton and semolina dishes, amongst others.

Taco seasoning is a commercial mixture of spices (including ingredients such as chilli powder, cumin, paprika, onion and garlic) used for flavouring a taco – the traditional Mexican dish consisting of a corn or wheat tortilla folded around a filling of meat or vegetables and often eaten with a topping of salsa, guacamole, cilantro, tomatoes, onions and lettuce.

Tandoori masala is an Indian, Pakistani and Afghan spice mixture used specifically with a tandoor (traditional clay oven). It varies from region to region but usually includes *garam masala*, lemon juice, cayenne pepper, garlic, ginger, onion and other spices. Pre-made *tandoori masala* is nowadays commonly sold in supermarkets.

Tea spice is a mixture of spices used to add flavour to tea (*chai*). It is especially popular in Ethiopia, the Middle East and countries around the Mediterranean region. The mixture varies but often contains cinnamon, cloves and cardamom, sometimes with herbs such as mint and thyme.

Tikka masala is a variable mixture of spices used to season chicken tikka, a popular dish of Indian (or British?) origin. Pieces of chicken are marinated in spices and yogurt, baked in a tandoor oven and then served in orange-coloured masala sauce made of cream, tomatoes, turmeric, paprika powder and other spices.

Vadouvan is a ready-to-use spice mixture, similar to Indian masala but with a French influence. This "French curry" has become popular in gourmet cooking. It is a combination of garlic, onion and shallots with typical masala spices such as chilli pepper, turmeric, cardamom and black pepper and also fresh herbs such as curry leaves.

Za'atar, zatar or **zahter** is a term that refers to 1: an individual herb (originally *Origanum syriacum* but nowadays also other Middle Eastern herbs with a similar flavour) and 2: a blended seasoning (condiment) made from dried herbs mixed with sesame seeds, salt and sometimes dried sumac (then resulting in red *za'atar*). *Za'atar* represents the Middle Eastern origins of the equally popular European oregano (e.g. as the main seasoning flavour of Italian pizzas).

33

Seasonings and condiments

Various products are used to enhance the taste, natural flavour or colour of food. These are usually classified as **seasonings** when they are added to the food while it is being prepared or especially sprinkled on afterwards to add more flavour, whereas **condiments** are served as accompaniments with the meal. They are often placed on the dinner table and applied by the diner, according to individual taste preference. Examples of well-known condiments include salt, black pepper, chilli pepper, horseradish sauce, hot sauce, brown sauce (e.g. HP sauce), chow-chow, achar, chutney, gherkins, ketchup, mint sauce, mustard sauce, olive oil, pickle, relish, sambal, soy sauce, Tabasco, teriyaki sauce, vinegar, wasabi and Worcestershire sauce. Ingredients such as truffles, dried fruits, alcohol, herbs and spices may also serve as condiments.

Seasonings are mixtures of natural substances (including herbs and spices) that are added to food to enhance the flavour and sometimes also to act as preservatives. There are five basic types of seasonings: saline (salt, seasoned salt and saltpetre), acid (vinegar, lemon juices), hot (black pepper, chilli pepper and other types of pepper), sweet (sugar, honey) and savoury or umami (fish sauce, soy sauce, hydrolyzed vegetable proteins, glutamate). Infused oils, flavoured with herbs and spices, are also used for seasoning. **Commercial seasonings** are blends of the five basic seasonings with various herbs and spices. They include "barbecue seasoning" (with celery seed, coriander, black pepper, red pepper, white pepper, paprika, salt, onion, garlic, brown sugar, cumin, and a natural smoky flavour), "Cajun seasoning" (salt and spices, including red pepper, paprika, garlic, and onion), "lemon and pepper seasoning" (with salt, black pepper, sugar, onion, citric acid, garlic, celery, lemon flavouring, and a sharp peppery taste and tart lemony flavour), "jerk seasoning" (sugar, salt and spices, including chilli pepper, allspice, thyme, onion and a strong allspice flavour), "Italian seasoning" (with basil, oregano, rosemary, sage, thyme and a pungent spicy flavour), "Chesapeake Bay-style seafood seasoning" (with celery salt, paprika and other spices, including mustard, chilli pepper, black pepper, bay leaves, cloves, allspice, ginger, mace, cardamom, cinnamon and a hot, spicy, pungent flavour), "Creole seasoning" (with salt, paprika, granulated garlic, cayenne pepper, ground black pepper, mustard, fresh lemon peel, ground bay leaf, filé powder and a salty, hot and spicy, peppery and garlic flavour), "Southwest seasoning" (with salt, paprika, chilli powder, cumin, coriander, cayenne pepper, ground black pepper, crushed red pepper, granulated garlic and a salty, spicy, chilli pepper flavour), "pizza seasoning" (with Parmesan cheese, garlic, onion, red pepper, thyme, basil, oregano and a cheesy, medium pungent, "Italian herb" flavour), "fajita seasoning" (with salt, garlic, onion, red pepper, natural lime flavour and spices, including black pepper, cumin and oregano, and a spice flavour and citrus aroma).

Seasoned salt is a mixture of salt and a diversity of powdered spices, which may include black pepper, paprika, garlic powder, onion powder, mustard, celery seeds, parsley, thyme, turmeric, marjoram and others, sometimes with cayenne pepper, asafoetida and even soup powders or meat (chicken extracts). It often contains monosodium glutamate (MSG) or soy sauce powder to give a savoury (umami) taste. Seasoned salt, often yellowish in colour, is sold in supermarkets under various brand names as seasoning salt, season salt or general purpose seasoning (e.g. the Swiss-made "Aromat", a well-known brand name in Europe and South Africa). These products are commonly used in fish-and-chips shops and other fast food or take-away food outlets. "Chip spice" is an English seasoned salt that contains spices and tomato powder. In Australia and New Zealand, seasoned salt is known as "chicken salt" because it usually (but not always) contains chicken extracts as the second main ingredient. **Celery salt** is a mixture of table salt or sea salt and ground fruits ("seeds") of celery or sometimes lovage. It is used in cocktails, hot dogs, salads, stews and as an ingredient of seasonings. **Garlic salt** is simply salt flavoured with one part dried garlic powder for every three parts of salt. It usually also contains an anti-caking agent such as calcium silicate. Garlic salt or ground garlic powder can be used as substitutes for fresh garlic. *Sharena sol* (literally "colourful salt") is a Bulgarian table condiment comprising salt, paprika and savory (*chubritsa*) instead of

Examples of popular condiments: pepper, salt, soy sauce, brown sauce, chutney, tomato sauce (ketchup), Worcestershire sauce and chilli pepper sauce

salt and pepper. ***Svanuri marili*** ("Svanetian salt") is a Georgian table condiment made from salt mixed with dried and powdered spices, including hot chilli peppers, garlic, coriander, dill and blue fenugreek.

Sauces are liquid or semi-liquid preparations that are added to food to improve flavour and appearance and to add moisture. Sauce-making has been developed to a fine art in France, where the sauce chef (*saucier*) is considered to be the solo artist in the orchestra of the kitchen. In the French tradition there are five primary types of sauces that give rise to endless variations (secondary sauces) when other ingredients are added. These are béchamel (milk-based, thickened with a white roux), *espagnole* (fortified brown veal stock), velouté (white stock, thickened with a roux or a *liaison*), hollandaise (emulsion of egg yolk, butter and lemon or vinegar) and *tomate* (tomato-based). Sauces may be served cold (mayonnaise, apple sauce, *pistou*), lukewarm (pesto) or warm (most of them). Pan sauces are obtained by deglazing a pan. Many commercial, ready-made sauces are available, including ketchup (tomato sauce) and soy sauce, or branded inventions ranging from classics such as HP sauce and Worcestershire sauce to modern derivatives, such as Japanese *okonomiyaki*, *tonkatsu* and *yakisoba*. Sauces made from stewed fruit (apple sauce, cranberry sauce) are often eaten with meat dishes (pork, poultry) or served as desserts. Traditional British sauces include "gravy" (eaten with roasts), spiced "bread sauce", apple sauce, mint sauce, horseradish sauce, English mustard and custard (on desserts). Ketchup, brown sauce (e.g. HP sauce) and Worcestershire sauce are commercial inventions that have found their way to the USA, Australia, South Africa and many other parts of the world.

Italian sauces are many and varied – some are used with meat and vegetable dishes (e.g. *bagna càuda*, *gremolata* and *salsa verde*) or with pizza (e.g. carbonara, pesto and *ragù alla Bolognese*). Famous dessert sauces are *crema pasticcera*, *crema al mascarpone* and *zabajobe*. Japanese sauces are mostly based on *shōyu* or soy sauce (e.g. *ponzu*, *yakitori*) or miso (e.g. *amamiso*, *gomamiso*). Wasabi or *wasabi-joyo* is used with sushi and sashimi. Examples of Chinese sauces include chilli sauces, *douban jiang*, hoisin sauce, oyster sauce, sweet bean sauce and sweet-and-sour sauce. Thai and Vietnamese fish sauces are made from fermented fish. Indian, Pakistani and other South Asian sauces are based on chilli peppers, tomatoes, tamarind, coconut milk and chutneys. Spanish salsa, such as *pico de gallo*, *salsa verde* and *salsa roja* are made from tomatoes, onions and spices. Mexican moles usually include chilli peppers and sometimes chocolate.

Green sauce is a term used for various sauces made from green herbs as the main ingredients. *Grüne Soße* (Green Sauce) is a German speciality mainly associated with the cities of Frankfurt am Main and Kassel and is made with a fixed combination of seven different fresh herbs (borage, chervil, chives, cress, parsley, salad burnet and sorrel) mixed with oil, hard-boiled egg yolks, vinegar, sour cream and white pepper. It is served cold with young potatoes, meat (typically *Frankfurter Schnitzel*), fish or vegetables. *Salsa verde* in Italy refers to a cold green sauce made from chopped (nowadays blended) parsley leaves with garlic, onion, capers, anchovies, vinegar and olive oil. It is served as a condiment or dipping sauce. Mexican *salsa verde* is a hot or cold green sauce made with tomatillos and chilli peppers. Mint sauce is a British and Irish sauce made from chopped spearmint leaves and traditionally eaten with roast lamb. *Gremolata* is an Italian (Milanese) green sauce made with parsley and traditionally served with braised veal shank. *Chimichurri* is an Argentinian and Uruguayan sauce made from parsley or cilantro and used with roast meat. Pesto originated in Genoa (northern Italian) and is made from pounded basil, garlic, pine nuts and olive oil. The French (Provençal) *pistou* resembles pesto but is made without pine nuts. Pesto seems to be based on *moretum*, an ancient Roman herb and cheese sauce eaten with bread. The herbs were typically pounded (as is still done today when making pesto), hence the names *pesto* and *moretum* (the origin of the words "pestle" and "mortar"). *Sauce verte* is a French mayonnaise made with a purée of mixed herbs.

Dips or **dipping sauces** are liquid condiments used in all parts of the world to add flavour to various finger foods such as bread, crackers, pita bread, *injera*, dumplings, falafel, crudités, seafood, pieces of meat and cheese, potato chips and tortilla chips. Examples of popular dips include southern Mediterranean aioli, Tex-Mex *chili con queso* (a dip for tortilla chips), Indian chutney, European fondue, American "French onion dip" or "California dip", Mexican guacamole, Levantine hummus, European and American ketchup or tomato sauce, Italian marinara sauce, European mayonnaise and mustard, Thai *nam chim*, Greek spiced olive oil (a dip for bread), Spanish/Mexican salsa (for tortilla chips), sour cream dip (for potato chips), East Asian soy sauce, North American spinach dip (for tortilla chips and vegetables), Thai and Vietnamese *sriracha* sauce (a type of hot sauce), East Asian sweet and sour sauce, European tartar(e) sauce, Japanese *tentsuyu*, Greek tzatziki or Turkish and Serbian *tarator* and Japanese wasabi sauce (for sushi and sashimi).

Pickles and preservatives were vital in the days before refrigeration became available and were important methods of preserving food, especially in hot climates. The use of salt, vinegar and other preservatives date back to ancient times. Pickles are often referred to as *achar*, the Hindi word for pickle. **Chutney** or *chatney* is a general term for a wide range of Indian condiments containing fruits, vegetables and spices in almost limitless combinations. Mango chutney, made from green mango fruits, is particularly popular. The word "chutney" is also used for Anglo-Indian commercial condiments containing chopped and cooked fruits and spices in vinegar. An example of a popular sweet and spicy chutney is "Major Grey's Chutney", made from mango, raisins, vinegar, lime juice, onion, tamarind, sugar and spices. The famous South African brand, Mrs HS Ball's chutney (locally called *blatjang*, dating from 1870) is made from apricots (or peaches), sugar, water, vinegar, starch, salt and caramel colourant. **Relish** refers to pickled or cooked fruits or vegetables used as a condiment. These include jams, chutneys and pickles (e.g. North American "relish", which is pickled cucumber jam). Famous British examples are "Gentleman's relish", a spiced anchovy paste dating from 1828 and "Branston pickle", a British commercial pickled chutney, first made in 1922 in the Branston suburb of Burton upon Trent. "Chow-chow" or "chowchow" is an American commercial pickled relish made from a combination of vegetables. "Piccalilli" is a Western-style Indian pickle or relish made from vegetables and spices. "Sambal" is the term used for a typical Indonesian condiment, a sauce that combines chilli peppers with various other ingredients, including shrimp paste, fish sauce, garlic, ginger, onions or shallots, palm sugar, lime juice and rice vinegar.

Sauerkraut (literally "sour cabbage") is a fermented condiment made from finely sliced cabbage. The unique flavour and texture are caused by lactic acid bacteria. Similar products include Korean *kimchi* or *kimchee*, Japanese *tsukemono* and Chinese *suan cai*. Many Chinese flavouring materials are made from pressed, fermented plants.

Essences are usually solutions of nature-identical flavour compounds in alcohol (e.g. vanilla essence is a solution of pure synthetic vanillin). **Extracts**, on the other hand, are complex mixtures of chemical compounds extracted from herbs or

Examples of sauces. Top row: wasabi paste, mayonnaise, English mustard, hummus, tartare sauce and chilli paste; bottom row: horseradish sauce, brown sauce, chutney, green mango chutney, tomato sauce and teriyaki sauce

spices (through distillation, solvent extraction or critical fluid extraction). Vanilla extract, for example, is a complex mixture of many flavour ingredients. Well-known essences are made to create the flavours of almond, citrus, coffee, lemon, orange blossoms, peppermint, pineapple, raspberry, strawberry and vanilla. Less well known are *kewra* (pandanus essence) and maraschino essence (a sweet liquid made from marasca cherries). The fragrant water left over after oil distillation is also popular as flavourants, such as *buchu* water, orange-flower water and rose water.

Vinegar is the most important and widely used condiment and flavourant. It is a mixture of water and acetic acid, produced by the fermentation of the alcohol (ethanol) in wine and other alcoholic beverages by acetic acid bacteria. Vinegars often retain some of the flavours of the grapes or other fruits from which the wine was made. The most widely used vinegars are wine vinegars (made from white or red wine), balsamic vinegar (made from the must of white trebbiano grapes and aged in wooden caskets for many years) and white vinegar or "distilled vinegar" (made by the fermentation of distilled alcohol, obtained from malted corn/maize). Vinegar is also made from beer in Austria, Germany and the Netherlands, from ale in the United Kingdom (where salt and malt vinegar are the traditional seasoning for fish and chips), sugarcane juice or palm sugar in Brazil and the Philippines, coconut water in Southeast Asia, dates or raisins in the Middle East, rice in Japan and rice, wheat, millet or sorghum in China. Chinese (Cantonese) sweetened vinegar

is made from rice wine, sugar, cloves, ginger, and other spices. Herb vinegars are popular in the Mediterranean region, flavoured with thyme, oregano or tarragon and easily made at home by adding a few springs of herbs to pale-coloured and mild-tasting white wine vinegar. Vinegar may also be flavoured with chilli peppers, garlic and other spices. Vinegar is not only a popular condiment but an important ingredient of marinades, pickling mixtures, vinaigrettes and other salad dressings, mustard, ketchup and mayonnaise. *Vinagrete*, for example, is the name given in Brazil to barbecue sauces made from vinegar and olive oil with tomatoes, onions and parsley.

Alcoholic beverages are widely used in cookery. **Wine** is not only an indispensable component of a good meal, but also an essential flavour ingredient in many dishes. Red wine is used in coq au vin, game ragouts, stews, marinades and cooked pears. It accompanies garlic, onions and mushrooms in thickened sauces. White wine (preferably dry and acidic) is used for cooking fish, seafood dishes and white meats. Mirin is a sweetened, viscous, sherry-like rice wine used in Japanese cooking. High-alcohol beverages such as clear spirits (aquavit, arrack, brandy, Calvados, cane spirit, gin, grappa, ouzo, rum, sake, sambuca, schnapps, slivovitz, tequila, vodka, whisky and others) are not only useful to flambé desserts but also to give extra flavour or character to a wide range of savoury and sweet dishes. Malted and fermented drinks such as stout, beer, cider, mead and palm wine all have their special functions in the kitchen. Absinth, cola, *geisho*, hops and quinine are examples of

Examples of different types of vinegar

plant-derived flavourants associated with these beverages. **Herbal extracts** (e.g. Angostura bitters) and **liqueurs** are useful sources of flavours in sweet desserts and confectionery because they are produced from many different herbs and spices (often in complicated and closely guarded secret combinations). Examples include advocaat, coffee liqueurs (Kahlúa, Tia Maria), crème liqueurs (crème de bananes, crème de cacao, crème de cassis, crème de menthe), fruit and nut liqueurs (amaretto, cherry brandy, Cointreau, Curaçao, Frangelico, Grand Marnier, kirsch, Malibu, maraschino, parfait amour, Van der Hum), herb and spice liqueurs (anise, anisette, Benedictine, chartreuse, Galliano, goldwasser, kümmel, pastis, Pernod, sambuca, Strega) and whisky-based liqueurs such as Drambuie, Irish Mist and Southern Comfort.

Non-alcoholic beverages contain water-soluble flavour compounds. Various types of tea (made from fresh or more often fermented leaves) are used in cooking, such as black tea or chai, yerba maté, rooibos tea, hibiscus tea and honeybush tea. The process of fermentation in this case is actually an enzymatic oxidation during which phenolic compounds are modified and various flavour components are produced. Coffee, guarana, cola and cacao are not only used for their stimulant action but they are also important sources of flavour. Fruit juices, vegetable juices, soft drinks and herbal infusions of bergamot, chamomile, dandelion, elderflower, lemon balm, lemon verbena, lovage, mint, rosehip, rosemary and yarrow can all be used in the kitchen to add flavour and/or colour to dishes.

Food colourings are used to make food more attractive or to replace colours lost during preparation. Negative perceptions about the health and safety of artificial colourants are leading to an increasing range of natural food dyes. Well-known examples include alkanet (*Alkanna tinctoria*) for red colour, annatto for red or yellow (in butter and cheese), saffron in paella, pandan in green sauces, beetroot in Russian and Polish soups, butterfly pea (*Clitoria ternatea*) for a blue colour in Thai desserts and drinks, elderberry (*Sambucus nigra*) formerly in red wine, as well as marigold (*Calendula officinalis*) and safflower (*Carthamus tinctorius*) in Georgian cuisine. In China, there are three important colourants: red fermented rice (as powder or water extract), fragrant herb (*xiang cao*, which is the dried and powdered leafy shoots of fenugreek – it gives a green colour to steamed pastries) and gardenia fruits (soaked in water, they give a yellow dye). Gardenia fruits are used to colour bean curd and pea meal (used to bake the well-known bright yellow mooncake of Beijing).

Garnish is the term used for a decoration or embellishment of a dish or drink, mainly to improve the visual impact by adding colour but often also to add extra or contrasting flavours, textures, fragrances (and sometimes even taste). Garnishes are not always edible but they often are. Borage flowers, nasturtium flowers, parsley and chives are examples of traditional garnishes for dishes, while slices of lemon, citrus peels, olives and cherries go with drinks and cocktails. The use of microgreens (seedlings of edible herbs) to embellish dishes and salads has become fashionable in recent years.

The chemistry of taste and flavour

Eating and drinking is a sensory experience – visual, tactile, olfactory and gustatory. The last two, smell and taste, are the most important, and come into play to detect the chemical compounds responsible for the flavour of the food or drink. Smell makes the most important contribution to flavour detection and the combination of aroma compounds in any given food or drink can lead to a potentially limitless number of different flavours. In contrast, there are only five basic taste sensations: sweet, salty, sour, bitter and savoury (umami). Tasting and smelling are two completely separate physiological processes. Taste is perceived in specialized taste buds on the tongue and other parts of the mouth, while smells are captured by specialized receptor cells located in the top of the nasal canal. Only volatile compounds with a small molecular weight (less than about 400) that are able to diffuse into the air can be detected as aromas. It is therefore surprising that 80% of what we call taste is actually produced by the aroma of food, which the brain associates with that food. The process of smelling is different from tasting, because smelling occurs during the process of inhalation, while flavour perception occurs during exhalation. The aroma of a food item outside of the mouth therefore differs from the aroma when it has already been taken into the mouth. The art of cooking is the creation of unique combinations of tastes and flavours, integrating and balancing the inherent tastes and flavours of the food items with the addition of seasonings, and aroma-rich and/or tasty herbs and spices.

Many culinary herbs contain essential oil as their main flavour compounds. Examples of plant families with a high frequency of occurrence of volatile oils are the celery family (Apiaceae), mint family (Lamiaceae) and citrus family (Rutaceae). The oils are typically formed in pellucid glands within the leaves (visible as translucent dots when the leaf is held against the light) or in glandular hairs on the leaf surface. In the case of the Apiaceae, the oil is formed in resin ducts that occur in the leaves and fruits. Specialized oil cells (called vittae) are a special feature of the small dry fruits in this family. These fruits (often wrongly referred to as "seeds") are called schizocarps because they usually separate into two halves, the mericarps, at maturity. Well-known examples include anise, caraway, celery, coriander, cumin, dill and fennel. In *Citrus* species (lemons and limes), oils are not only formed in the leaves (petitgrain oils) but also in the glandular coloured part of the fruit rind (peel oils). Fresh peels are an essential component of many dishes and are used in the form of zest – the coloured part of the peel (*flavedo*) is grated, leaving behind the spongy white layer (*albido*). The dried peel of tangerine is an important spice in southern Chinese and Cape cooking (used with sweet potato).

Essential oils are often complex mixtures of up to 60 or more individual chemical compounds and it may be hard to predict which component (or combination of compounds) is responsible for the

characteristic aroma and flavour of the herb. In some cases it is easy: **limonene** gives the typical lemon flavour, (−)-**carvone** spearmint flavour, **anethole** anise flavour, **menthol** peppermint flavour and **camphor** the typical camphor smell.

limonene   (−)-carvone   menthol   anethole   camphor

Of interest are volatile sulphur-containing compounds such as **allicin**, found in garlic, **dimethyl sulphide** in the aroma of both white and black truffles (although the typical mushroom flavour is mainly due to **1-octen-3-ol**, also known as mushroom alcohol). Another group of flavour compounds is represented by **allyl isothiocyanate** and other volatile mustard oils, produced by plants such as mustard, horseradish and wasabi (family Brassicaceae). Headspace analysis and other chromatographic techniques are continuously expanding our understanding of the aroma impact

allicin   dimethyl suphide

1-octen-3-ol
(mushroom alcohol)   allyl isothiocyanate

of individual essential oil compounds. In some cases the aroma impact is produced by a minor constituent that occurs only in trace amounts but has a very low threshold of detection. In other cases the unique aroma may be the result of an additive or even synergistic effect of several volatile compounds.

When heat is applied to some foods, a non-enzymatic browning occurs, and hundreds of new and pleasant flavour compounds are created. Two different and as yet poorly understood processes may be involved, depending on the level of heating and the substances that are heated. The one is caramelization, also called toasting, a partial burning process (pyrolysis) through which sugar turns brown and releases volatile compounds with a typical caramel flavour. Many of us make use of this process every day when we make toast for breakfast. The sugars in the bread are caramelized, resulting in the familiar brown, crisp and tasty outer layer. The other process is roasting (called the Maillard reaction), a chemical reaction between amino acids and reducing sugars caused by heat. The most familiar examples are the roasting of coffee beans and the tempering (brief roasting) of herbs and spices during the initial phases of cooking to release the flavours. Complex mixtures of aroma and flavour compounds are formed when the nucleophilic amino groups of amino acids react with the carbonyl groups of the sugars. The type of compounds that form depends on the type of amino acids that are present. The Maillard reaction (often in combination with caramelization) is responsible for the flavour and taste of toasted bread, French fries, malted barley, roasted coffee, roasted meat and maple syrup. One of the common flavour compounds created by the Maillard reaction when baked goods such as biscuits, bagels and tortillas turn brown or when popcorn is heated is **6-acetyl-2,3,4,5-tetrahydropyridine**. A similar flavour is found in **2-acetylpyrroline**, a substance that occurs naturally in fragrant rice and in the wilted leaves of the fragrant pandan (*Pandanus amaryllifolius*). The minimum level of detection of both these aroma compounds is less than 0.06 nanograms per litre. (An average human cell weights 1 ng, a grain of maize pollen 250 ng). Enzymatic browning of food is generally unwanted (e.g. the browning of bananas or the surface of cut apples) but it plays a crucial role in the colour and flavour of tea and dried fruits such as raisins and figs.

Another important and complicated method commonly used to create new foods and food flavours is through controlled fermentation. During the process, sugars are converted to alcohol or organic acids and proteins are broken down to amino acids through the action of microorganisms (yeasts and bacteria). It is used in a very wide range of foods and is often responsible for the umami taste (e.g. in fish and soy sauce). Examples include products based on fermented beans (miso, *nato*, soy sauce, soy paste, tofu), grains (beer, kvass, bread, *injera*, rice wine, whisky, vodka), fruits (wine, vinegar, cider, brandy), vegetables (sauerkraut, *kimchi*, pickles), honey (mead, *tej*), tea (kombucha), milk (yogurt, quark, cheese), fish (*garum*, fish sauce, shrimp paste) and meat (chorizo, *jamón ibérico*, pepperoni, salami). Vegetable proteins may also be broken down to amino acids and associated umami flavour compounds through direct acid hydrolysis, resulting in so-called acid-hydrolysed vegetable protein or HVP. The dark brown liquid that is formed contains glutamic acid (the well-known flavour enhancer) and other natural amino acids and is nowadays commonly used in the manufacture of commercial soy sauce. A similar result is achieved by processing yeast proteins, resulting in dark brown savoury products well known under brand names such as Marmite and Vegemite (see notes under umami, the fifth taste).

The senses of touch (tactile or somatosensory detection) and vision are also relevant. Pungency, the "hot" or "spicy" taste of chilli peppers, black pepper, ginger and horseradish is traditionally viewed as a sixth taste in India and China but it is technically not a taste because the burning sensation does not arise from taste buds but from a direct stimulation of somatosensory nerve fibres in the mouth (those responsible for the detection of pain and temperature). There are several different somatosensory effects that contribute to the overall impression of a food item. A pungent and burning sensation is caused by **capsaicin** in chilli pepper, allyl isothiocyanate in mustard oils and **piperine** in black pepper. An imaginary cool sensation is typical of menthol in peppermint and camphor in

6-acetyl-2,3,4,5
-tetrahydropyridine

2-acetylpyrroline

camphor oil, while a tingling or "tingling numbness" sensation is created by substances such as **spilanthol** (in spilanthes, *Spilanthes acmella*) and **hydroxy-α-sanshool** in Sichuan pepper (*Zanthoxylum piperitum*). An astringent or "puckering" sensation is commonly caused by tannin-rich foods and wine. A "metallic" taste (usually unacceptable) is caused by certain foods. The Scoville scale is a measurement of the pungency of chilli peppers, created by Wilbur Scoville in 1912. The number of times a measured amount of pepper extract has to be diluted to make the capsaicin undetectable gives the Scoville heat units (SHU). Pure capsaicin, for example, has a Scoville rating of 16 000 000, while the rating for extremely hot chilli peppers exceeds 200 000. The method is imprecise and has largely been replaced by direct measurement of the level of pungent chemical(s) using high-performance liquid chromatography.

capsaicin

piperine

spilanthol

hydroxy-α-sanshool

Two cultural factors may result in the same taste experience being described as pleasant or unpleasant. Firstly, the temperature at which the food or drink is served must match the expectation or custom. Secondly, the appearance and colour of the food items must be attractive. Red or orange-coloured drinks, for example, tend to be perceived as sweeter than green-coloured drinks even if they are otherwise identical. The principle is also used in charcuterie (a reddish purple or pinkish colour is desirable in processed meat) and in dairy products (where a yellow colour is called for in butter and cheese).

The ability to taste, known as gustatory perception or gustation, has an important survival value for humans and animals. A sweet taste indicates energy-rich food; an umami taste indicates the presence of important amino acids (the building blocks of proteins); a salty taste helps to regulate the electrolyte balance in the body; a sour taste may give a signal that the food is fermented and perhaps no longer safe; a bitter taste gives a warning of the presence of natural toxins – the majority of poisonous substances are intensely bitter. There are genetic differences in the ability to taste. Persons with an exceptionally sensitive sense of taste are called supertasters. Children are much better at tasting than adults, because about half of our taste buds are lost before the age of twenty. Some foods or drinks may have an aftertaste that can be different to that of the food or drink itself. An acquired taste is a liking for a food or drink that develops over time after an initial phase of dislike, caused by some unfamiliar aspect such as a strange odour, taste or appearance.

Humans and animals can taste food because of the presence of a complicated sensory system comprising thousands of taste buds in the mouth and throat, with nerves connecting them to the brain where the chemical signals coming from the taste buds (each with 50 to 100 or more taste cells) are interpreted. The taste buds are the main sensory organs and are highly concentrated on the top of the tongue, where they are located in thousands of small papillae, visible to the naked eye. Each papilla contains hundreds of taste buds. Taste buds are also present on the roof, sides and back of the mouth and in the throat. The idea that certain parts of the tongue have different taste buds for detecting sweet, sour, bitter or sour taste is a common misconception. The density of taste buds varies in different parts of the mouth and tongue but the taste buds themselves are all the same.

The chemical compounds responsible for taste dissolve in saliva and can then enter the taste buds through pores. Once inside the pore, they interact with either ion channels or with proteins on the surfaces of taste receptors. This interaction triggers chemical signals which are transmitted to the brain through the central nervous system. Sweet-tasting molecules trigger a response by binding to T1R2 and T1R3 proteins in the taste cells; bitter molecules bind to T2R receptors, umami molecules bind to T1R1 and T1R3 receptors and both sour and salt molecules enter the taste cells and flow through ion channels. The various receptors in the taste cells and the flow rate of the active compounds allow the brain to distinguish between different tastes and also the magnitude of each taste.

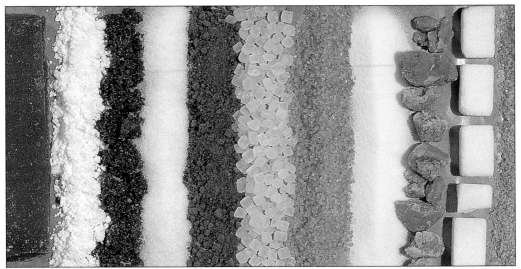

Various forms of sugar: Chinese lump sugar, icing sugar, treacle, cane sugar, Chinese brown sugar, sugar crystals, demerara sugar, castor sugar, palm sugar, sugar cubes and yellow sugar

Sweetness

Sweet tastes are triggered by sugars and a few other natural and artificial compounds. Nowadays we use pure sugars such as sucrose (table sugar) but the original source of sweetness was foods rich in carbohydrates such as honey, sweet-tasting fruits and sucrose-rich phloem sap from sugar cane and palm stems. Sugar and honey are important sources of energy. Sugars differ in their sweetness, usually measured relative to the sweetness of sucrose, which is taken as 1. Glucose is less sweet than sucrose (sweetness value of ca. 0.75) while fructose is sweeter (sweetness value of ca. 1.5). Natural sources of sugar such as molasses and palm sugar are not only used for the sugar they contain but also for a diversity of natural flavour compounds that are often essential for the success of certain dishes. **Sucrose, glucose, fructose** and

lactose (found in milk) are amongst the most important natural sugars in the human diet, while starch is the most important carbohydrate. It forms the bulk of staple foods such as wheat, maize (corn), rice, potatoes and cassava and the processed foods derived from them, including bread, pasta, noodles, porridge, pancakes and tortillas.

Starch is a polysaccharide consisting of a large number of glucose molecules arranged into **amylose** and **amylopectin** units. Amylopectin usually represents about 80% of the total starch, but some cultivars of maize and rice (known as waxy maize and glutinous rice) have only amylopectin and no amylose. The enzymes that hydrolyze starch (known as amylases) are present in human saliva and are also produced by the pancreas. Amylase from the surface of grains such as barley breaks starch down into **maltose** units during the process of malting (mashing). Cooking is necessary to improve the digestibility of starch, so that the use of fire was a prerequisite for the switch in the human diet from low to high starch. Human populations with starch-rich diets tend to have more amylase genes. Starch is commercially hydrolyzed by acids and/or enzymes into fragments known as dextrins. Examples are maltodextrin, an almost tasteless filler and thickener, and glucose syrup (known as corn syrup in the USA), used as sweetener and thickener, in processed food. A complete breakdown gives dextrose (commercial

sucrose

glucose

fructose

maltose

lactose

glucose). Dextrose solutions can be enzymatically modified (using isomerase) so that many of the glucose molecules are converted to fructose, thus producing so-called "high fructose corn syrup", which is the main sweetener used in the USA for processed foods and beverages.

amylose

maltose

amylopectin

The long-term intake of too many calories and lack of exercise lead to weight gain and poor health but eating too few calories can be equally harmful. The recommended daily intake for an adult is 2 000 calories and the recommended daily sugar intake is 90 grams. Each gram of sugar adds four calories to the diet (fat adds nine, alcohol seven) but it should be taken into account that fruits, vegetables, milk products and soft drinks all add to the daily sugar intake.

Natural and artificial non-sugar sweet substitutes such as **xylitol**, **saccharin** and **aspartame** have become popular to control weight and health, especially for diabetics. However, the excessive use of artificial sweeteners may disturb the body's natural ability to regulate carbohydrate intake and thus lead to weight gain rather than weight loss.

xylitol

saccharin

aspartame

Some plant glycosides are much sweeter than sugar, such as glycyrrhizin from liquorice root (30 times sweeter than sucrose) (see *Glycyrrhiza glabra*) and stevioside from stevia leaves (250 times sweeter than sucrose) (see *Stevia rebaudiana*). Thaumatin, a sweet protein from the West African katemfe fruit (*Thaumatococcus daniellii*) is 2 000 times sweeter than sucrose. The sweetest known chemical compound is a synthetic substance called **lugduname**, with a sweetness value up to 300 000. It has not yet been approved for human use. Some natural compounds are able to inhibit sweet taste perception, including gymnemic acid from the leaves of gymnema or gurmari vine (*Gymnema sylvestre*) and ziziphin from the leaves of the Chinese jujube (*Ziziphus jujube*). **Lactisole** is an artificial sweetness inhibitor used in some sweet jellies and fruit preserves to suppress the sweetness and bring out the fruit flavours. In contrast, miraculin, a tasteless glycoprotein from the miracle fruit (*Synsepalum dulciferum*) and curculin, a sweet-tasting protein from the lumbah plant (*Curculigo latifolia*) modify taste buds so that sour foods are perceived as sweet. The effect lasts up to one hour.

lugduname

lactisole

Sugar is a natural preservative, widely used in jams, jellies, syrups and sauces. White or brown sugar also has an important role in cooking because it acts like a seasoning to enhance and balance flavours. A pinch of sugar added to carrots, peas and sweet corn improves the taste. Sugar is often added to tomato dishes, soups and stews to reduce the acidity and improve the flavour. Chinese chefs use sugar to produce the typical sweet and sour tastes associated with meat dishes. Sugar is used in baking and candy-making, not only to ensure the desired texture but to contribute to taste and colour. When heated, sugar caramelizes – an effect seen (and tasted!) when toast turns brown. Sweet (and sour) tastes are often added to dishes in the form of dried fruits such as dates, figs, prunes and apricots.

Examples of sour-tasting flavourants. Top row: citric acid, white grape vinegar, mayonnaise, tamarind and lemon; Middle row: red and white grapes, lime, tamarind pulp, tamarind fruits and dried apricots. Front row: hibiscus, sour figs, kokum, opened tamarind fruit and dried apricots

Sourness

Sour taste or acidity is triggered by natural acids such as acetic acid, citric acid and tartaric acid. The sourness of a food or substance is measured by comparison to dilute **hydrochloric acid**, which has a sourness index of 1. Citric acid, present in lemons and limes, for example, has a sourness index of 0.46. The most common food sources containing acids are fruit such as apricots, lemons, limes, oranges, grapes, tamarind, sour figs and kokum. The choice of acid can influence the taste of food because they are slightly different in taste. Vinegar is the most widely used sour condiment. It contains **acetic acid**, which gives it a sour taste

and distinctive smell. **Ascorbic acid**, better known as vitamin C, is a useful food preservative present in lemons and other fruits and has a crisp and slightly sour taste. **Citric acid** is responsible for the acid taste of citrus fruits, while **oxalic acid** is found in rhubarb and sorrel. **Lactic acid** is present in fermented milk products (e.g. sour cream) and gives these products a distinctive and pleasant sour taste. **Malic acid** gives apples and apple cider its delicious sour taste, while **tartaric acid** does the same in grapes and wine (and **phosphoric acid** in carbonated soft drinks).

The balance between sweetness and acidity is an important aspect of cooking and the contrast is exploited by skilled chefs to create unforgettable dishes. The desired effect can be achieved by combining sour and sweet food items on the same plate or by incorporating them into a single dish, such as sugar and vinegar with sweet and sour-tasting fruits, as used in Chinese (Cantonese) sweet and sour pork. This dish demonstrates the Chinese philosophical principle of yin and yang – opposites that complement each other. Another example is the popular "sour candy", sweets in which sugar is combined with citric acid.

acetic acid

ascorbic acid
(vitamin C)

citric acid

oxalic acid

lactic acid

malic acid

tartaric acid

phosphoric acid

Saltiness

Salt or table salt is a crystalline mineral composed mainly of **sodium chloride** ($NaCl$). It is one of the oldest food seasonings. The word "salary" comes from the Latin word *salarium*, the money given to a Roman soldier to buy salt. Salzburg and Hallstatt in Austria are traditional centres of salt production (named after the German and Celtic words for salt – *Zalze* and *Hall*). Salt is provided in salt shakers or salt mills on most dinner tables around the world. Salt may be unrefined (e.g. sea salt) and therefore add some aroma compounds when sprinkled on food. Refined salt (e.g. table salt) may be 100% pure but sometimes contains small amounts of additives such as iodine, fluoride, iron or folic acid to counteract the symptoms of mineral deficiencies. Seasoned salts contain herbs, spices and flavour enhancers. In Chinese and other East Asian cuisines, salty soy sauce, fish sauce or oyster sauce are traditionally used for seasoning (while cooking) and not salt. Salt is essential to regulate the fluid balance of the body but excessive intake may increase the risk of high blood pressure. The recommended daily intake of salt in healthy adults is 4 000 mg (ca. 0.14 oz) per day (UK) and the upper limit is set at 5 750 mg (ca. 0.2 oz) in the USA, Canada, Australia and New Zealand. When a comparative measure of the saltiness of food is required, then sodium chloride is used as reference, with an arbitrary saltiness value of 1.

Salting is a method to preserve food using dry edible salt, similar to pickling with brine (salty water). Most microorganisms that cause food decay are unable to survive in high salt environments because of the high osmotic pressure. The brine often contains nitrites (saltpetre) because it gives otherwise grey-coloured pickled meat products such as salami and corned beef an attractive reddish colour. **Potassium nitrate** (KNO_3) is not only the original ingredient for making gunpowder and fertilizer but also one of several nitrogen-containing compounds referred to as saltpetre. **Sodium nitrate** ($NaNO_3$) and especially the more effective **sodium nitrite** ($NaNO_2$) are nowadays preferred for pickling and as a food ingredient to ensure food safety.

Bitterness

The ability of humans to detect bitter-tasting, toxic substances at very low levels must have been an important survival mechanism for our species. Bitterness is the most sensitive of the tastes and is usually regarded as unpleasant but many foods and drinks are inherently bitter or are intentionally spiked with bittering agents such as hops or various alcoholic preparations known as bitters. Examples include ale, beer, bitter gourd, chicory leaves, cocoa, coffee, citrus peel, dandelion leaves, marmalade, maté, olives and tonic water. **Quinine** is the bitter-tasting alkaloid in tonic water where it is added at a level of about 2.7 g per litre. This alkaloid is used as a reference standard to compare the bitterness of substances and has been assigned a bitterness value of 1. It has a detection limit of 0.0026 g per litre. The most bitter of all known substances is denatonium, which has a bitterness value of 1 000. **Denatonium benzoate** is added to industrial liquids as a bitterant and aversive agent to prevent people (especially children) from accidentally ingesting toxic liquids such as methanol and antifreeze (ethylene glycol). The ability to taste bitterness varies from person to person. Two synthetic substances (phenylthiocarbamide and 6-*n*-propylthiouracil) are tasteless to some persons but bitter to others and are commonly used to study the genetics of bitter perception.

In recent years there has been a revival of the concept of the *amarum* (bitter-tasting substances) or *aromaticum amarum* (aromatic bitters, such as "Angostura" tonic) that enhances appetite and good digestion. Bitter compounds not only stimulate the flow of saliva but the bitter taste in the mouth sends a signal (via the *nervous vagus*) to the stomach to increase the excretion of gastric juices. The ancient practice of eating bitter herbs or drinking bitter tonics has continued to this day and the customary gin and tonic before dinner makes perfect sense from a digestive point of view. The

sodium chloride

sodium nitrite

sodium nitrate

potassium nitrate

quinine

denatonium benzoate

Examples of different types of salt: English sea salt flakes, Himalayan pink rock salt, Swedish smoked sea salt flakes, Pakistani black rock salt (*kala namak*) and Hawaiian black lava salt (sea salt blended with activated charcoal)

art of balancing the bitterness of food items such as olives and chicory leaves with salty, sweet, sour or umami tastes is at the heart of many culinary customs and traditional recipes.

Umami

Umami is a savoury or "meaty" taste caused by amino acids, which results in a mouth-watering effect and a long-lasting aftertaste that balances and integrates all the flavours of a dish. It is one of the five basic tastes which also include sweet, sour, salty and bitter. Umami was first discovered by the Japanese scientist Kikunae Ikeda in 1908, who created the term from the Japanese words *umai* ("delicious") and *mi* ("taste").

glutamic acid

monosodium glutamate (MSG)

γ-aminobutyric acid (GABA)

inosine-5'-monophosphate

guanosine-5'-monophosphate

The taste is caused by **L-glutamate**, an abundant naturally occurring non-essential amino acid, and also by 5'-ribonucleotides such as **guanosine-5'-monophosphate** (GMP) and **inosine-5'-monophosphate** (IMP).

Umami was scientifically recognized as a fifth taste, distinct from saltiness, as recently as 1985. The ability to detect the umami taste is due to the presence of specialized receptors on human and animal tongues. Glutamate in its acid form has almost no effect, but the salts of **glutamic acid**, known as **glutamates**, ionize to give the distinctive taste of umami, detected on the tongue as the carboxylate anion of glutamate. For this reason the sodium salt of glutamate (**monosodium glutamate or MSG**) is used in commercial flavour enhancers. The overall taste-enhancing effect is strengthened by the simultaneous use of GMP and IMP which act with MSG in a synergistic way – the taste intensity is increased beyond that of the sum of the two ingredients.

Yeast extracts or **yeast spreads** are the common names for different types of yeast products used as flavour enhancers and food additives. They are made by natural enzymatic processes (autolysed yeast) or by acid hydrolysis (hydrolysed yeast), during which the proteins of yeast cells (mostly brewer's yeast, *Saccharomyces cerevisiae*) are broken down to amino acids and peptides, in much the same way as soy sauce and soy pastes are made. The products are in the form of liquids, pastes and powders and contain guanosine monophosphate

Well-known bitter tasting products | Condiments used for an umami (savoury) taste

(GMP) as a major umami flavour compound. Yeast autolysates are well known by their brand names in various countries, including Marmite and Oxo (UK, Ireland, South Africa and New Zealand), Vegemite (Australia), Cenovis (Switzerland) and Vitam-R (Germany). These products are almost unknown in the USA and other countries.

There is some controversy around the use of MSG and the addition of yeast extracts as natural flavourants. Some people are sensitive to MSG and numerous anecdotes of headaches and other side effects have been reported since 1968 by people who have eaten a Chinese meal, so that the syndrome became known as the Chinese Restaurant Syndrome. There is as yet no convincing scientific evidence that the levels normally used in food or naturally present in food are harmful except when large amounts are taken without food. It is interesting to observe the link between MSG and **GABA (gamma-aminobutyric acid)**, a well-known inhibitory neurotransmitter that can be formed from MSG by a single decarboxylation step. The mind- and mood-enhancing effects of substances such as capsaicin (see chilli), myristicin (see nutmeg) and MSG, and their role in the pleasure of the dining experience remain as a fascinating topic for future research.

The umami taste has been used since ancient times because it occurs naturally in many different food items. A fermented fish sauce known as *garum* was popular in Roman times. This traditional source

of glutamate is also employed in Chinese and other Asian cuisines where soy sauce, fish sauce, shrimp paste and other fermentation products are essential ingredients. In Western countries, ripe tomatoes, cheese, mushrooms and cured meats, all rich in glutamates and inosinates, serve the same purpose. Recent inventions of black sauces and pastes (e.g. Worcestershire sauce, Maggi sauce, Marmite and Vegemite) were inspired by the ancient custom of fermenting fish and soy proteins. The process is nowadays accomplished in three days by acid hydrolysis (using hydrochloric acid) of vegetable proteins (often wheat protein) to produce hydrolysed vegetable protein (HVP), rich in glutamates. The synergistic effects brought about by the umami taste explain why certain food combinations have become classics. Umami was first discovered in the basic Japanese broth known as *dashi*, in which there is a synergistic interaction between the glutamate in fermented *kombu* seaweed and the ribonucleotides in dried bonito (fish) flakes. Similar food parings of other culinary traditions come to mind, where vegetables and cheese, both rich in glutamates are combined with beef, pork, chicken or fish, all rich in inosinates. Examples are Parmesan cheese with tomato and mushrooms in Italian pastas and pizzas, leeks and chicken meat in Chinese chicken soup and Scottish cock-a-leekie, and cheese and beef in an American cheeseburger. An excellent dish also requires the correct choice and amount of herbs and spices that will create balance, add complexity or round out the overall flavour even more.

References and further reading

There are literally hundreds of excellent general books on the history, cultivation and culinary uses of herbs and spices. These can easily be located on publishers' websites. In terms of books on culinary traditions, the *Culinaria* series comes to mind as a spectacular and excellent source of information (first published by Könemann but now by H.F. Ulmann, an imprint of Tandem Verlag). The list of titles is continually expanding and includes The Caribbean, China, European Specialities, France, Germany, Greece, Hungary, Italy, Russia, Southeast Asian Specialities and Spain.

The following is a short list of reviews and other useful additional references.

Andrew, D. 2002. *Dangerous tastes: The story of spices.* University of California Press, Berkeley.

Blank, I., Milo, C., Lin, J., Fay, F.B. 1999. Quantification of aroma-impact components by isotope dilution assay recent developments, pp. 63–74. In: Teranishi, R., Wick, E.L., Hornstein, I. (Eds), *Flavor chemistry: 30 years of progress.* Plenum Publishers, New York.

Borget, M. 1993. *Spice plants.* The Tropical Agriculturalist Series. Macmillan, London.

Brown, D. 2008. *RHS encyclopedia of herbs.* Dorling Kindersley, London.

Burkill, I.H. 1966. *A dictionary of the economic products of the Malay Peninsula*, Vol. 1 & 2. Crown Agents for the Colonies, London.

Couplan, F., Duke, J. 1998. *The encyclopedia of edible plants of North America.* McGraw-Hill, New York.

De Guzman, C.C., Siemonsma, J.S. (Eds). 1999. *PROSEA Handbook 13. Spices.* PROSEA, Bogor.

Delwiche, J. 2004. The impact of perceptual interactions on perceived flavor. *Food Quality and Preference* 15: 137–146.

Farrel, K.T. 1999. *Spices, condiments and seasonings.* Aspen Publishers, Gaithersburg, USA.

Franke, W. 1997. *Nutzpflanzenkunde. Nutzbare Gewächse der gemäßigten Breiten, Subtropen und Tropen.* George Thieme Verlag, Stuttgart.

Geha, R.S., Beiser, A., Ren, C., Patterson, R., Greenberger, P.A., Grammer, L.C., Ditto, A.M., Harris, K.E., Shaughnessy, M.A., Yarnold, P.R., Corren, J., Saxon, A. 2000. Review of alleged reaction to monosodium glutamate and outcome of a multicenter double-blind placebo-controlled study. *The Journal of Nutrition* 130 (4 S Supplement): 1058S–1062S.

Gernot Katzer's Spice Pages (http://gernot-katzers-spice-pages.com).

Harborne, J.B., Baxter, H. 2001. *Chemical dictionary of economic plants.* Wiley, New York.

Hill, T. 2005. *The spice lover's guide to herbs and spices.* John Wiley & Sons, Hoboken.

Hu, S.-Y. 2005. *Food plants of China.* The Chinese University Press, Hong Kong.

Hutton, W. 1997. *Tropical herbs and spices.* Periplus Editions, Singapore.

Janick, J., Simon, J.E. (Eds). 1993. *New crops.* Wiley, New York.

Kaye, J. 2000. *Symbolic olfactory display.* M.Sc. thesis, Massachusetts Institute of Technology.

Kiple, K.F., Ornelas, K.C. (Eds). 2000. *The Cambridge world history of food.* Cambridge University Press, Cambridge.

Larousse. 1999. *The concise Larousse gastronomique.* Hamlyn, London.

Margvelashvili, J. 1991. *The classic cuisine of Soviet Georgia.* Prentice Hall Press, New York.

McVicar, J. 2007. *Jekka's complete herb book.* Kyle Cathie Limited, London.

National Geographic. 2008. *Edible. An illustrated guide to the world's food plants.* National Geographic, Washington.

Ortiz, E.L. 1992. *The encyclopedia of herbs, spices, & flavourings.* Dorling Kindersley, London.

Phillips, B. 2013. *The book of herbs.* Hobble Creek Press, Springville.

Porcher, M.H. et al. 1995–2020. *Multilingual multiscript plant name database.* University of Melbourne. (http://www.plantnames.unimelb.edu.au).

Rogers, J. 1990. *What food is that? And how healthy is it?* Weldon Publishing, Sydney.

Seideman, J. 2005. *World spice plants. Economic usage, botany, taxonomy.* Springer, Heidelberg.

Simpson, B.B., Ogorzaly, M.C. 2001. *Economic botany: Plants in our world*, Chapter 8. Spices, Herbs, and Perfumes (3rd ed.). McGraw-Hill International Edition. New York.

Smartt, J., Simmonds, N.W. (Eds). 1995. *Evolution of crop plants* (2nd ed.). Longman Scientific and Technical Publishers, London.

Swahn, J.O. 1999. *The lore of spices. Their history, nature and uses around the world.* Senate Publishing, London.

Toth, S., Mussinan, C. (Eds). 2012. *Recent advances in the analysis of food and flavors.* ACS Symposium Series Vol. 1098. American Chemical Society.

Vaughan, J.G., Geissler, C.A. 1997. *The New Oxford Book of Food Plants.* Oxford University Press, Oxford.

Van Wyk, B.-E. 2005. *Food plants of the world.* Timber Press, Portland.

Westphal, E., Jansen, P.C.M. (Eds). 1989. *Plant resources of South East Asia: A selection.* Pudoc, Wageningen.

Wiersema, J.H., Leon, B. 2013. *World economic plants. A standard reference* (2nd ed.). CRC Press, Boca Raton.

Zeven, A.C., De Wet, J.M.J. 1982. *Dictionary of cultivated plants and their regions of diversity.* (2nd ed.). Centre for Agricultural Publishing and Documentation, Wageningen.

Zohary, D., Hopf, M. 2000. *Domestication of plants in the Old World* (3rd ed.). Clarendon Press, Oxford.

The Herbs and Spices
(in alphabetical order by plant name)

The following icons are used to indicate the main use or uses of each species:

 herb

 spice

 both herb and spice

 food colourant (natural dye)
or flavourant

Aframomum corrorima
korarima • Ethiopian cardamom

Aframomum corrorima (A. Br.) Jansen (Zingiberaceae); *Ethiopiese kardamom* (Afrikaans); *korarima* (Amharic); *korarima, cardamome d'Ethiopie, poivre d'Ethiopie* (French)

DESCRIPTION Ethiopian cardamom or korarima is the ripe or near-ripe seeds of the korarima plant, usually sold enclosed in the dried fruit. These are smooth, shiny brown and a few (3–5) mm (ca. ⅙ in.) in diameter. They have a distinctive sweet smell (when crushed) and a delicious mild sweet spicy taste very similar to cardamom.

THE PLANT A slender, leafy perennial of up to 2 m (ca. 6 ft) tall, with thin, fleshy stems and broad, oblong leaves arising from a creeping rhizome.[1] Attractive pink flowers are borne near the ground and are followed by bright red, fleshy fruits containing numerous seeds.

ORIGIN The use of korarima is well known only from Ethiopia and Eritrea, where it is an important spice and an essential component of the traditional cuisine.[2] The plant is thought to be endemic to Ethiopia (and perhaps Sudan) but it may also be present in northern Kenya, western Uganda and Tanzania.[1] The spice is exported to a limited extent, especially to Sudan, Egypt, Iran and India.

CULTIVATION Plantations are limited to Ethiopia (especially rain forests in the highlands of the southwest) and Eritrea but the spice is also wild-harvested to some extent. Plants can be grown from seeds but a quicker option is to divide the rhizomes.[2] High humidity, mild temperatures and some shading are required.[2] Flowering and fruiting occurs almost throughout the year.

HARVESTING The fruits turn bright red when they are ripe and ready for harvesting and drying. They are hand-picked and dried in the shade, resulting in a pale grey, bottle-shaped product that is sold throughout Ethiopia. These fruits closely resemble those of Melegueta pepper but they are much larger and the seeds are smooth.

CULINARY USES Korarima is an important ingredient of *berbere, awaze, mitmita* and other Ethiopian spice mixtures.[2,3] Traditional *berbere* comprises 1½ cups korarima, 2½ cups dried chili, ½ cup each of salt, red onion and garlic, 4 teaspoons each of nigella, ajowan and ginger and 1 teaspoon each of cloves, black pepper, cinnamon, coriander, savory, cumin, fenugreek seeds, sesame seeds and basil (and/or rue). Korarima is traditionally used to flavour coffee[2,3] and can be used in the same way as cardamom (or in combination with cardamom to increase the complexity of the flavour).

FLAVOUR COMPOUNDS The essential oil contains cineole, terpinyl acetate and nerolidol as the main chemical compounds but also sabinene, neryl acetate, geraniol, β-pinene, α-terpineol and terpin-4-ol.[4,5] The seeds are chemically similar to those of cardamom (*Elettaria cardamomum*) but the aroma-active compounds do not appear to be known.

NOTES The seeds are used in Ethiopian traditional medicine as tonics and carminatives.[2,3]

nerolidol

1,8-cineole

terpinyl acetate

1. Lock, J.M. 1997. Zingiberaceae. In: Edwards, S., Demissew, S., Hedberg, I. (Eds). *Flora of Ethiopia and Eritrea*, Vol. 6, pp. 324–329. National Herbarium, Addis Ababa and University of Uppsala, Uppsala.

2. Jansen, P.C.M. 1981. Spices, condiments and medicinal plants in Ethiopia, their taxonomy and agricultural significance. *Agricultural Research Reports* 906. Centre for Agricultural Publishing and Documentation, Wageningen.

3. Demissew, S. 1993. A description of some essential oil bearing plants in Ethiopia and their indigenous uses. *Journal of Essential Oil Research* 5: 465–479.

4. Abegaz, B., Asfaw, N., Lwande, W. 1994. Chemical constituents of the essential oil of *Aframomum corrorima* from Ethiopia. Sinet: *Ethiopian Journal of Science* 17: 145–148.

5. Başer, K.H.C., Kürkçüoglu, M. 2001. The essential oils of *Aframomum corrorima* (Braun) Jansen and *A. angustifolium* K.Schum. from Africa. *Journal of Essential Oil Research* 13: 208–209.

Fruits and seeds of Ethiopian cardamom or korarima (*Aframomum corrorima*)

Ethiopian cardamom: ripe or near-ripe seeds of *Aframomum corrorima*

3 mm

Flowering plant (*Aframomum corrorima*)   Flowers (*Aframomum corrorima*)   Leaves (*Aframomum corrorima*)

Aframomum melegueta
Melegueta pepper • grains of paradise

Aframomum melegueta (Roscoe) K. Schum. (Zingiberaceae); *malegueta-peper* (Afrikaans); *khayrbûâ* (Arabic); *paradijskorrels* (Dutch); *malaguette, maniguette, poivre de Guinée, graines de paradis* (French); *Malagettapfeffer, Meleguetapfeffer, Paradieskörner, Guineapfeffer* (German); *grani de melegueta, maniguetta* (Italian); *rajskiye zyorna, malagvet* (Russian); *malagueta* (Spanish)

DESCRIPTION Small reddish-brown seeds of about 3 mm (⅛ in.) in diameter, resembling cardamom but with a distinctive warty surface. When crushed they have an aromatic smell and a sharp and pungent but spicy taste (once described as "dense fragrance underlined with heat"). The ground seeds have a greyish colour. The product is also known as grains of paradise and alligator pepper but sometimes called Guinea grains or Guinea pepper[1] (see notes).

THE PLANT A leafy, clump-forming perennial herb, up to 1.4 m (ca. 5 ft) high, with attractive orange flowers and ovoid fruit capsules borne near the ground.

ORIGIN West coast of Africa, including Cameroon, Ghana (the main source of exports), Ivory Coast, Liberia, Nigeria and Togo. This area was once known as Melegueta or the "pepper coast". The name is derived from the ancient empire of Melle or Meleguetta in the upper headwaters of the Guinea River, which was the original centre of commercial exploitation.[1]

CULTIVATION It is easily cultivated by planting the rhizomes in warm, humid and partially shaded places.

HARVESTING The ripe fruits (apparently edible) are dried to a pale grey colour and then tied in small bundles.

CULINARY USES This was a valuable substitute for pepper in the days before the oceanic spice route and is still important in West and North African cookery (e.g. as ingredient of spice powders, such as Moroccan *ras el hanout* and Tunisian five-spice mixture called *gâlat dagga*). This unique spice is making a comeback, not only as an ingredient of old recipes (to flavour beers, spiced wines and sausages) but as a (less pungent) substitute for pepper in stews and vegetable dishes. It should be ground and added in liberal amounts to the dish just before serving.

FLAVOUR COMPOUNDS The essential oil contains caryophyllene and humulene, with smaller amounts of their oxides, as well as α-cardinol. The pungent taste can be ascribed to (6)-paradol and other hydroxyarylalkanones.

NOTES Strictly speaking, "grains of paradise" refers to *Aframomum granum-paradisi*, while "Melegueta pepper" is the correct name for *A. melegueta*. Most authorities regard the former as a synonym of the latter. Another species, *A. angustifolium*, is known as great cardamom, Cameroon cardamom or Madagascar cardamom. Melegueta pepper is used in West Africa to freshen the breath, warm the body and to treat dyspepsia. *Xylopia aethiopica* (Guinea pepper) is another West African spice used in much the same way as pepper.[3] The smoked and dried fruits have a peppery flavour ascribed mainly to the presence of linalool in the essential oil.[4]

[6]-paradol

α–humulene

β–caryophyllene

1. **Burkill, H.M. 2000.** *The useful plants of West Tropical Africa* (2nd ed.), Vol. 5, pp. 312–314. Royal Botanic Gardens, Kew.
2. **Fernandez, X., Pintaric, C., Lizzani-Cuvelier, L., Loiseau, A.-M., Morello, A., Pellerin, P. 2006.** Chemical composition of absolute and supercritical carbon dioxide extract of *Aframomum melegueta*. *Flavour and Fragrance Journal* 21: 162–165.
3. **Burkill, H.M., 1985.** *The useful plants of West Tropical Africa* (2nd ed.), Vol. 1, pp. 130–132. Royal Botanic Gardens, Kew.
4. **Tairu, A.O., Hofmann, T., Schieberle, P. 1999.** Identification of the key aroma compounds in dried fruit of *Xylopia aethiopica*. In: Janick, J. (Ed.), *Perspectives on new crops and new uses*, pp. 474–478. ASHS Press, Alexandria.

Melegueta pepper (*Aframomum melegueta*)

Seeds of *Aframomum melegueta*

Plant of *Aframomum melegueta*

Guinea pepper (*Xylopia aethiopica*), showing seeds

53

Agathosma betulina
buchu • round leaf buchu

Agathosma betulina (Berg.) Pillans [= *Barosma betulina* (Berg.) Bartl.& H.L. Wendl.] (Rutaceae); *boegoe*, *rondeblaarboegoe* (Afrikaans); *buchu* (French); *Bucco* (German); *buchu* (Italian); *buchu* (Spanish)

DESCRIPTION Small, rounded, gland-dotted leaves with a distinctive aromatic smell and a characteristic blackcurrant flavour. The leaves of round leaf buchu have a length to width ratio of less than 2:1 (more oblong in shape in the less desirable *A. crenulata* known as oval leaf buchu).[1]

THE PLANT A much-branched woody shrub of about 1 m (ca. 3 ft) in height with white or pale purple star-shaped flowers that develop into characteristic segmented capsules with shiny black seeds. The species resprouts after fire, unlike *A. crenulata* which regenerates from seeds only. The latter is an erect, sparsely branched shrub of up to 1.8 m (ca. 6 ft) high.[1]

ORIGIN The species is highly localised in nature and is endemic to the Cederberg region of the Western Cape Province of South Africa. *Agathosma crenulata* has a more southern distribution.[1]

CULTIVATION Buchu is propagated from seeds or cuttings and is grown in low phosphate (24 ppm maximum), sandy, low pH soils (pH 2.5–4.5) with ample watering (but in winter only).[2] Plantations are located in the fynbos region of South Africa with its Mediterranean climate, and recently also in Australia. Buchu is relatively difficult to maintain in a herb garden but can be successfully grown in a pot, taken under cover to avoid summer rain.

HARVESTING The product has been wild-harvested until cultivation became a viable option. Bundles of leafy stems are air-dried in the shade and processed to separate the leaves from the stems. A large part of the harvest is steam-distilled to obtain the essential oil.

CULINARY USES Buchu leaf and buchu oil are popular as flavour components in alcoholic beverages (brandies and liqueurs), mineral water, cool drinks, herbal teas and ice teas. The oil is used in the food industry to improve the taste of beverages, sweets and confectionery. Good quality buchu brandy makes an excellent aperitif (with lemonade and/or tonic water). It is used sparingly to enhance the flavour of sweets, jams, jellies and confectionery.

FLAVOUR COMPOUNDS The leaves and essential oil have a very characteristic minty odour and flavour. The blackcurrant-like tones are ascribed to sulphur-containing minor compounds in the essential oil (mainly 8-mercapto-*p*-menthan-3-one),[3] while the mint-like flavours are due to diosphenol (so-called buchu camphor) and other volatiles such as isomenthone.[3,4] Oval leaf buchu (*A. crenulata*) is considered to be inferior because of its high pulegone content and near absence of diosphenol. The sulphur compounds give the typical "catty" notes that are also found in blackcurrant and hopped beer.

NOTES Buchu is a traditional aromatic bitters and tonic, used to treat indigestion and urinary ailments.

isomenthone ψ-diosphenol pulegone 8-mercapto-*p*-menthan-3-one

1. **Spreeth, A.D. 1976.** A revision of the commercially important *Agathosma* species. *Journal of South African Botany* 42: 109–119.
2. **Blommerus, L. 2007.** *Buchu* Agathosma – *cultivation, economics and uses*. Agricultural Research Council, Elsenburg.
3. **Kaiser, R., Lamparsky, D., Schudel, P. 1975.** Analysis of buchu leaf oil. *Journal of Agricultural Food Chemistry* 23: 943–950.
4. **Posthumus, M.A., Van Beek, T.A., Collins, N.F., Graven, E.H. 1996.** Chemical composition of the essential oils of *Agathosma betulina, A. crenulata* and an *A. betulina* × *A. crenulata* hybrid (buchu). *Journal of Essential Oil Research* 8: 223–228.

Buchu: left, round leaf buchu (*Agathosma betulina*); right, oval leaf buchu (*A. crenulata*)

Round leaf buchu (*Agathosma betulina*)

Leaves (*Agathosma betulina*)

Fruits and seeds (*Agathosma betulina*)

Oval leaf buchu (*Agathosma crenulata*)

Cultivated plant (*Agathosma betulina*)

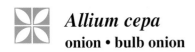

Allium cepa
onion • bulb onion

Allium cepa L. (Alliaceae); *ui* (Afrikaans); *cong tou, yang cong* (Chinese); *oignon* (French); *Küchenzwiebel* (German); *cipolla* (Italian); *tamanegi* (Japanese); *bawang* (Malay); *cebola* (Portuguese); *cebolla* (Spanish)

DESCRIPTION A rounded bulb covered with a brown or purplish, papery skin. It has a strong flavour and cutting releases pungent volatile compounds that cause weeping. Onions are used fresh, dried or pickled.

THE PLANT A biennial plant with hollow leaves and rounded clusters of small white flowers. Shallot (*A. cepa* var. *ascalonicum*) differs in forming clusters of bulbs.[1,2] The name "scallion" is the original name for the shallot but is now often used for spring onions, which may also include leeks or Welsh onion (in the USA) or young *Allium cepa* plants.[1,2] The tree onion (*A. ×proliferum*) has small bulbs that develop within the inflorescence. *Allium chinense* (Chinese onion) is an Asian species known as *chiao tou* in China or *rakkyo* in Japan. The small bulbs are fried or they are pickled in vinegar or brine.

ORIGIN An ancient cultigen (first recorded 3000 BC), thought to be derived from wild species in the Middle East (possibly Iran and Afghanistan).[2]

CULTIVATION Onions are grown as an annual crop. Critical for success are the choice of cultivar, the time of sowing, long daylight hours during bulb formation and light soils. Seedlings are set out 10 cm (4 in.) apart in rows of 30 cm (1 ft) wide. Direct sowing (10 mm / ⅖ in. deep) requires thinning, thus providing young onions for the kitchen.

HARVESTING Drying is encouraged by reducing irrigation and by bending over the tops. Select thin-necked bulbs for storage because they keep better. Dry onions should be stored in a shaded, well-ventilated place.

CULINARY USES Few culinary herbs rival the onion in popularity and diversity of uses.[3] The bulbs are eaten raw as an ingredient of salads, soups, sauces, stews and curries or may be processed for use in pickles and chutneys. Adding chopped onions as an ingredient to a sour fruit sauce turns it into a chutney. The small bulbs of *rakkyo* (*chiao tou*) may be fried or are commonly pickled in brine or vinegar. Examples of popular dishes include French onion soup (*soupe à l'oignon*) and *slaphakskeentjies* (an old Cape favourite of small onions cooked in a sweet-sour sauce).

FLAVOUR COMPOUNDS Onions are rich in sugar, which caramelizes during the frying process. This, together with the enzymatic breakdown of sulphur compounds, results in the delicious savoury taste. Four non-volatile, odourless sulphoxides occur in *Allium* species as the main flavour and odour precursors, of which isoalliin is characteristic of *A. cepa*.[4] When cells are damaged, isoalliin is converted by alliinase to form isoallicin, the main thiosulphinate intermediate in *A. cepa* (0.14–0.35 µmol per g wet weight).[4] Isoallicin is further converted to volatile sulphur compounds. The weeping reaction caused by cutting onions with a blunt knife is due to another volatile sulphur compound known as lachrymatory factor (syn-propanethial S-oxide).[4]

NOTES Onions are known to inhibit blood-clotting and hence thrombosis.

1. Gregory, M., Fritsch, R.M., Friesen, N.W., Khassanov, F.O., McNeal, D.W. 1998. Nomenclator Alliorum. Allium *names and synonyms – a world guide*. Royal Botanic Gardens, Kew.

2. Fritsch, R.M., Friesen, N. 2002. Evolution, domestication and taxonomy. In: Rabinowitch, H.D., Currah, L. (Eds), Allium *crop science: recent advances*, pp. 5–30. CABI Publishing, New York.

3. Farrel, K.T. 1999. *Spices, condiments and seasonings*. Aspen Publishers, Gaithersburg, USA.

4. Benkeblia, N., Lanzotti, V. 2007. *Allium* thiosulfinates: chemistry, biological properties and their potential utilization in food preservation. *Food*1: 193–201.

Onions, pickle onions and spring onions (*Allium cepa*)

Onion plants (*Allium cepa*)

Flowers of *Allium cepa*

Shallots (*Allium cepa* var. *ascalonicum*)

Chinese onion (*Allium chinense*)

Tree onion (*Allium ×proliferum*)

57

Allium fistulosum
Welsh onion • Japanese bunching onion

Allium fistulosum L. (Alliaceae); *Walliese ui, groenui* (Afrikaans); *cong, da cong* (Chinese); *duan bawang* (Indonesian); *ciboule* (French); *Winterzwiebel* (German); *cipolo d'inverno, cipoletta* (Italian); *negi* (Japanese); *daun bawang* (Malay); *cebolinha commun* (Portuguese); *ceboletta comun* (Spanish); *ton horm* (Thai)

DESCRIPTION The fleshy stems (about the thickness of a finger) are without distinct bulbs and bear hollow leaves with an inflated appearance (*fistulosum* means "hollow"). It is variously known as scallion, green onion, bunching onion or salad onion (USA), Welsh onion (England), *ciboule* (France) and *negi* (Japan).[1] It is superficially similar to leeks (*A. ameloprasum*) but the latter have flat leaves.

THE PLANT A robust perennial herb, easily distinguished by the tufted habit, absence of bulbs, hollow leaves and rounded clusters of small white flowers.

ORIGIN A cultigen of Chinese origin. This is the original Chinese garden onion (recorded 100 BC) which spread to Japan (recorded AD 720). "Welsh onion" is derived from the German or old English word *welsche* (meaning *foreign*). "Spring onion" may refer to this species or to young onions (*A. cepa*) that are used in the same way. Scallion or shallot (*A. cepa* var. *ascalonicum*) is also a type of bunching onion that can be confused with Welsh onion. The name "Chinese onion" is best avoided due to a possible confusion with *A. chinense*. "Chinese chives" is the name of yet another species (see *A. tuberosum*).[1,2]

CULTIVATION This species (unlike onion) can be sown almost any time of the year or the tufts can be divided. Welsh onion and various hybrids with *A. cepa* (such as 'Beltsville Bunching' in the USA) have become popular as annual crops. They are best grown in full sun but will tolerate some shade. Aphids may be troublesome. Cultivars can be short and thick (e.g. 'Shimonita'), red-skinned (e.g. 'Red Beard' and 'Santa Clause') or have slightly bulbous bases (e.g. 'Yoshima').

HARVESTING Stems with leaves are cut any time of the year – one of few vegetables that can be harvested all year round.

CULINARY USES Welsh onion is a traditional ingredient of oriental cooking, used in stir-fries and in many other dishes. The stems and leaves are eaten fresh as an ingredient of salads and are often seen as garnishes, shaped into distinctive brushes or tassels.

FLAVOUR COMPOUNDS The mild, onion-like flavour is derived from sulphur-containing compounds such as isoalliin and methiin that are enzymatically modified to form volatile sulphides via thiosulphinate intermediates (only 0.08 µmol thiosulphinates per g fresh weight, hence the mild flavour).[3] The main volatile compounds in *A. fistulosum* are dipropyl disulphide (30%), methyl propyl trisulphide and dipropyl trisulphide.[4]

NOTES Leeks (*Allium ameloprasum*, formerly known as *A. porrum*) also have a mild flavour (0.15 µmol thiosulphinates per g),[3] making them ideal for flavouring soups (French *vichyssoise* and Scottish "cock-a-leekie") and stews. There are three groups: ordinary leek (var. *porrum*), great-headed or Levant garlic (var. *holmense*) and kurrat (var. *kurrat*).[1] Leeks and kurrat leaves are eaten as vegetables while Levant garlic bulbs are used as a mild seasoning.

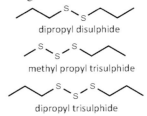

dipropyl disulphide

methyl propyl trisulphide

dipropyl trisulphide

1. Fritsch, R.M., Friesen, N. 2002. Evolution, domestication and taxonomy. In: Rabinowitch, H.D., Currah, L. (Eds), Allium *crop science: recent advances*, 5–30. CABI Publishing, New York.

2. Harvey, M.J. 1995. Onion and other cultivated alliums. In: Smartt, J., Simmonds, N.W. (Eds), *Evolution of crop plants* (2nd ed.), pp. 344–350. Longman, London.

3. Benkeblia, N., Lanzotti, V. 2007. *Allium* thiosulfinates: chemistry, biological properties and their potential utilization in food preservation. *Food* 1: 193–201.

4. Pino, J.A., Rosado, A., Fuentes, V. 2000. Volatile flavor compounds from *Allium fistulosum* L. *Journal of Essential Oil Research* 12: 553–555.

Welsh onion (*Allium fistulosum*)

Flowers of Welsh onion (*Allium fistulosum*)

Flowers of Welsh onion (*Allium fistulosum*)

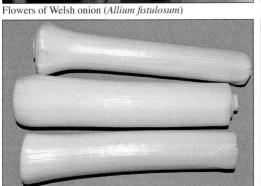

Leeks (*Allium ameloprasum*)

Flowers of leek (*Allium ameloprasum*)

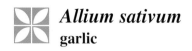

Allium sativum
garlic

Allium sativum L. (Alliaceae); *knoffel* (Afrikaans); *suan* (Chinese); *ail, ail blanc* (French); *Knoblauch* (German); *aglio* (Italian); *gaarikku* (Japanese); *bawang puteh* (Malay); *czosnek* (Polish); *alho* (Portuguese); *ajo* (Spanish)

DESCRIPTION Garlic bulb segments (known as "cloves") are axillary buds enclosed within a communal papery shell. They are mostly used fresh, less often dried and powdered or steeped in oil. Garlic scapes (*suan tai* in Sichuan) are commonly used in Chinese cuisine.

THE PLANT A perennial herb with flat leaves arising from an underground bulb. The rounded flower cluster is surrounded by a sheath-like bract. Elephant garlic is a hybrid between garlic and onion with very large bulb segments and a mild flavour.[1-3]

ORIGIN Garlic is a cultigen of unknown origin dating back to at least 3000 BC. It is perhaps derived from *Allium altaicum*, a central Asian species.[1-3]

CULTIVATION Garlic hardly ever produces viable seeds and is therefore propagated by simply planting the cloves. The best planting time is from late summer to autumn. They are spaced 100 to 150 mm (4 to 6 in.) apart and planted at a depth of 50 mm (2 in.). Unlike other species, the bulbs are formed entirely below the soil surface. Light soil and regular feeding with nitrogen fertilizer is recommended. Irrigation should be discontinued when the leaves start to whither.

HARVESTING Bulbs are harvested when the leaves have completely died. They are left in the sun for one day and then brought inside for storage in a cool, well-ventilated place. The bulbs can be sliced and dried to produce garlic flakes or garlic powder.

CULINARY USES The flavour of garlic is characteristic of Italian, French, Spanish and Portuguese cooking but garlic has become popular in most parts of the world. It is typically fried in a little oil as an overture to preparing an excellent dish. It may be pounded and mixed with other ingredients to create seasoning paste (garlic purée), aioli (garlic mayonnaise), pesto, garlic butter and garlic oil. Whole or sliced cloves can be added to soups, stews, fish and meat dishes. Roasted garlic (especially elephant garlic) is sometimes eaten as a side dish.

FLAVOUR COMPOUNDS The strong flavour is due to di-2-propenyl disulphide (diallyl disulphide) as main volatile compound. It is formed from alliin, a non-volatile, odourless sulphoxide characteristic of intact *A. sativum* bulbs, via allicin, a thiosulphinate intermediate. The levels of thiosulphinate in garlic can be quite high (up to 36 µmol per g wet weight).[3,4]

NOTES Garlic is used in herbal medicine for its antibiotic action and lipid-lowering effects. The sulphur is excreted mainly through the lungs, hence the noticeable effect on the breath. In Europe and Asia, wild-harvested *Allium ursinum* (bear's garlic, wood garlic or ramsons) is making a comeback as culinary herb. It produces about 21 µmol thiosulphinates per g wet weight).[4] Care should be taken not to confuse it with the poisonous but superficially similar autumn crocus (*Colchicum autumnale*). Other garlic-like herbs include *A. schorodoprasum* (sandleek) from Southeast Asia and *A. tricoccum* (ramp) from the United States.

1. Mathew, B. 1996. *A review of* Allium *section* Allium. Royal Botanic Gardens, Kew.
2. Fritsch, R.M., Friesen, N. 2002. Evolution, domestication and taxonomy. In: Rabinowitch, H.D., Currah, L. (Eds), Allium *crop science: recent advances*, 5–30. CABI Publishing, New York.
3. Block, E. 2010. *Garlic and other alliums. The lore and the science*. The Royal Society of Chemistry, Cambridge.
4. Benkeblia, N., Lanzotti, V. 2007. *Allium* thiosulfinates: chemistry, biological properties and their potential utilization in food preservation. *Food* 1: 193–201.

Garlic bulbs, flakes and powder (*Allium sativum*)

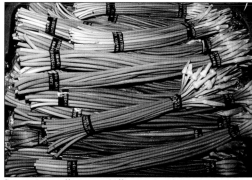
Suan tai or garlic scapes (*Allium sativum*)

Garlic plants (*Allium sativum*)

Bear's garlic (*Allium ursinum*)

Allium schoenoprasum
chive • chives

Allium schoenoprasum L. (Alliaceae); *grasui* (Afrikaans); *bei cong* (Chinese); *ciboulette*, *civette* (French); *Schnittlauch* (German); *cipoletta* (Italian); *cebolinha* (Portuguese); *cebollino* (Spanish)

DESCRIPTION The slender, hollow leaves are grass-like, up to 30 cm (ca. 1 ft) long and up to 6 mm (¼ in.) in diameter. They have a mild, onion-like smell and taste. Chives are best used fresh but can also be freeze-dried or frozen in ice cubes.

THE PLANT An attractive perennial tuft with thin, tubular leaves growing from a cluster of slender, cylindrical white bulbs and bearing decorative purple flowers. Chive (onion chive) is sometimes confused with garlic chive (Chinese chive) but the latter has flat leaves and white flowers (see *A. tuberosum*).

ORIGIN Chive occurs naturally in Europe and Asia and the cultivated plant closely resembles the wild species.[1] Domestication has apparently occurred (more than once) in the Mediterranean region, with the earliest records dating back to the 16th century.

CULTIVATION Temperate conditions are required but chives also grow well at high elevations in the tropics. Commercial growers propagate plants from seeds (sown shallowly, 6 mm / ¼ in. deep) but the home grower can simply divide the clumps. To maintain vigour, mature plants should be divided every second year.[2] Chive is an attractive edging plant.

HARVESTING Leaves may be cut every six to eight weeks during the growing season, before the plants flower and become dormant in winter.[2] They are best used fresh but chopped and dried chives are produced on a commercial scale.[2]

CULINARY USES Chives are one of the most widely utilized and popular garnishes, used (whole or finely chopped) to add a mild onion flavour to omelettes, scrambled eggs, sauces, mashed potatoes, cream cheese, cottage cheese, salads, soups, savoury dishes and dips.[3] They are traditionally used in French cookery as a key ingredient of *fines herbes*, an aromatic mixture of chives with parsley, chervil and tarragon, in various proportions (and sometimes with other ingredients added).[3] *Fines herbes* are used in sauces, cream cheese, sautéed vegetables, omelettes and meat dishes. Chives are also an ingredient of the famous Frankfurt green sauce.

FLAVOUR COMPOUNDS Sulphur-containing volatile compounds are released through enzymatic action from methiin, propiin and other so-called CSOs [(+)-*S*-alk(en)yl cysteine sulphoxides] when the leaves are bruised and chopped (see *A. cepa* and *A. sativum*).[4] The mild taste is due to the low levels of thiosulphinates (only about 0.19 µmol per g wet weight).[4] A recent study reported nearly 1% allicin, as well as numerous phenolic compounds, including isoquercetrin and kaempferol.[5]

NOTES Chives are sometimes confused with spring onions but the latter are young onion (*A. cepa*) plants harvested before the bulbs have formed. The name spring onion is also used for shallots or scallions, a variety of onion with multiple stems.

methiin   propiin

1. **Fritsch, R.M., Friesen, N. 2002.** Evolution, domestication and taxonomy. In: Rabinowitch, H.D., Currah, L. (Eds), Allium *crop science: recent advances*, pp. 5–30. CABI Publishing, New York.
2. **Farrel, K.T. 1999.** *Spices, condiments and seasonings.* Aspen Publishers, Gaithersburg, USA.
3. **Larousse. 1999.** *The concise Larousse gastronomique.* Hamlyn, London.
4. **Benkeblia, N., Lanzotti, V. 2007.** *Allium* thiosulfinates: chemistry, biological properties and their potential utilization in food preservation. *Food* 1: 193–201.
5. **Vlasa, L., Parvu, M., Parvu, E.A., Toin, A. 2013.** Chemical constituents of three *Allium* species from Romania. *Molecules* 18: 114–127.

Chive plant (*Allium schoenoprasum*)

Leaves (*Allium schoenoprasum*)

Flowers of chives (*Allium schoenoprasum*)

Leaves of chive (left) and garlic chive

63

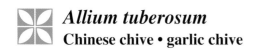

Allium tuberosum
Chinese chive • garlic chive

Allium tuberosum Rottler ex Sprengel (Alliaceae); *Chinese grasui, knoffel-grasui* (Afrikaans); *jiu cai* (Chinese); *ciboule de chine à feuilles larges* (French); *Chinesischer Schnittlauch* (German); *nira* (Japanese); *bawang kuchai* (Malay); *cive chino* (Spanish); *kui chaai* (Thai)

DESCRIPTION The leaves are grass-like, thin and flat (neither tubular and hollow as in chive and onion, nor folded lengthwise as in garlic and leek). Chopped leaves have a more pungent taste than spring onions or chives. The young inflorescences (flowering buds) may be green but are often blanched to produce white or yellow chives. These characteristic bunches are a common sight on oriental markets and may be presented as "oriental garlic", "Chinese leeks" or "flowering chives". They should not be confused with *suan tai* or garlic scapes, distinguished by their club-shaped and elongated (acuminate) buds.

THE PLANT The plant forms dense clumps of grass-like, flat, somewhat drooping leaves and erect clusters of white flowers.

ORIGIN East Asia and northeastern India. It is an ancient crop but the exact origin or site of domestication is not clear.[1,2] It has become naturalized in many parts of the world.

CULTIVATION Propagation is achieved by seeds or by dividing the mature clumps. Both methods can yield more than 50 tons of green leaves per hectare per annum. The plant tolerates a wide range of climates and soils and has become popular for use in home herb gardens. Leaves or flowering stalks are harvested just before the flowers open. White or yellow chives are produced by growing the plants in darkness, using clay pots, plastic sheeting or straw to cut out sunlight and reduce photosynthesis. In commercial plantations, the production of blanched chives weakens the plants so that it is alternated with green chive production.[3]

HARVESTING Leaves (or the small, narrow bulbs) can be harvested at any time during the growing season. The most prized part is the flowering stalk, harvested before the buds emerge from the sheaths. Green chives can be stored in a cold room for more than two weeks but blanched chives have a shelf life of less than five days.

CULINARY USES Green or blanched leaves are used as a culinary herb for garnishing and flavouring soups and rice or noodle dishes. They are also enjoyed as a vegetable and an ingredient of stir-fries. Prolonged cooking should be avoided as it results in a bitter taste. The distinctive, usually blanched flower buds are a popular (but expensive) vegetable dish with an onion-like taste.[4]

FLAVOUR COMPOUNDS The flavour is due to sulphur-containing compounds that are released enzymatically when the plants are bruised or cooked. The mild garlic-like flavour originates from rather low levels of thiosulphinates (ca. 2 μmol per g wet weight).[5] A complex mixture of volatile compounds is produced, of which dimethyl disulphide and dimethyl trisulphide are the main compounds.

NOTES Chinese chive is a remedy for fatigue in traditional Chinese medicine. The leaves and the bulbs are applied to insect bites, cuts and wounds. The whole plant is said to repel insects and moles.

dimethyl disulphide   dimethyl trisulphide

1. **Fritsch, R.M., Friesen, N. 2002.** Evolution, domestication and taxonomy. In: Rabinowitch, H.D., Currah, L. (Eds), Allium *crop science: recent advances*, pp. 5–30. CABI Publishing, New York.
2. **Block, E. 2010.** *Garlic and other alliums. The lore and the science*. The Royal Society of Chemistry, Cambridge.
3. **Van Wyk, B.-E. 2005.** *Food plants of the world*. Briza Publications, Pretoria.
4. **Hu, S.-Y. 2005.** *Food plants of China*. The Chinese University Press, Hong Kong.
5. **Benkeblia, N., Lanzotti, V. 2007.** *Allium* thiosulfinates: chemistry, biological properties and their potential utilization in food preservation. *Food* 1(2): 193–201.

Garlic chives, the flowering stems of *Allium tuberosum*

Plants (*Allium tuberosum*)

Leaves (*Allium tuberosum*)

Flowers and seeds (*Allium tuberosum*)

Aloysia citrodora
lemon verbena • vervain

Aloysia citrodora Palau[= *A. triphylla* (L'Herit.) Britton; *Lippia citriodora* H.B.K.; *L. triphylla* (L'Herit.) Kunze]
(Verbenaceae); *sitroenverbena* (Afrikaans); *verveine odorante* (French); *Zitronenstrauch* (German); *limoncina*,
erba luigia, *cedrina* (Italian); *verbena olorosa*, *hierba luisa* (Spanish)

DESCRIPTION The leaves are somewhat rough in texture and have a very strong smell of lemon when crushed. They are used (fresh or dried) in cookery, while the essential oil or solvent extracts are used in aromatherapy, perfumery, soaps and cosmetics. The correct scientific name is *Aloysia citrodora*[1] and synonyms include *Lippia citriodora* and *Aloysia citriodora*.[1,2]

THE PLANT A deciduous, woody shrub of up to 3 m (ca. 10 ft) high with aromatic leaves in whorls of three or four and small, pale mauve flowers in loose panicles. The leaves are highly aromatic and have a sandpapery texture.

ORIGIN South America (indigenous to Argentina and Chile). The Spaniards brought it to Europe in the 18th century.

CULTIVATION This is one of the most rewarding flavour plants to grow in the herb garden and it is useful to have a regular supply of fresh leaves. Full sun (or at least morning sun) is required, as well as a light loam soil and regular watering. It can withstand mild frost but in northern regions it is kept indoors during the winter. The plant is easily propagated from cuttings but can also be grown from seeds. Regular pruning is beneficial as the shrub often tends to become untidy. Commercial plantings are found in many parts of the world, including France and Spain.

HARVESTING Commercial harvesting is done in early summer at full bloom and again in autumn. The leafy flowering tops are cut and dried in the shade. Steam distillation for the valuable essential oil (known as lemon verbena oil or verbena oil) should follow soon after harvesting to ensure better quality and higher yields (up to 1% dry weight but more often less than 0.5%).

CULINARY USES Lemon verbena may be used as a substitute for lemongrass and as a refreshing lemon flavour ingredient in stews, soups and stir-fries (remove the leaves before serving as they are leathery and not edible). It is excellent for flavouring summer drinks and can be included in salads, fruit salads, baked custard, rice puddings and jams. Fresh leaves may be used to flavour stocks for seafood dishes, while dried and powdered leaves can be added to fish and meat stuffings.[3] The main commercial uses are for health teas, tisanes, various liqueurs and fruit juice drinks.

FLAVOUR COMPOUNDS Oil of verbena contains limonene (hence the lemon smell) and citral as main compounds.[4,5] Citral is a mixture of geranial (also known as citral A) and neral (also known as citral B). The presence of many other components, including carvone, geraniol, linalool and others add to the intensity and complexity of the flavour.

NOTES Lemon verbena tea is believed to have digestive, antispasmodic and sedative properties. Dried leaves are commonly used as an ingredient of potpourri.[2]

limonene   geranial   neral

1. **Armada, J., Barra, A. 1992.** On *Aloysia* Palau (Verbenaceae). *Taxon* 41(1): 88–90.
2. **Mabberley, D.J. 2008.** *Mabberley's plant-book* (3rd ed.). Cambridge University Press, Cambridge.
3. **Larousse. 1999.** *The concise Larousse gastronomique*. Hamlyn, London.
4. **Harborne, J.B., Baxter, H. 2001.** *Chemical dictionary of economic plants*. Wiley, New York.
5. **Argyropoulou, C., Daferera, D., Tarantilis, P.A., Fasseas, C., Polissiou, M., 2007.** Chemical composition of the essential oil from leaves of *Lippia citriodora* H.B.K. (Verbenaceae) at two developmental stages. *Biochemical Systematics and Ecology* 35(12): 831–837.

Lemon verbena, the leaves of *Aloysia citrodora*

Leaves (*Aloysia citrodora*)

Leaves and flowers (*Aloysia citrodora*)

Flowers (*Aloysia citrodora*)

Alpinia galanga
greater galangal • galangal

Alpinia galanga (L.) Sw. (Zingiberaceae); *groot galanga* (Afrikaans); *dà gao liang jiang* (Chinese); *galanga* (French); *Große Galgant* (German); *laos* (Indonesian); *galanga* (Italian); *lengkuas* (Malay); *galang* (Spanish); *kha* (Thai)

DESCRIPTION The rhizomes superficially resemble fresh ginger but are more robust (about 50–100 mm / 2–4 in. in diameter), less flattened and have a pinkish colour. It is also known in English as galanga, laos or Siamese ginger[1] (but galingale is *Cyperus esculentus*). Rhizomes are sold fresh, sliced and dried or as a powder. It has a very pungent taste and a somewhat peppery flavour.

THE PLANT Galangal is a leafy perennial, up to 3.5 m (nearly 12 ft) high, similar to ginger but much more robust.[1] The plants rarely flower in cultivation. Lesser galangal (*Alpinia officinarum*) is superficially similar but much smaller (up to 1.5 m / 5 ft in height) with attractive (but infrequent) white and purple flowers. It is in turn larger than the poorly known small galangal (*Kaempferia galanga*), also called Chinese galangal or *shan nai*.

ORIGIN Greater galangal is indigenous to Indonesia and other parts of tropical Asia. It is commonly cultivated as a spice in Thailand, Malaysia and Laos. Galangal was commonly used in European cooking during the Middle Ages and is still used as a flavour ingredient in digestive bitters and liqueurs. The galangal mentioned in old literature may be of Chinese origin and is then more likely to refer to lesser galangal or *gao liang jiang* (*A. officinarum*). This species has been used since the Middle Ages in eastern Europe and Russia to flavour teas and alcoholic beverages. Yet another species is small galangal (*Kaempferia galanga*) believed to be of Southeast Asian origin. It is the *kencur* of Indonesian cuisine (used fresh, e.g. in Malaysian *nasi ulam*) and the dried *sha jiang* (sand ginger) of Chinese (Sichuan) cookery (the plant is called *shan nai*).[2]

CULTIVATION Greater galangal is widely cultivated in tropical regions. Hot, humid conditions and light soils are required. Propagation is achieved by dividing the rhizomes.

HARVESTING There is no definite season and rhizomes can be harvested at any time of the year, while they are actively growing.

CULINARY USES Galangal is characteristic of Malaysian and Thai cooking. Slices of the tough rhizome (or large chunks, vigorously bruised to release the flavours) are added to curry dishes and soups. The young shoots and unopened flower buds are steamed and eaten as vegetables.

FLAVOUR COMPOUNDS Essential oil and several unusual chemical compounds contribute the strong flavour of galangal.[3] The main components of the essential oil are reported to be 1,8-cineole, β-pinene, myrcene, chavicol and methyl cinnamate, while the pungent taste is due to 1'S-1'-acetoxychavicol acetate and several isomers[4] which seem to occur in the plant as glycosides.

NOTES Galangal and related plants are widely used in Chinese, Indian and Indonesian traditional medicine and the dried rhizomes are nowadays freely available.

methyl cinnamate

1'S-1'-acetoxy-chavicol acetate

1,8-cineole

1. Burkill, I.H. 1966. *A dictionary of the economic products of the Malay Peninsula*, Vol. 2, pp. 1327–1332. Crown Agents for the Colonies, London.
2. Hu, S.-Y. 2005. *Food plants of China*. The Chinese University Press, Hong Kong.
3. Harborne, J.B., Baxter, H. 2001. *Chemical dictionary of economic plants*. Wiley, New York.
4. Kubota, K., Someya, Y., Kurobayashi, Y., Kobayashi, A. 1999. Flavour characteristics and stereochemistry of the volatile constituents of greater galangal (*Alpinia galanga* Willd.). In: Shahidi, F., Ho, C.-T. (Eds), *Flavour chemistry of ethnic foods*, pp. 97–104. Springer, New York.

Greater galangal (*Alpinia galanga*): mature (left) and young rhizomes

Greater galangal (*Alpinia galanga*)

Lesser galangal (*Alpinia officinarum*) and greater galangal (right)

Plants of greater galangal (*Alpinia galanga*)

Small galangal (*Kaempferia galanga*)

Anethum graveolens
dill

Anethum graveolens L. (= *A. sowa* Roxb.) (Apiaceae); *dille* (Afrikaans); *bazrul shibbat, shibit* (Arabian); *aneth* (French); *Dill* (German); *anithi, sotapa, sowa* (Hindi); *aneto* (Italian); *koper* (Polish); *ukrop* (Russian); *eneldo* (Spanish); *dill* (Swedish)

DESCRIPTION Dill weed is the whole plant with finely divided aromatic leaves, small flowers and immature fruits ("seeds"), used fresh or dried.[1] Dill fruits ("dill seeds") are the small dry schizocarps, up to 6 mm (¼ in.) long that split into two flat, elliptic, single-seeded mericarps at maturity. They are brown with a lighter-coloured edge.

THE PLANT An erect, fennel-like annual or biennial herb, up to 1 m (ca. 3 ft) with blue-green, deeply divided (pinnately compound) leaves and small yellow flowers borne in umbels.[1]

ORIGIN Southwestern Asia and southern Europe, but is naturalized in many parts of Eurasia and North America. It was regarded as a symbol of vitality in Roman times.

CULTIVATION Dill is grown as an annual crop and requires cool temperatures, long days, a regular supply of water and deep, fertile, slightly acidic to neutral, loam soils. It is sensitive to extreme temperatures, drought and strong winds. Commercial production is centred in India and Pakistan.[1] It is cultivated in kitchen gardens all over the world for culinary use.

HARVESTING Dill leaves are best used fresh but dried leaves ("dill weed") are also available. Harvesting for the leaves (or for distilling the essential oil) is done just before the plants flower. Yields of about 25 tons fresh leaves per hectare can be expected (or 0.5 tons of dry fruits per hectare).[1]

CULINARY USES Dill herb is more aromatic than fennel. Twice as much fresh herb is used to equal the required amount of dry herb.[1] The most famous applications are for the seasoning and garnishing of fish dishes, potato salad and pickled gherkins (e.g. Russian pickled cucumbers).[2]

The aromatic fruits are used in dill pickles, dill butter, breads, cheeses, sauces, vegetables and especially vinegars.[2] Dill fruits are used as a spice in North African meat dishes and in traditional Scandinavian crayfish or salmon. The essential oil has numerous commercial uses as an ingredient of confectionery, desserts, condiments, beverages and meat seasonings (especially for sausages).

FLAVOUR COMPOUNDS The essential oil of dill weed (dill herb) has α-phellandrene and (+)-dill ether (anethofuran) as main compounds, giving it the typical dill flavour.[3,4] The fruits are rich in (+)-carvone (up to 70%) and (+)-limonene (30 to 40%), with smaller amounts of α-phellandrene and dillapiole. Dill weed oils with higher levels (10 to 20%) of α-phellandrene and less than 20% carvone are therefore preferred. Fruit oil of Indian dill (a chemotype previously known as *Anethum sowa*) is rich in dillapiole (up to 40%).[3,4]

NOTES Dill or dill essential oil is used in traditional medicine for digestive ailments and infant flatulence. Long known for its carminative powers, its name is derived from the Old Norse word "dilla", meaning to lull or soothe.

(+)-dill ether   dillapiole

(−)-α-phellandrene   (+)-carvone   (+)-limonene

1. **Farrel, K.T. 1999.** *Spices, condiments and seasonings.* Aspen Publishers, Gaithersburg, USA.
2. **Kiple, K.F., Ornelas, K.C. (Eds). 2000.** *The Cambridge world history of food.* Cambridge University Press, Cambridge.
3. **Reichert, S., Wüst, M., Beck, T., Mosandl, A. 1998.** Stereoisomeric flavor compounds LXXXI: dill ether and its *cis*-stereoisomers: synthesis and enantioselective analysis. *Journal of High Resolution Chromatography* 21: 185–188.
4. **Rădulescu, V., Popescu, M.L., Ilies, D.-C. 2010.** Chemical composition of the volatile oil from different plant parts of *Anethum graveolens* L. (Umbelliferae) cultivated in Romania. *Farmacia* 58: 594–600.

Leaves ("dill weed") and fruits ("dill seeds") of *Anethum graveolens*

Fruits of dill (*Anethum graveolens*)

3 mm

Flowering plant (*Anethum graveolens*)   Flowers (*Anethum graveolens*)

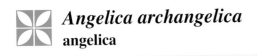

Angelica archangelica
angelica

Angelica archangelica L. (Apiaceae); *angelika* (Afrikaans); *angélique* (French); *Engelwurz* (German); *archangelica* (Italian); *angélica* (Spanish); *kvan* (Swedish)

DESCRIPTION The thick, hollow young stems and leaf stalks and the dry, flat and ribbed fruits (schizocarps) are most often used in cookery. The spindle-shaped roots are the source of a valuable essential oil.[1]

THE PLANT Angelica is a robust biennial herb of about 2 m (ca. 6 ft) high when in flower. The hollow and fluted stems bear large leaves with toothed margins. Especially characteristic are the very large umbels of green flowers. Cultivated angelica may be confused with European wild angelica (*A. sylvestris*) or American angelica (*A. atropurpurea*), a purple-stemmed plant indigenous to North America) with similar uses.

ORIGIN Europe and Asia. The Vikings allegedly spread the herb to central and southern Europe from Iceland, Scandinavia and Russia. It has been cultivated in Europe and elsewhere as a culinary and medicinal herb for centuries, especially in monasteries. *Angelica* means "angel" and suggests divine healing power.[1]

CULTIVATION Angelica thrives in rich, well-drained soil with a slightly acidic pH. It is a hardy plant that can withstand even severe frost. It dies after flowering in the second year.[1]

HARVESTING Stems and leaves are usually harvested in the spring of the second year while roots intended for flavour use are dug up in the autumn of the first year.[1] Flowering is sometimes prevented by pruning to prolong root growth. The fruits (seeds) are harvested when ripe.

CULINARY USES Stems, leaf stalks, roots, fruits and essential oil are used to flavour herbal teas and famous alcoholic beverages such as Benedictine, chartreuse, gin, vermouth and vespétro. The fresh leaves or leaf stalks are sometimes added to cream cheese, salads, sauces, soups, vegetables and meat dishes. Candied stems (traditionally produced in Niort in France)[2] are used to decorate cakes and desserts. To make candied angelica, mature but green stem pieces are soaked in water, softened by cooking in sugar syrup and cleaned by removing the fibrous outer layer. The stem pieces become glassy and translucent when soaked for three days in boiling syrup. They are then dried in an oven and dusted with castor sugar.[2]

FLAVOUR COMPOUNDS The valuable essential oil obtained from the roots or the fruits has a musky, earthy odour and a pungent bittersweet flavour.[1,4] It contains numerous mono- and sesquiterpenoids, especially α-pinene, limonene, β-phellandrene and carene (δ-3-carene), as well as coumarins (osthenole, osthole and angelicin)[3] and macrolides (especially 15-pentadecanolide and 13-tridecanolide).[4] The musky scent is ascribed to the macrolides.

NOTES Angelica has appetite stimulant, stomachic and spasmolytic properties. Chinese angelica or *dang-gui* (*A. polymorpha* var. *sinensis*) is important in Chinese traditional medicine.

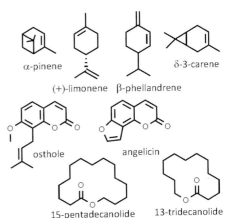

1. **Farrel, K.T. 1999.** *Spices, condiments and seasonings.* Aspen Publishers, Gaithersburg, USA.
2. **Larousse. 1999.** *The concise Larousse gastronomique.* Hamlyn, London.
3. **Harborne, J.B., Baxter, H. 2001.** *Chemical dictionary of economic plants.* Wiley, New York.
4. **Nivinskiene, O., Butkiene, R., Mockute, D. 2003.** Changes in the chemical composition of essential oil of *Angelica archangelica* L. roots during storage. *Chemija (Vilnius)* 14: 52–56.

Leaf, petiole and candied stem of *Angelica archangelica*

10 mm

Fruits (*Angelica archangelica*)

Fruits (*Angelica archangelica*)

Plant (*Angelica archangelica*)

Swollen leaf bases (*Angelica archangelica*)

Anthriscus cerefolium
chervil • garden chervil • French parsley

Anthriscus cerefolium (L.) Hoffm. (Apiaceae); *kerwel* (Afrikaans); *cerfeuil* (French); *Gartenkerbel* (German); *cerfoglio* (Italian); *chaviru* (Japanese); *chabil* (Korean); *cerefólio* (Portuguese); *perifollo, cerafolio* (Spanish)

DESCRIPTION The fresh or dried leaves are used. They are bright green, pinnately compound (divided into numerous narrow segments) and superficially resemble flat-leaf parsley.

THE PLANT Chervil or garden chervil is a sparsely hairy annual herb with bright green leaves that are divided into numerous flat segments. The plant bears small white flowers in characteristic umbels. These are followed by small dry fruits (sometimes called seeds) that are oblong in shape and about 10 mm (just under ½ in.) long.

ORIGIN Chervil originated from central Asia (Iran to southern Russia). It is an ancient culinary herb with a long recorded history of use by the Greeks and Romans. Chervil is mentioned in the famous *Historia naturalis* of the Roman writer Pliny the Elder (AD 23–79).[1] The herb was introduced from the eastern Mediterranean to the rest of Europe. It has become naturalized in the United States.

CULTIVATION Chervil is easily grown in the kitchen garden from seeds sown at regular intervals, from spring onwards. The plant thrives in rich, well-drained soil and prefers cool, partially shaded conditions so that it is often grown in spring and autumn but not mid-summer.[2] In large urban centres it is readily available from commercial plantings.

HARVESTING Leaves are harvested during vegetative growth, before the plants start to flower (60 to 90 days after sowing). They are immediately taken to the cool room and kept moist until use.

CULINARY USES Chervil should be used fresh. The aromatic and spicy leaves with their delicate aniseed flavour are a favourite among French chefs because it is one of the essential components of *fines herbes* (together with parsley, tarragon and chives).[3] The parsley-like leaves are widely used as a condiment and garnish with fish or egg dishes, soups and omelettes. Chopped leaves can be sprinkled over any salad, vegetable (especially green peas), soup or meat dishes and included in dips and stuffings. Chervil is an important ingredient of *vinaigrette* ("French dressing"), a vinegar-based dressing for salads or cold dishes. Other French sauces that typically contain chervil include *béarnaise* (a hot creamy sauce served with fish or meat dishes) and *gribiche* (a cold mayonnaise-like sauce served with cold fish or calf's head).[3] It can be used as a substitute for French tarragon, which is chemically similar. Chervil is also an ingredient of Frankfurt green sauce.

FLAVOUR COMPOUNDS Chervil fruits contain about 1% essential oil with estragole (methylchavicol) and 1-allyl-2,4-dimethoxybenzene as major compounds and smaller amounts of anethole (aniseed flavour) and other constituents.[4] The oil is used commercially as a food ingredient to flavour beverages, condiments, confectionery and meat products. Estragole and anethole have irritant properties so that the oil should be used in small amounts only.

NOTES *Chaerophyllum bulbosum* is a similar but unrelated plant, known as turnip-rooted chervil. It has edible roots that are eaten as a vegetable in central and northern Europe.

estragole (methylchavicol)   1-allyl-2,4-di-methoxybenzene   anethole

1. **Mabberley, D.J. 2008.** *Mabberley's plant-book* (3rd ed.). Cambridge University Press, Cambridge.
2. **Farrel, K.T. 1999.** *Spices, condiments and seasonings.* Aspen Publishers, Gaithersburg, USA.
3. **Larousse. 1999.** *The concise Larousse gastronomique.* Hamlyn, London.
4. **Harborne, J.B., Baxter, H. 2001.** *Chemical dictionary of economic plants.* Wiley, New York.

Chervil, the leaves of *Anthriscus cerefolium*

Plant (*Anthriscus cerefolium*)

Fruits (*Anthriscus cerefolium*)

Young fruits (*Anthriscus cerefolium*)

Flowers (*Anthriscus cerefolium*)

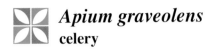

Apium graveolens
celery

Apium graveolens L. (Apiaceae); *seldery* (Afrikaans); *qin cài, sai kan choi* (Chinese); *célerei* (French); *Sellerie* (German); *sèdano* (Italian); *oranda mitsuba* (Japanese); *daun sop* (Malay); *apío* (Spanish); *khuen chai* (Thai)

DESCRIPTION The small dry fruits ("celery seeds") are about 1 mm (0.04 in.) in diameter and have minute corky ridges visible on the fruit surface. The flat leaves of Chinese celery (var. *secalinum*) are used as a condiment, while stalk celery (var. *dulce*) and celeriac (var. *rapaceum*) are vegetables but nevertheless add a lot of flavour when used in soups and stews.

THE PLANT The celery plant is an erect, biennial aromatic herb with bright green foliage and umbels of tiny cream-coloured flowers followed by small dry fruits.

ORIGIN The wild form of celery occurs over large parts of Europe, Asia and Africa and has a long history of use as a culinary herb by the ancient Chinese, Egyptians, Greeks and Romans. Chinese celery (*qin cài*) is derived from the highly aromatic wild Asian form of the species. Vegetable uses (stalk celery, celeriac) developed later.[1]

CULTIVATION Seeds are directly sown or container-grown seedlings may be planted. Celery requires rich organic soil and thrives even under saline conditions. Celery is a marsh plant and therefore requires moist conditions and regular watering.

HARVESTING Ripe fruits are gathered in the second year from mature plants. Chinese celery leaves are cut at regular intervals throughout the season. Stalk celery has to be blanched but self-blanching cultivars have been developed that spontaneously lose the chlorophyll in their leaf stalks.[1] Yields of 40 to 60 tons per hectare can be expected.

CULINARY USES The Malay name for Chinese celery is *daun sop* or "soup leaf" because it is mainly used in soup, to which it adds a strong, bitter taste. In Asian cooking it is also used in noodle or rice dishes and in stir-frying. Stalk celery can be used raw in salads but is best known for adding a rich spicy flavour to soups and stews. Celery seeds are used as a spice in Indian cooking and are an essential ingredient of many spice mixtures and recipes in French, English and American cookery. These include meat dishes, sausages, salami, corned beef, soups, gravies, beverages, confectionery, ice cream and chewing gum.[2]

FLAVOUR COMPOUNDS Celery leaves and petioles owe their distinctive flavour to a combination of four compounds, namely 3-butyltetrahydrophthalide, 3-butylphthalide, apiole and myristicin.[3,4] The fruits contain essential oil with limonene as main compound and smaller amounts of β-selinene, apiole, santalol, α- and β-eudesmol and dihydrocarvone. Also present in the seed oil are small amounts (1% each) of 3-butylphthalide and sedanolide, to which the characteristic odour is ascribed.[3]

NOTES Celery salt has become popular in low-salt diets. It is made from dried and powdered celeriac mixed with sea salt.

apiole

myristicin

3-butyl-4,5-dihydrophthalide

3-butyl-phthalide

1. **Riggs, T.J. 1995.** Umbelliferous minor crops (Umbelliferae). In: Smartt, J., Simmonds, N.W. (Eds), *Evolution of crop plants* (2nd ed.), pp. 481–484. Longman, London.
2. **Farrel, K.T. 1999.** *Spices, condiments and seasonings.* Aspen Publishers, Gaithersburg, USA.
3. **Harborne, J.B., Baxter, H. 2001.** *Chemical dictionary of economic plants.* Wiley, New York.
4. **Sellami, I.H., Bettaieb, I., Bourgou, S., Dahmani, R., Limam, F., Marzouk, B. 2012.** Essential oil and aroma composition of leaves, stalks and roots of celery (*Apium graveolens* var. *dulce*) from Tunisia. *Journal of Essential Oil Research* 24: 513–521.

Leaves and fruits of *Apium graveolens*

Stem celery (*Apium graveolens*)

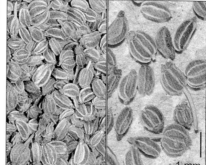
Fruits (*Apium graveolens*)

Celeriac (*Apium graveolens*)

Flowers (*Apium graveolens*)

Armoracia rusticana
horseradish

Armoracia rusticana P.Gaertn., Mey. & Scherb. (Brassicaceae); *peperwortel* (Afrikaans); *la gen* (Chinese); *křen* (Czech); *piparjuuri* (Finnish); *raifort, grand raifort* (French); *Meerrettich, Kren* (German); *torma* (Hungarian); *cren* (Italian); *hoosuradiishu, seiyou wasabi* (Japanese); *chrzan* (Polish); *khren* (Russian); *rábano picante* (Spanish); *pepparot* (Swedish)

DESCRIPTION The thick fleshy roots are yellowish brown in colour with white flesh. Intact roots have no odour, but a hot, biting, pungent taste and a sharp smell develop when they are grated, crushed or powdered.[1]

THE PLANT A perennial leafy herb of up to 1.2 m (4 ft) in height. It bears large, soft, bright green leaves, small white flowers and oblong seed capsules. The fruits only rarely form viable seeds.[1]

ORIGIN Probably southeastern Europe and western Asia.[2,3] It appears to be a sterile cultigen, of hybrid origin, that has been cultivated only for the last 2 000 years.[2,3] The plant has been particularly popular in central Europe (sometimes called "German mustard") but is now grown commercially in many parts of the world. It has become invasive in parts of the United States.[1]

CULTIVATION Propagation is from root crowns or root cuttings. It prefers deep, moist, rich loam soil and grows best is semi-shaded places. Horseradish is a perennial crop in Europe but is grown as an annual in the United States.[1]

HARVESTING Roots are dug up in autumn and the lateral roots removed and stored in a dark place for planting in the following season. At the factory, the roots are cleaned, trimmed, grated and processed. Larger roots are preferred as they allow easier and more efficient processing.

CULINARY USES Horseradish sauce is made by mixing grated root with vinegar and salt and blending it with milk or cream into a ready-to-use cream sauce. The root can be dried and powdered or mixed with mustard (Creole mustard), vinegar (horseradish vinegar), mayonnaise or red beet juice (red horseradish).[4] Famous variations include the English "Tewkesbury mustard" (mentioned by Shakespeare), Austrian *Krensenf*, German *Tafelmeerrettich* and American "horseradish sauce". The sauce is used sparingly as a condiment, commonly with meat and fish dishes. It is particularly popular in the Alsace, Germany, eastern Europe, Russia and Scandinavia but has also become a favourite in the United States as an ingredient of hot and cold sauces, including mustards, dips, relishes and salad dressings. It is an ideal condiment for beef, pork, sausages and potato salad. Scandinavian dishes such as herring, smoked reindeer and smoked trout are traditionally served with horseradish sauce. Horseradish loses its pungency when it is exposed to the heat of cooking. It is therefore best used fresh or added to a warm dish just before serving.

FLAVOUR COMPOUNDS The pungent aroma is due to mustard oils (glucosinolates). When horseradish roots are damaged by grating or grinding, the glucosinolate (sinigrin) that is stored in the vacuoles of the root cells comes into contact with enzymes (thioglucosidase or myrosinase). The sinigrin is hydrolysed and further chemically modified to produce the highly reactive and pungent allyl isothiocyanate.[1]

NOTES Horseradish can be dangerous when taken in large quantities. The mustard oils may irritate the lining of the mouth, throat, nose, digestive system and urinary tract.

sinigrin → allyl isothiocyanate

1. Farrel, K.T. 1999. *Spices, condiments and seasonings.* Aspen Publishers, Gaithersburg, USA.

2. Courter, J.W., Rhodes, A.M. 1969. Historical notes on horseradish. *Economic Botany* 23: 156–164.

3. Wedelsbäck Bladha, K., Olsson, K.M. 2011. Introduction and use of horseradish (*Armoracia rusticana*) as food and medicine from antiquity to the present: emphasis on the Nordic countries. *Journal of Herbs, Spices & Medicinal Plants* 17: 197–213.

4. Kiple, K.F., Ornelas, K.C. (Eds) 2000. *The Cambridge world history of food.* Cambridge University Press, Cambridge.

Horseradish, the roots of *Armoracia rusticana*

Horseradish sauce

Plant (*Armoracia rusticana*)

Flowers (*Armoracia rusticana*)

Artemisia dracunculus
tarragon • French tarragon • estragon

Artemisia dracunculus L. (Asteraceae); *dragon* (Afrikaans); *tarkhūn* (Arabic); *estragon* (French); *Esdragon*, *Estragon* (German); *estragon* (Italian); *estragon* (Spanish)

DESCRIPTION The fresh, dried or frozen leaves are narrowly oblong and hairless, with an anise-like aroma and liquorice taste. Russian (false) tarragon is sometimes used as a substitute.[1]

THE PLANT It is a branched and spreading perennial herb, about 0.2 m (ca. 8 in.) high, with inconspicuous off-white flower heads. Russian tarragon, a cultivar or a separate species (then called *A. dracunculoides*), is very similar but more robust (and less aromatic).

ORIGIN Southeastern Europe and central Asia. The early culinary history is obscure but the name "tarragon" is derived from *tarkhūn*, the Arabic name.[1] Tarragon only became popular as a salad green and flavouring agent in the 16th century.

CULTIVATION French tarragon rarely produces seeds and is propagated by division or by stem or root cuttings. Commercial production occurs mainly in southern Europe, western Asia and North America. It requires warm, sunny conditions and slightly alkaline, well-drained soils. Plantations are replaced every three years. Russian tarragon is much easier to grow.

HARVESTING The thin leafy stems are harvested twice a year, potentially yielding about 15 tons fresh leaves or 30 kg (66 lbs) essential oil per hectare. Leaves are carefully dried in the shade to retain as much of the flavour (and green colour) as possible.

CULINARY USES Tarragon or estragon is one of the most sought after herbs amongst gourmet chefs because of its delicate anise flavour, reminiscent of liquorice.[1] It is one of the four ingredients of the French *fines herbes* (together with parsley, chervil and chives). Tarragon vinegar is made by steeping a few fresh leafy twigs in a bottle of white wine vinegar. It is an essential ingredient of famous sauces such as béarnaise, hollandaise and tartare that classically accompany asparagus, green beans, peas and other vegetables (as tarragon cream) or chicken, meat and eggs. Leaves (preferably fresh) are used to flavour meat dishes, stews, fish dishes, salads, pickles and mustard sauces. Extracted oleoresin is an ingredient of commercial food products – one part is equivalent to four parts of dried leaves.

FLAVOUR COMPOUNDS Leaves contain essential oil with up to 70% estragole (methylchavicol) as main ingredient, accompanied by several minor volatile constituents.[2,3] Estragole is an isomer of *trans*-anethole and is responsible for the anise flavour. The pungent taste is due to *cis*-pellitorin (an isobutyramide).[4]

NOTES Beverages and food items traditionally flavoured with other *Artemisia* species include absinthe and the original Pernod and pastis (*A. absinthium*, wormwood or absinth), vermouth (*A. pontica*, Roman wormwood), beer and herbal teas (*A. abrotanum*, southernwood) and roasted goose (*A. vulgaris*, mugwort).

estragole
(methylchavicol)

anethole

1. **Larousse. 1999.** *The concise Larousse gastronomique*. Hamlyn, London.
2. **Lopes-Lutz, D.S., Alviano, D.S., Alviano, C.S., Kolodziejczyk, P.P. 2008.** Screening of chemical composition, antimicrobial and antioxidant activities of *Artemisia* essential oils. *Phytochemistry* 69: 1732–1738.
3. **Sayyah, M., Nadjafnia, L., Kamalinejad, M. 2004.** Anticonvulsant activity and chemical composition of *Artemisia dracunculus* L. essential oil. *Journal of Ethnopharmacology* 94: 283–287.
4. **Gatfield, I.L., Ley, J., Foerstner, J., Krammer, G., Machinek, A. 2003.** Production of *cis*-pellitorin and use as a flavoring. Patent: WO04/000787/ PCT/EP03/06545.

Tarragon leaves and tarragon vinegar (*Artemisia dracunculus*)

French tarragon (*Artemisia dracunculus*)

Russian tarragon (*Artemisia dracunculoides*)

Southernwood (*Artemisia abrotanum*)

Absinth, wormwood (*Artemisia absinthium*)

Roman wormwood (*Artemisia pontica*)

Mugwort (*Artemisia vulgaris*)

Bixa orellana
annatto • achiote • roucou

Bixa orellana L. (Bixaceae); *annato* (Afrikaans); *yan zhi shu* (Chinese); *rocou, roucou, annatto* (French); *annatou* (German); *kesumba* (Indonesian); *annatto* (Italian); *jarak belanda, kesumba* (Malay); *atsuete* (Philippines); *urucú, urucum* (Portuguese); *achiote, achiotillo* (Spanish); *kam tai* (Thai)

DESCRIPTION The seeds are cone-shaped and about 4 mm (just over ⅛ in.) long, with a bright red seed coat containing a natural pigment (*annatto* or *roucou*).[1] They have practically no taste or aroma.

THE PLANT It is a distinctive shrub or small tree bearing heart-shaped leaves and clusters of pink flowers. Between 10 and 50 seeds are borne in each of the oblong, spiny pods.

ORIGIN *Bixa* is indigenous to tropical America. It has been used as ritual body paint by Native Americans and is known as *achiote, annatto, bijol* and *roucou* (*achiotl* is the original Aztec name). In the 17th century, the Spanish introduced it to almost all warm regions of the world.[1]

CULTIVATION Any tropical or frost-free subtropical area with a high rainfall and well-drained soil is suitable for commercial production, preferably with a dry period during seed ripening.[2,3] High-yielding clones are selected and grown from cuttings but most commercial trees are propagated from seeds even though seeds and bixin yields may then be quite variable.[2,3] Plantations can be found in many Central American countries and in Africa and Asia. Brazil is a major producer for local consumption but Peru and Kenya are responsible for most of the annual ca. 7 000 tons in international trade. Consumption has increased in recent years because of bans on synthetic food dyes (it is a natural alternative to tartrazine) and various applications in confectionery, soaps and cosmetics.

HARVESTING Fruits are harvested (by knife or secateurs) when they start to split open and before the seeds are exposed to sunlight and rain. Seed yields of around 1 000 to 2 000 kg (2 200 to 4 400 lbs) per hectare (dry weight) are typical.[3] The seeds are cleaned by mechanical sieving and winnowing, taking care to avoid abrasion damage.

CULINARY USES Annatto is used mainly to give an attractive yellow colour to margarine, butter, cheese, processed salad oils and smoked fish.[3] The whole or ground seeds have almost no flavour but are nevertheless important because of the colour it adds to Meso-American and Philippine dishes. It is an essential ingredient of a spice mixture called *recado rojo* (achiote paste) that is used in marinades and rubs for meat, Mexican *cochinita pibil* (pork), *tascalate* (a drink), *empanadas* and *tamales*. It takes the place of saffron in Jamaican akee-fish-and-rice and in many traditional dishes in the Philippines (where it is called *atsuete*).

FLAVOUR COMPOUNDS The main pigment is a carotenoid called bixin (2 to 3% yield).[3,4] It occurs as a mixture of the *cis* and *trans* isomers. Seeds are extracted with dilute sodium or potassium hydroxide, resulting in the natural bixin being converted to water-soluble norbixin salt. An alternative is extraction with vegetable oil, giving a suspension of *cis*- and *trans*-bixin.[3,4]

NOTES Annatto very rarely causes food-related allergies in sensitive persons.

cis-bixin

trans-bixin

cis-norbixin

trans-norbixin

1. Ingram, J.S., Francis, B.J. 1969. The annatto tree (*Bixa orellana* L.) – a guide to its occurrence, cultivation, preparation and uses. *Tropical Science* 11: 97–102.
2. Venter, M.W. 1981. The cultivation of *Bixa orellana* L. *Crop Production* (South Africa) 5: 87–89.
3. Henry, B.S. 1992. Natural food colours. In: Hendry, G.A.F., Houghton, J.D. (Eds), *Natural Food Colourants* (2nd ed.), pp. 40–79. Blackie Academic & Professional, Glasgow.
4. Preston, H.D., Rickard, M.D. 1980. Extraction and chemistry of annatto. *Food Chemistry* 5: 47–56.

Annatto fruits and seeds (*Bixa orellana*)

Flowers and leaves (*Bixa orellana*)

Seed extract

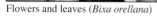

Seeds (*Bixa orellana*)

Fruits (*Bixa orellana*)

Flower (*Bixa orellana*)

83

Boesenbergia rotunda
fingerroot • Chinese keys • krachai • suo shi

Boesenbergia rotunda (L.) Mansf. [= *B. pandurata* (Roxb.) Schltr.] (Zingiberaceae); *vingerwortel* (Afrikaans); *ao chun jiang*, *suo shi* (Chinese); *Chinesischer Ingwer* (German); *temu kunchi* (Malay); *kunci* (Indonesian); *krachai* (Thai); *bong nga truat*, *cu ngai*, *ngai num kho* (Vietnamese)

DESCRIPTION Fingerroot comprises the short rhizome with several finger-like roots attached. The roots are orange-brown, yellow on cut surfaces with a highly aromatic, spicy flavour. They are best used fresh but are sometimes available in pickled or frozen form.

THE PLANT This is a small, ginger-like perennial herb with broad leaves and attractive pink flowers with a characteristic purple lip.[1,2] The correct name of the plant is *B. rotunda*; the name *B. pandurata* is still widely used but is regarded as a synonym by most authorities.[1,2]

ORIGIN Chinese keys are widely distributed in southern and eastern Asia, including India, Sri Lanka, Thailand, Vietnam, Malaysia, Indonesia, Indo-China and southern China.[1] The exact origin is not clear, as the plant is an important crop, grown as a spice in Cambodia, Thailand, Vietnam and Indonesia and as a traditional medicine in central Asian countries, extending to Russia and Hungary.

CULTIVATION A tropical climate, wet conditions and shade are important requirements but the plant is relatively easy to grow. It is propagated by division of the rhizomes.

HARVESTING Fresh roots are dug up at any time of the year. They are sold fresh on the markets, often alongside their close relatives, ginger and galangal.

CULINARY USES Finely chopped or sliced roots are an important ingredient of the cuisines of Indonesia, Cambodia, Thailand and Vietnam. In Thailand, fresh rhizomes and roots of krachai are eaten raw in salads, or they are added to fish curries, soups, stews, stir-fries, pickles, vegetables and other dishes. It is considered a necessary ingredient to balance strong fish or meat odours in dishes such as *kaeng tai pla* (a fish curry) and *gaeng pa* (jungle curry, made with pork or chicken). In Cambodia, *k'cheay* (the Khmer name) is one of eight components of the traditional *kroeung* spice pastes. It is also widely used in Javanese cuisine. The medicinal value (as a functional food) is utilized in modern-day instant soups produced in China.

FLAVOUR COMPOUNDS The distinctive flavour is usually ascribed to the presence of camphor and geraniol as major components in the essential oil of the roots but it is likely that several other compounds also contribute to the aroma.[3] A complex mixture of chemical compounds is present, of which chalcones (e.g. panduratin A and boesenbergin A) have been the focus of many pharmacological studies.[4]

NOTES Fresh or dried fingerroot is an important traditional medicine, used to treat all types of digestive disturbances.[4]

camphor

geraniol

boesenbergin A

panduratin A

1. **Wu, D., Larsen, K. 2000.** Zingiberaceae. In: Wu, Z.Y., Raven, P.H. (Eds), *Flora of China*, Vol. 24, pp. 322–377. Science Press, Beijing.
2. **Sirirugsa, P. 1992.** A revision of the genus *Boesenbergia* Kuntze (Zingiberaceae) in Thailand. *Natural History Bulletin of the Siam Society* 40: 67–90.
3. **Sukari, M.A., Mohd Sharif, N.W., Yap, A.L.C. et al. 2008.** Chemical constituents variations of essential oils from rhizomes of four Zingiberaceae species. *The Malaysian Journal of Analytical Sciences* 12: 638–644.
4. **Tan, E.-C., Lee, Y.-K., Chee, C.-F. et. al. 2012.** *Boesenbergia rotunda*: from ethnomedicine to drug discovery. *Evidence-Based Complementary and Alternative Medicine* 12, Article ID 473637, 25 pages. doi:10.1155/2012/473637

Fingerroot or Chinese keys, the rhizome and roots of *Boesenbergia rotunda*

Plants (*Boesenbergia rotunda*)

Flower (*Boesenbergia rotunda*)

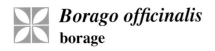

Borago officinalis
borage

Borago officinalis L. (Boraginaceae); ***komkommerkruid*** (Afrikaans); ***bourrache*** (French); ***Boretsch*** (German); ***boragine*** (Italian); ***borraja*** (Spanish)

DESCRIPTION Borage leaves and stems are covered in bristly hairs and the flowers are typically bright blue but may also be pink or rarely white. The taste and flavour of borage are similar to that of cucumber.

THE PLANT This is a robust, leafy annual with thick, soft stems bearing large leaves and clusters of drooping, star-shaped flowers.

ORIGIN The plant is believed to be of Middle Eastern origin but it occurs naturally in the western Mediterranean region and southern Europe.[1] According to beliefs dating back to ancient Greece, borage gives courage and eliminates sadness. It has spread to most temperate regions of the world where it is often grown in herb gardens.

CULTIVATION Propagation is from seeds – borage may seed itself and can become somewhat weedy. Nowadays it is often grown more for decorative than culinary purposes. Large-scale production is aimed at producing seeds for the extraction of the valuable seed oil.

HARVESTING Leaves and flowers may be sporadically harvested throughout the season. Seeds are mechanically harvested for oil extraction.

CULINARY USES Freshly picked young leaves have a mild, cucumber-like taste and are traditionally used as a potherb and ingredient of salads, puréed soups and cucumber pickles. The characteristic taste is quickly lost when the leaves are cooked or boiled. Borage is still popular in Germany, Spain, France, northern Italy and Poland despite the fact that it contains trace amounts of poisonous alkaloids. In Germany, borage is one of the essential ingredients of the popular Frankfurt green sauce (*Frankfurter Grüne Sauce*) together with six other herbs (burnet, chervil, chives, cress, parsley and sorrel), vegetable oil, boiled egg yolks, vinegar and white pepper. The borage may be replaced with cucumber. Green sauce is typically served with meat, fish, vegetables or young potatoes. In Britain, borage leaves and flowers have traditionally been added to gin drinks and cocktails (such as Pimm's No. 1, Pimm's cup and claret cup). Nowadays, a sliver of cucumber peel is often used instead. The crystallized (candied) flowers are one of only a few food items with a blue colour and are a popular garnish on cakes, desserts and fruit salads.

FLAVOUR COMPOUNDS The main volatile compound responsible for the cucumber-like flavour of borage is (*E,E*)-2,4-decadienal.[2] It is structurally closely related to 2,6-nonadienal (so-called "cucumber aldehyde"), the main aroma compound of cucumbers.[3]

NOTES Pyrrolizidine alkaloids occur in trace amounts in leaves and flowers,[4] so that regular culinary use may not be entirely safe. The same is true for comfrey (*Symphytum officinalis*) formerly much used for making herbal tea. Borage seed oil ("starflower oil") has become a popular functional food item and dietary supplement to alleviate the symptoms of eczema and stress-related ailments. It contains gamma-linoleic acid (GLA), an essential fatty acid.

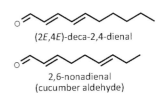

(2*E*,4*E*)-deca-2,4-dienal

2,6-nonadienal
(cucumber aldehyde)

1. **Mabberley, D.J. 2008.** *Mabberley's plant-book* (3rd ed.). Cambridge University Press, Cambridge.
2. **Mhamdi, B., Wannes, W.A., Dhifi, W., Marzouk, B. 2009.** Volatiles from leaves and flowers of borage (*Borago officinalis* L.). *Journal of Essential Oil Research* 21: 504–506.
3. **Schieberle, P., Ofner, S., Grosch, W. 1990.** Evaluation of potent odorants in cucumbers (*Cucumis sativus*) and muskmelons (*Cucumis melo*) by aroma extract dilution analysis. *Journal of Food Science* 55: 193–195.
4. **Martina, H., Holger, J., Gerhard, S. 2002.** Thesinine-4'-O-β-glucoside – the first glycosylated plant pyrrolizidine alkaloid from *Borago officinalis*. *Phytochemistry* 60: 399–402.

Leaf and flowers of borage (*Borago officinalis*)

Flowers (*Borago officinalis*)

Seeds (*Borago officinalis*)

3 mm

Comfrey (*Symphytum officinalis*)

Flowers (*Borago officinalis*)

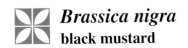

Brassica nigra
black mustard

Brassica nigra (L.) Koch (Brassicaceae); *swart mosterd* (Afrikaans); *senafitch* (Amharic); *hei jie* (Chinese); *moutarde noire* (French); *Schwarzer Senf* (German); *kali rai* (Hindi); *senape nera* (Italian); *kuro garashi* (Japanese); *mostarda negra* (Portuguese); *mostaza negra* (Spanish)

DESCRIPTION Black mustard is the small, reddish brown or blackish, spherical ripe seeds harvested from *B. nigra* plants. They are about 1 mm (0.04 in.) in diameter and have a nutty flavour (especially when briefly heated in oil) but no aroma. A pungent taste develops after chewing the seeds for some time. "Black", "White" or "Brown" are often used interchangeably but mostly refer to *B. nigra*, *Sinapis alba* and *B. juncea*, respectively. Brown mustard or Indian mustard (*B. juncea*) originated as a hybrid between *B. nigra* and *B. campestris* and is used as an oilseed in India or as Dijon mustard in France (Dijon is the main production area of mustard paste).[1] The seeds are slightly larger (up to 2 mm / 0.08 in. in diameter) and paler in colour. Many different cultivars of brown mustard have been developed as vegetables in China, in the same way as cabbage (*B. oleracea*) was developed in Europe.[2] *Mizuna* or Japanese greens are a cultivated variety of turnip (*B. rapa*).

THE PLANT Black mustard is a robust annual with broad leaves, large, branched inflorescences bearing small yellow flowers and smooth, beaked capsules that split open when they ripen (those of brown mustard remain closed at maturity). White mustard (see *Sinapis alba*) has hairy fruits and yellowish seeds.[3]

ORIGIN Black mustard originated in the Middle East and has been an important spice since ancient times.[3,4] It was rapidly replaced by brown mustard around the 1950s, because the latter has indehiscent (non-shattering) capsules making it suitable for large-scale harvesting. More than 160 000 tons of mustard seeds are sold each year, making it the most important spice by volume and second only to black pepper in monetary terms.[3]

CULTIVATION Black and brown mustard are both hardy annual crops grown from seeds sown in spring.

HARVESTING Black mustard capsules have to be harvested by hand at regular intervals while those of brown mustard are non-shattering and therefore ideal for large-scale mechanical harvesting.

CULINARY USES The seeds are widely used as a spice and ingredient of spice mixtures and curry powders. Various types of mustard sauces are among the most widely used condiments in Western countries, served since the 14th century in a special mustard pot.[1] The typical mustard taste ("spicy and piquant") develops about 15 minutes after the powdered seeds have been mixed with some form of liquid, thus allowing for enzyme activity. Water is used in the case of "English mustard", while vinegar or white wine is the traditional choice in "Continental mustard". Egg yolk or grape must (*mustum ardens* in Latin or "piquant must", hence "mustard") are also used.[1] Mustard sauces traditionally accompany hot or cold meats and sausages, including the famous American hot dog. It is often available from special dispensers at fast food outlets.

FLAVOUR COMPOUNDS Volatile mustard oil with its pungent taste is released through enzymatic action (by myrosinase) when the mustard-oil glycoside sinigrin (ca. 1% in the seeds) is converted to allyl isothiocyanate.

NOTES Black mustard has antibacterial activity.

sinigrin   allyl isothiocyanate

1. Larousse. 1999. *The concise Larousse gastronomique.* Hamlyn, London.
2. Van Wyk, B.-E. 2005. *Food plants of the world.* Briza Publications, Pretoria.
3. Hemingway, J.S. 1995. Umbelliferous minor crops (Umbelliferae). In: Smartt, J., Simmonds, N.W. (Eds), *Evolution of crop plants* (2nd ed.), pp. 82–86. Longman, London.
4. Vaughan, J.G., Hemingway, J.S. 1959. The utilization of mustards. *Economic Botany* 13: 196–204.

Black mustard (*Brassica nigra*) with brown mustard (*B. juncea*) and white mustard (*Sinapis alba*)

Mustard powder and mustard pot

Seeds (*B. nigra*)

1 mm

Flowers and fruits (*Brassica nigra*)

Plants (*Brassica nigra*)

Plants (*Brassica juncea*)

Camellia sinensis
tea • chai

Camellia sinensis (L.) O. Kuntze (= *Thea sinensis* L.) (Theaceae); *tee* (Afrikaans); *cha* (Chinese – Cantonese, Mandarin); *te* (Chinese – Min dialect); *thé* (French); *chay* (Hindi); *Tee* (German); *tè* (Italian); *da, cha, sa* (Japanese); *da, cha* (Korean); *teh* (Malay); *chá, chai* (Portuguese); *chay* (Russian); *té* (Spanish); *te* (Swedish); *cha* (Thai)

DESCRIPTION Tea is typically made from a "flush", which comprises the terminal bud ("pekoe tip") together with two to four leaves immediately below it (respectively called "orange pekoe", "pekoe", "pekoe souchong" and "souchong").[1] The stems can be used for twig tea (*kukicha*). The three basic types are green tea (unfermented), oolong tea (semi-fermented) and black tea (fully fermented).[1,2]

THE PLANT Chinese tea (var. *sinensis*) is a much-branched shrub of about 3 m (10 ft) high with relatively small leaves, while Assam tea (var. *assamica*) is a tree (10 m / 33 ft or more when not trimmed) with larger and more pointed leaves.

ORIGIN Tea occurs naturally in southern China, Cambodia, Laos, Myanmar and Thailand, where it has been an important part of local cultures for thousands of years. During the colonial era, plantations of Chinese and Assam tea were established in Java, India and later Sri Lanka and Kenya to satisfy a rapidly increasing demand, which currently exceeds 4 million tons per year. Nearly 80% is black tea – green tea is mainly used in China, Japan, India and parts of Arabia. Oolong tea is favoured in China and Taiwan.[3]

CULTIVATION Seedlings or cuttings are planted in tropical or subtropical climates with an annual rainfall of at least 1 270 mm (50 in.). Slow growth at high elevations results in more flavour and superior quality.

HARVESTING Tea leaves are harvested by (weekly or fortnightly) hand-plucking of the flushes (to ensure uniformity and quality), or by mechanical plucking. Black tea is produced by a carefully controlled process of withering, rolling (to break cell walls), "fermentation" (enzymatic oxidation and polymerization of phenolic compounds), firing (to inactivate enzymes and to dry the product) and finally sorting, grading and packing. Flowers of jasmine or *mo li* (*Jasminum sambac*) and *gui hua* (*Osmanthus fragrans*) are used in China to flavour tea.

CULINARY USES Tea is an essential ingredient of oriental cooking (especially in China, Japan and Vietnam) and is used in meat, fish and egg dishes. Well-known examples are tea eggs, tea-smoked duck and Chinese stir-fried shrimps. The use of tea in food and desserts is gaining popularity in Western countries such as France and the United States. Food-pairing with speciality teas (rather than wine) is another new trend.

FLAVOUR COMPOUNDS The taste and flavour of tea is determined by a complicated mixture of volatile terpenoids, organic acids, astringent phenolics (e.g. epigallocatechin) and bitter-tasting alkaloids (mainly caffeine, at 2 to 4%).[3,4] Non-terpenoid hexenols (especially *cis*-3-hexen-1-ol) give tea its fresh green flavour; monoterpene alcohols (mainly linalool and geraniol, but also other compounds such as indole, *cis*-jasmone and others)[3,4] are responsible for the floral fragrances.

NOTES Roman chamomile (*Chamaemelum nobile*) and German chamomile (*Matricaria recutita*) are popular herbal teas and flavour components of health drinks.

epigallocatechin

caffeine

cis-3-hexen-1-ol

indole

linalool   geraniol   *cis*-jasmone

1. Kiple, K.F., Ornelas, K.C. (Eds). 2000. *The Cambridge world history of food*. Cambridge University Press, Cambridge.
2. Stella, A., Brochard, G., Beauthéac, N., Donzel, C., Walter, M. 1992. *The book of tea*. Flammarion, Paris.
3. Chaturvedula, V.S.P., Prakash, I. 2011. The aroma, taste, color and bioactive constituents of tea. *Journal of Medicinal Plants Research* 5: 2110–2124.
4. Borse, B.B., Rao, L.J.M., Nagalakshmi, S., Krishnamurthy, N. 2002. Fingerprint of black teas from India: Identification of the regio-specific characteristics. *Food Chemistry* 79: 419–424.

Tea types: oolong, orange pekoe, Earl Grey, keemun, Russian black, silver and gunpowder

Leaves and flower (*Camellia sinensis*)

3 mm

Ethiopian tea spice

Jasmine, *mo li* (*Jasminum sambac*)

Osmanthus, *gui hua* (*Osmanthus fragrans*)

Roman chamomile (*Chamaemelum nobile*)

German chamomile (*Matricaria recutita*)

91

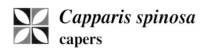

Capparis spinosa
capers

Capparis spinosa L. (Capparidaceae); *kappers* (Afrikaans); *ci shan gan* (Chinese); *câpres* (French); *Kapern* (German); *kiari* (Hindi); *cappero* (Italian); *keepaa* (Japanese); *melada* (Malay); *alcaparras* (Portuguese); *alcaparro* (Spanish); *kebere* (Turkish)

DESCRIPTION Capers are the small unopened flower buds, usually preserved in brine. The young fruits (caperberries) are also used. Both the buds and the fruits have a pungent, slightly bitter, sour and salty taste and a fragrant and fruity aroma.

THE PLANT It is thought to be a hybrid between *C. sicula* and *C. orientalis* that has been maintained as a cultigen for many centuries.[1] The spreading to somewhat erect woody shrub has slightly thorny stipules and fleshy leaves. The attractive but short-lived flowers have a central stalked ovary surrounded by numerous stamens that develops into a fleshy berry.[1]

ORIGIN The caper bush is indigenous to the Mediterranean region, including Spain, France, Italy, Greece and Turkey.[1] Remains of capers have been found in several prehistoric archaeological sites (dating back to 9000 BC) and the product has been recorded in ancient texts (e.g. Greek and Roman) as a popular condiment and food item.[2]

CULTIVATION Capers are collected from wild plants but are mostly grown commercially from seeds or cuttings. Germination rates are low and cuttings are not easy to root.[2] Plants survive extreme summer aridity and can be productive for 25 or more years. Regular pruning is essential and irrigation and fertilizer application may result in improved yields, from as low as 1 kg up to 9 kg per plant per year.[2] Spain, Italy, France, Turkey and Morocco are the main producers.[2]

HARVESTING Buds are hand-picked before sunrise, before they open – a very labour-intensive process. They are pickled in white wine vinegar or in brine (Sicilian capers), not only to preserve them but also to remove the bitter taste.[2] The salt is washed off before use. The ripe berries are picked, stalks and all. These are later convenient when the fruits are eaten by hand.

CULINARY USES Capers are an important spice and condiment with a sharp, piquant and salty-sour flavour. They are used for caper sauce and many other sauces (*tartare, remoulade, ravigote, vinaigrette,* sauce *gribiche* and tarragon sauce) that accompany salads, rice, pasta, eggs, seafood and meat dishes (especially mutton and lamb).[2,3] Capers have become popular as a garnish for pizzas, canapés and cheeses, and as a cocktail snack. Sicilian *pasta colle sarde* has capers as important ingredient (with sardines, parsley, pine nuts, raisins and tomatoes).

FLAVOUR COMPOUNDS The pungent taste is due to sulphur compounds, mainly methylglucosinolate (glucocapperin) that is enzymatically converted (by myrosinase) to methyl isothiocyanate.[4] The flavour is also ascribed to numerous volatile compounds, including cinnamaldehyde, benzaldehyde, ethyl hexadecanoate, *trans*-nerolidol, 4-terpineol and linalool.[4] Compounds with a raspberry-like flavour have also been reported, including α- and β-ionone, frambinone, frambinylalcohol and zingerone.[4]

NOTES Nasturtium, marigold and other types of flower buds are sometimes used as inferior substitutes for capers.

methylglucosinolate   methyl isothiocyanate
(glucocapperin)

1. **Inocencio, C., Rivera, D., Obón, M.C., Alcaraz, F., Barreña, J.-A. 2006.** A systematic revision of *Capparis* section *Capparis* (Capparaceae). *Annals of the Missouri Botanical Garden* 93: 122–149.
2. **Sozzi, G.O. 2001.** Cape bush: Botany and horticulture. *Horticultural Reviews* 27: 125–185.
3. **Rivera, D., Inocencio, C., Obón, M.C., Alcaraz, F. 2003.** Review of food and medicinal uses of *Capparis* L. subgenus *Capparis* (Capparidaceae). *Economic Botany* 57: 515–534.
4. **Romeo, V., Ziino, M., Giuffrida, D., Condurso, C., Verzera, A. 2007.** Flavour profile of capers (*Capparis spinosa* L.) from the Eolian Archipelago by HS-SPME/GC–MS. *Food Chemistry* 101: 1272–1278.

Capers – flower buds and young fruits of *Capparis spinosa*

Capers

Flower (*Capparis spinosa*)

Flowering plant (*Capparis spinosa*)

Ripe fruit (*Capparis spinosa*)

Capsicum annuum
paprika • chilli • cayenne pepper

Capsicum annuum L. (Solanaceae); *rissie* (Afrikaans); *tian jiao* (Chinese); *poivron* (French); *Spanischer Pfeffer,*
Gewürzpaprika (German); *hara mirch* (Hindi); *paprika* (Hungarian); *pimento, pepperone* (Italian); *peppaa* (Japanese);
cabai (Malay); *pimento* (Portuguese); *pimiento picante* (Spanish); *phrik* (Thai); *biber* (Turkish)

DESCRIPTION The fruits are fleshy, hollow berries, very variable in shape and size as a result of domestication and breeding over many centuries. They may be sweet and non-pungent ("sweet peppers" or "capsicum") or hot (known as "chili peppers" in the USA and as "chillies" or "chilli" in British English). The fruits are used fresh or dry, often as flakes or powder (red pepper), ranging from paprika (mild) to cayenne (moderately hot).

THE PLANT *Capsicum annuum* typically has solitary white flowers and pendulous (drooping) fruits, while *C. frutescens* usually has clustered flowers and very small red fruits borne in an upright position.[1] It has been suggested that *C. annuum*, *C. frutescens* and *C. chinense* should be combined into one species (then called *C. annuum*) because they are almost impossible to tell apart.[1,2] There are many forms of *C. annuum*, ranging from sweet (e.g. bell pepper) to hot (e.g. cayenne), including the famous Mexican 'Jalapeño' and 'Serrano', but also 'Cascabel', 'Catarina', 'Chilhuacle', 'Costeño', 'De Agua', 'Fresno', 'Guajillo', 'Pasilla', 'Pequin', 'Poblano' and 'Pulla', as well as New Mexico chillies such as the 'Anaheim'.

ORIGIN Southern United States through Mexico to northern Peru, Bolivia and the West Indies.[2] The presumed area of origin is Meso-America.[3] The oldest evidence of domesticated chillies came from a cave in Mexico, dated to 5000–6500 BC.[1–3] Columbus introduced *C. annuum* to Europe at the end of the 15th century, from where this new "pepper" was enthusiastically spread to almost all corners of the world.[1–3]

CULTIVATION The plants are perennial subshrubs that thrive in warm, dry climates. In temperate and cold regions, however, they are killed by frost and therefore mostly grown from seeds as an annual crop.[1] Success depends on high temperatures, rich, well-drained soil, regular feeding with compost or liquid fertilizer and regular but sparse watering.

HARVESTING Fruits are best picked at full maturity when maximum flavour and pungency have developed.

CULINARY USES Chilli peppers and paprika are popular all over the world for adding pungency, colour and flavour to a wide range of dishes (see *C. frutescens*). Chilli pepper has largely replaced black pepper as the main pungent principle in Indian and Indonesian food, from which it also takes the name "pepper".

FLAVOUR COMPOUNDS The hot (pungent, piquant) sensation is caused by six natural capsaicinoids, the main ones being capsaicin and dihydrocapsaicin (respectively ca. 70% and 20% of total capsaicinoids). They are the laurate esters of the carotenoids capsanthin and capsorubin.[4] Pure capsaicin has a pungency of 16 000 000 Scoville Heat Units (SHU).

NOTES The Scoville Scale is explained in the section on THE CHEMISTRY OF TASTE AND FLAVOUR on page 40.

capsaicin

dihydrocapsaicin

1. Eshbaugh, W.H. 1993. History and exploitation of a serendipitous new crop discovery. In: Janick, J., Simon, J.E. (Eds), *New crops*, pp. 132–139. Wiley, New York.
2. Walsh, B.M., Hoot, S.B. 2001. Phylogenetic relationships of *Capsicum* (Solanaceae) using DNA sequences from two noncoding regions: the chloroplast *atpB-rbcL* spacer region and nuclear *waxy* introns. *International Journal of Plant Science* 162: 1409–1418.
3. McLeod, M.J., Guttman, S.I., Eshbaugh, W.H. 1982. Early evolution of chili peppers (*Capsicum*). *Economic Botany* 36: 361–368.
4. Harborne, J.B., Baxter, H. 2001. *Chemical dictionary of economic plants.* Wiley, New York.

Paprika peppers and cayenne peppers (*Capsicum annuum*)

'Serrano' (*Capsicum annuum*)

Dried cayenne peppers (*Capsicum annuum*)

Cayenne pepper and 'Serrano'

'Pablano' (*Capsicum annuum*)

Paprika pepper (*Capsicum annuum*)

95

Capsicum frutescens
chilli • bird chilli • Tabasco pepper

Capsicum frutescens L. (Solanaceae); *brandrissie* (Afrikaans); *la jiao* (Chinese); *piment, cayenne* (French); *Tabasco*, *Chili Pfeffer* (German); *mirch* (Hindi); *peperoncino* (Italian); *kidachi tougarashi* (Japanese); *lada merah, cabai merah* (Malay); *peri-peri, pimentão* (Portuguese); *guindilla, ají* (Spanish); *pilipili, piri piri, peri peri* (Swahili); *phrik kheenuu* (Thai); *dar biber* (Turkish); *öt* (Vietnamese)

DESCRIPTION Fruits are typically small (less than 25 mm or 1 in. long), bright red when ripe, oval to oblong, invariably very hot and resemble the wild forms of the species.[1]

THE PLANT A frost-sensitive perennial with green-ish-white flowers and small, red fruits typically borne in an upright position.[1,2] The best-known forms[3,4] are the African bird's eye chilli (*piri piri* or *peri peri, kambuzi* pepper), Indonesian *tjabe rawit*, Philippine *siling labuyo*, Brazilian *malagueta* and especially American 'Tabasco'. There are three more sources of hot chillies. The South American *C. chinense* ("yellow lantern chili") has dull white flowers and variously coloured and shaped fruits,[2] including 'Habanero' (once considered to be the hottest chilli in the world) and 'Trinidad Moruga Scorpion' (the current champion, measured at over 2 million SHU on the Scoville scale). The Central American and Bolivian *aji* (*C. baccatum* – *aji* is the Caribbean word for chilli) is a distinct species with the petals spotted at the base[2] and includes the popular 'Aji Amarillo', 'Bishop's Crown' and 'Peppadew'. The Andean *rocoto* or tree chilli (*C. pubescens*) has purple flowers, hairy leaves and globose red fruits with black seeds (cultivars include 'Canario' and 'Manzano').[2]

ORIGIN Central and South America.[1-3] It is naturalized in Africa – *piri piri* is the southern African and later Portuguese name given to the "new pepper" introduced by the Portuguese.[1]

CULTIVATION See *C. annuum*. *C. frutescens* has apparently not been domesticated before colonial times[1] and is still much less widely cultivated than *C. annuum*. 'Tabasco' is grown on a large scale in Louisiana, USA.[4] *Capsicum pubescens* is somewhat more cold tolerant.

HARVESTING Chilli fruits are hand-harvested from cultivated plants or from naturalized wild plants (bird pepper).

CULINARY USES The hot, spicy flavour is an essential part of Mexican, West Indian, Indian and Indonesian dishes and has become very popular in most parts of the world. The source of the pungency is mostly *C. annuum* (often 'Cayenne' or 'Serrano') and not *C. frutescens*. Examples of spicy sauces and hot spice blends include American Tabasco sauce, Chinese *öt*, Ethiopian *berbere*, North African *harissa*, Indian *masala*, and many more. Fresh, dried or powdered chillies (red pepper, cayenne pepper or paprika powder) are widely used to flavour stews, meat dishes, potatoes, cheese and eggs. Famous dishes include Mexican *chilli con carne* (beef and red beans), Portuguese *peri-peri* chicken, Hungarian *goulash* (stew with paprika) and various Mediterranean specialities such as *gazpacho*, *piperade* and *ratatouille*. *Capsicum baccatum* has been used in Yucatan and Caribbean-style cooking for centuries and is a popular condiment in Peruvian cuisine (together with red onion and garlic), where it is mainly used fresh in sauces (e.g. *huancaina*) and with meat (e.g. *aji de gallina*). It is usually dried and ground for Bolivian dishes (e.g. *fricase paceno*). The mild fruits of 'Peppadew' are a popular condiment in South Africa.

FLAVOUR COMPOUNDS Capsaicin is the main pungent principle with smaller amounts of dihydrocapsaicin.[4] (See *Capsicum annuum* for the chemical structures of the two main compounds.)

NOTES Capsaicin is the active ingredient in pepper sprays, used in riot control and personal self-defence products.

1. **McLeod, M.J., Guttman, S.I., Eshbaugh, W.H. 1982.** Early evolution of chili peppers (*Capsicum*). *Economic Botany* 36: 361–368.
2. **Walsh, B.M., Hoot, S.B. 2001.** Phylogenetic relationships of *Capsicum* (Solanaceae) using DNA sequences from two noncoding regions: the chloroplast *atpB-rbcL* spacer region and nuclear *waxy* introns. *International Journal of Plant Science* 162: 1409–1418. 2001.
3. **DeWitt, D., Bosland, P.W. 2009.** *The complete chile pepper book*. Timber Press, Portland.
4. **Farrel, K.T. 1999.** *Spices, condiments and seasonings*. Aspen Publishers, Gaithersburg, USA.

Selection of chilli peppers (*Capsicum frutescens* and *C. annuum*)

Bird's eye pepper (*Capsicum frutescens*)

5 mm

Dried chilli peppers (*Capsicum frutescens*)

'Scotch Bonnet' (*Capsicum chinense*)

'Habanero' (*Capsicum chinense*)

'Bishop's Crown' (*Capsicum baccatum*)

'Peppadew' (*Capsicum baccatum*)

Carthamus tinctorius
safflower

Carthamus tinctorius L. (Asteraceae); *saffloer* (Afrikaans); *hong hua* (Chinese); *carthame* (French); *Färberdistel, Saflor* (German); *cartamo* (Italian); *cártamo* (Spanish)

DESCRIPTION The dried florets are about 6 mm (¼ in.) long and have a bright orange colour.

THE PLANT A thistle-like annual with spiny leaves and attractive flower heads surrounded by spiny bracts.

ORIGIN Safflower is a cultigen (no wild plants have ever been found). It was probably developed in the Middle East, from where it spread to ancient India, Egypt, Sudan, Ethiopia and later to Europe and China.[1] Garlands of safflower were found in the tomb of Tutankhamen in Egypt.[1] Since ancient times, safflower has been a source of orange-red dye, used to colour cloth (largely replaced by synthetic aniline dyes that are cheaper). The seed oil is now the most important product.

CULTIVATION Safflower is cultivated in several countries in the Middle East for the production of dried flowers that are used as a saffron substitute. It is susceptible to diseases and therefore requires a short, wet season and long, dry season.[2]

HARVESTING The florets are harvested by hand. Spineless forms have been developed to ease harvesting.[2] The fruits (achenes or "seeds") are non-shedding and harvested mechanically.

CULINARY USES Dried flowers (florets) are used as a saffron substitute for colouring and flavouring rice dishes and as ingredient of spice mixtures and herbal teas. It is sometimes sold as true saffron to unsuspecting or uninformed clients. Another member of the family used in the same way is Mexican marigold (*Tagetes erecta* or *T. lemmonii*). Powdered florets ("petals") are an essential ingredient of the Georgian spice mixtures *khmeli*

ajika (based on hot chillies) and *khmeli suneli* (based on blue fenugreek, *Trigonella caerulea*), not only to add colour but also to enhance the flavours of other ingredients.[3] Marigold powder is a commercial source of yellow/orange food colouring and is an additive to chicken feed to improve the colour of egg yolks.[4] Leaves of the related *huacataya* or tagette (*Tagetes minuta*) are used to make a pesto-like bottled paste called "black mint sauce" which is popular in South America (e.g. as an ingredient of Peruvian *ocopa*, a potato dish). Florets of pot marigold (*Calendula officinalis*) are also used as saffron substitute and as edible garnish to add colour to salads.

FLAVOUR COMPOUNDS Saffron and other sources of edible florets are used mainly as food colourings and not so much for flavour. The pigment in safflower is a benzoquinone called carthamin (safflower yellow). Mexican marigold and *Chrysanthemum* species produce mixtures of carotenoids, of which lutein is often the main compound.

NOTES Cold-pressed safflower seed oil (popular in the health food industry) is polyunsaturated and contains 75% linoleic acid.

carthamin (safflower yellow)

lutein

1. **Knowles, P.F., Ashri, A. 1995.** Safflower. In: Smartt, J., Simmonds, N.W. (Eds), *Evolution of crop plants* (2nd ed.), pp. 47–49. Longman, London.
2. **Singh, V., Nimbkar, N. 2007.** Safflower (*Carthamus tinctorius* L.). In: Singh, R.J. (Ed.), *Genetic resources, chromosome engineering and crop improvement: Oilseed crops*, Vol. 4, pp. 167–194. CRC Press, Boca Raton.
3. **Margvelashvili, J. 1991.** *The classic cuisine of Soviet Georgia*. Prentice Hall Press, New York.
4. **Hadden, W.L. et al. 1999.** Carotenoid composition of marigold (*Tagetes erecta*) flower extract used as nutritional supplement. *Journal of Agricultural and Food Chemistry* 47: 4189–4194.

Safflower or false saffron (*Carthamus tinctorius*) with marigold and Mexican marigold

Safflower (*C. tinctorius*)

Flower heads (*Carthamus tinctorius*)

Fruits (*C. tinctorius*)

Marigold (*Calendula officinalis*)

Mexican marigold (*Tagetes lemmonii*)

Tagette or *huacataya* (*Tagetes minuta*)

99

Carum carvi
caraway

Carum carvi L. (Apiaceae); *karwy* (Afrikaans); *fang feng, se lu zi* (Chinese); *carvi* (French); *vilayati jeera* (Hindi); *Kümmel* (German); *carvi, cumino Tedesco, caro* (Italian); *karahe, kyarawei* (Japanese); *kaereowei, kaerowei* (Korean); *kminek* (Polish); *tmin* (Russian); *alcaravea* (Spanish)

DESCRIPTION The small, ripe fruits ("seeds") are pale brown, and each comprise two narrowly oblong and often slightly curved halves (mericarps), with prominent corky ribs along their length. They have a mild, spicy aroma and a bitter, somewhat peppery taste. Caraway has often been confused with the similar-looking cumin, as can be seen in vernacular names that originally referred to cumin (derived from the Latin *Cuminum*), such as *Kümmel* (German). Similarly, the Hindi name *jeera* is sometimes used for caraway but actually refers to cumin and black cumin (see *Cuminum cyminum*).

THE PLANT Caraway is a biennial herb with feathery, bipinnately compound leaves and umbels of small white flowers that appear in the second year. It grows to a height of 20 cm (8 in.) in the first year and about 60 cm (2 ft) in the second year.

ORIGIN The plant is indigenous to central Europe, the Mediterranean region and western and central Asia.[1,2] The fruits have been found in Neolithic villages, indicating that its culinary uses may date back to ancient times.[2] The Romans used the roots as a vegetable. Caraway is now a popular spice in many parts of the world.

CULTIVATION Plants are easily grown from seeds. They require full sun, rich, well-drained soil with a neutral pH and regular watering. In cold regions, the one-year-old plants are protected by mulching them in winter. Commercial cultivation occurs in eastern Europe, Germany, Finland and the Netherlands but most of the spice now originates from North Africa (especially Egypt).[3]

HARVESTING Fruits ripen in the second year and are harvested when they ripen and turn brown.

CULINARY USES Caraway is a typical spice of German-speaking countries and is an important flavour ingredient of rye bread, sauerkraut, *Schweinsbraten* (roast pork), goulashes, vegetables, potatoes, cheeses (e.g. Munster), cakes and biscuits as well as alcoholic beverages such as kümmel, schnapps and vespétro.[4] In the Netherlands, it is an ingredient of Leyden cheese, the most common type of *kominjekaas*. It is also a popular spice in Scandinavia, used to flavour food, confectionery, cheeses (e.g. Swedish *bondost* and Danish *harvati*) and beverages (*akvavit*). In France, caraway is used to flavour *dragées* (sugar-coated almonds).[4] In Britain it is often included in pickles and cabbage and traditionally served as a condiment (in a small dish) along with baked apples.[2] It is popular in confectionery (e.g. British caraway seedcake and Serbian scones) and desserts (e.g. Middle Eastern caraway pudding).

FLAVOUR COMPOUNDS The warm, aromatic flavour of caraway fruits is due to essential oil rich in R-(+)-carvone (60%) and limonene (40%)[5]. S-(−) carvone smells like spearmint and is the main compound of spearmint oil (see *Mentha spicata*). Only the R-enantiomer of limonene occurs in nature.

NOTES The fruits and essential oil are used to alleviate flatulence and stomach complaints.

R-(+)-carvone   S-(−)-carvone   (+)-limonene

1. **Mabberley, D.J. 2008.** *Mabberley's plant-book* (3rd ed.). Cambridge University Press, Cambridge.
2. **Kiple, K.F., Ornelas, K.C. (Eds). 2000.** *The Cambridge world history of food.* Cambridge University Press, Cambridge.
3. **Farrel, K.T. 1999.** *Spices, condiments and seasonings.* Aspen Publishers, Gaithersburg, USA.
4. **Larousse. 1999.** *The concise Larousse gastronomique.* Hamlyn, London.
5. **Harborne, J.B., Baxter, H. 2001.** *Chemical dictionary of economic plants.* Wiley, New York.

Leaf and flowers of caraway (*Carum carvi*)

Plants in flower (*Carum carvi*)

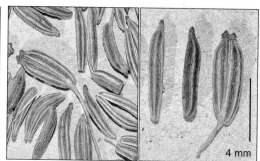

Caraway, the fruits of *Carum carvi*

Flowers (*Carum carvi*)

Chrysanthemum coronarium
chop suey greens • garland chrysanthemum • tangho • Japanese greens

Chrysanthemum coronarium L. (Asteraceae); *tangho, eetbare krisant* (Afrikaans); *tong hao, tong-mo* (Chinese); *tangho* (French); *Tangho* (German); *shungiku, shun giku* (Japanese); *chrysanthemo, tangho* (Italian)

DESCRIPTION The young plants (seedlings) with their soft, pale green and lobed or dissected leaves are used. They are strongly aromatic.

THE PLANT An annual herb with much-branched stems of about 1 m (3 ft) high bearing toothed or deeply dissected leaves and yellow to orange flower heads.

ORIGIN The plant is indigenous to the Mediterranean region but culinary uses have developed in China and Japan.[1] The plant also called crown daisy. Related flavour plants include the common or florist chrysanthemum (*C.* ×*morifolium*) (*ju hua* in Chinese), which has special cultural and symbolic significance in China and Japan[1] and costmary or alecost (*C. balsamita*), formerly used in Europe to flavour beer and ale.[1]

CULTIVATION In warm regions, chop suey greens can be grown throughout the year. Seeds germinate very easily and the plants are tolerant of cold and drought but do best when grown in well-drained fertile soil and watered regularly. The herb is produced by sowing seeds at regular intervals to ensure a continuous supply of seedlings.

HARVESTING Young plants are harvested before they are 20 cm (8 in.) high.

CULINARY USES Young shoots with leaves are popular as a culinary herb and vegetable in China, Japan and Korea. They add a strong and distinctive flavour to stir-fry dishes, soups and stews.[2] The leaves may also be cooked and eaten as a vegetable or as a side dish (*banchan*) in Korean cuisine. They are considered to be an essential ingredient of oyster omelettes in Taiwan and *nabemono* or *nabe* (hotpot dishes) in Japan. Chop suey greens are easily overcooked and are therefore very briefly stir-fried or added at the last minute to stews and hotpots. Cultivars of *C.* ×*morifolium* with small white or yellow flower heads are the source of "chrysanthemum tea" (*ju hua chá*, in Chinese) and are used as a spice in Chinese snake meat soup. *Gukhwaju* is a type of Korean rice wine flavoured with the flowers and Japanese sashimi may be garnished with them.

FLAVOUR COMPOUNDS The aroma of *C. coronarium* is not determined by a single chemical compound but by a combination of compounds with different aroma properties.[3] These include *cis*-3-hexenal (green grass aroma), 2-hexenal (green and fruity), methional (boiled potato/roasty), myrcene (described as peppery, spicy, balsamic and plastic), nonanal (green/waxy) and (*E, Z*)-2,6-nonadienal (cucumber-like). Flowers of *C.* ×*morifolium* appear to vary chemically depending on the cultivar but chrysanthenone and verbenyl acetate are often among the main essential oil compounds.[4]

NOTES In Northern Europe and Britain, the leaves of tansy (*C. vulgare*) are sometimes used to flavour pies, pastries and marinades while feverfew (*C. parthenium*) is used medicinally (as a migraine prophylactic).

(Z)-3-hexenal   (E)-2-hexenal   methional

chrysanthenone   verbenyl acetate

1. **Mabberley, D.J. 2008.** *Mabberley's plant-book* (3rd ed.). Cambridge University Press, Cambridge.
2. **Kiple, K.F., Ornelas, K.C. (Eds). 2000.** *The Cambridge world history of food.* Cambridge University Press, Cambridge.
3. **Zheng, C.H., Kim, T.H., Kim, K.H., Leem, Y.H., Lee, H.J. 2004.** Characterization of potent aroma compounds in *Chrysanthemum coronarium* L. (Garland) using aroma extract dilution analysis. *Flavour and Fragrance Journal* 19: 401–405.
4. **Ito, T., Tada, S., Sato, S. 1990.** Aroma constituents of edible chrysanthemum. *Journal of the Faculty of Agriculture, Iwate University* 20: 35–42.

Chop suey greens (*Chrysanthemum coronarium*)

Ju hua (*Chrysanthemum ×morifolium*)

Dried flower heads (*C. ×morifolium*)

10 mm

Alecost or costmary (*Chrysanthemum balsamita*)

Tansy (*Chrysanthemum vulgare*)

103

Cinnamomum verum
cinnamon • Ceylon cinnamon

Cinnamomum verum J. Presl (= *C. zeylanicum* Blume) (Lauraceae); *kaneel* (Afrikaans); *dar cini* (Arabic); *xi lan rou gui* (Chinese); *canelle de Ceylan* (French); *kanéla* (Greek); *darchini* (Hindi); *Ceylon-Zimt, Echter Zimt* (German); *canella* (Italian); *seiron nikkei* (Japanese); *kayu manis* (Malay); *canela* (Portuguese); *canelo de Ceilán* (Spanish)

DESCRIPTION Cinnamon is the pale brown (tan-coloured) pieces of inner bark or more often tightly rolled pieces of bark (quills). The closely related and very similar cassia (*C. aromaticum*) tends to be slightly more reddish brown. Both are also sold in powdered form. True cinnamon is more expensive but the preferred spice in Europe, Mexico, Australia and New Zealand, while cassia is the main product used in China, Southeast Asia, Canada and the United States, often sold under the culinary name of cinnamon but actually derived from *C. aromaticum* and *C. burmannii*.[1,2] Cassia buds are the dried immature fruits, harvested when they reach about one-fourth of their full size.

THE PLANT True cinnamon comes from *C. verum*, a medium-sized, evergreen tree with glossy three-veined leaves and small white flowers. The fruits are oblong and dark purple, resembling small acorns. Cinnamon is often confused with cassia (*C. aromaticum*), a tree with alternate leaves and smaller, more rounded fruits. Other commercial types of cassia bark include Indonesian cassia (*C. burmannii*), Saigon cassia (*C. loureiroi*) and Indian cassia (*C. tamala*).[1]

ORIGIN Cinnamon is indigenous to Sri Lanka and southwest India. It is one of the oldest of all spices and is mentioned in the Bible and in Sanskrit texts.[1,2] Trees are nowadays widely cultivated in tropical countries, with major commercial production centred in Indonesia, Sri Lanka, Seychelles and Madagascar. Cassia is thought to have originated in Myanmar (Burma) but is grown commercially on a large scale in Vietnam and especially China.[1,2]

CULTIVATION Trees are propagated from seeds or cuttings and are grown in humid tropical regions.

Plants are multi-stemmed and only about 2.5 m (ca. 8 ft) high because of repeated cutting and harvesting and subsequent coppicing.

HARVESTING The inner bark of branches and coppice shoots is stripped (every two years), scraped to remove the outer bark, folded into each other and then dried to produce quills.

CULINARY USES Cinnamon has a sweet aroma and a hot, spicy flavour.[3] The bark or the essential oil distilled from it are widely used in confectionery, puddings, custards, desserts and mulled wine.[3] The spice is also used in food processing, spice mixtures, sauces, meat dishes, stews, poultry, pickles and soups.[2,3] It is commonly used in Moroccan, Greek and Middle Eastern beef, lamb and chicken dishes.[2] Cassia has a similar flavour and aroma and is used in much the same way.

FLAVOUR COMPOUNDS The flavour of both cinnamon and cassia bark is due to cinnamaldehyde, the main component (ca. 40 to 80% in cinnamon oil, 70 to 90% in cassia oil).[4] The aroma of cassia is stronger and considered to be less delicate. The buds (young fruits) and leaves of *C. verum* contain eugenol as the main volatile compound[4] and the buds are still used in India and other countries as a spice, often in confectionery.

NOTES Camphor, obtained from the wood of the camphor tree (*C. camphora*) is used to a limited extent to flavour sweets and desserts.

cinnamaldehyde   eugenol   camphor

1. **Nayar, N.M., Ravindran, P.N. 1995.** Tree spices. In: Smartt, J., Simmonds, N.W. (Eds), *Evolution of crop plants* (2nd ed.), pp. 495–497. Longman, London.
2. **Kiple, K.F., Ornelas, K.C. (Eds). 2000.** *The Cambridge world history of food.* Cambridge University Press, Cambridge.
3. **Larousse. 1999.** *The concise Larousse gastronomique.* Hamlyn, London.
4. **Harborne, J.B., Baxter, H. 2001.** *Chemical dictionary of economic plants.* Wiley, New York.

Cinnamon – bark and powdered bark of *Cinnamomum verum*

Leaves (*C. verum*)

Flowers and fruits (*C. verum*)

Cassia bark (*C. aromaticum*)

Cinnamomum camphora

Cinnamomum loureiroi

Cinnamomum sieboldii

Citrus aurantiifolia
lime

Citrus aurantiifolia (Christm.) Swingle (Rutaceae); *lemmetjie* (Afrikaans); *lai meng* (Chinese); *limette acide* (French); *Saure Limette* (German); *lenuu, nimbu* (Hindi); *lima* (Italian); *raimu* (Japanese); *limau asam, jeruk nipis* (Malay); *lima ácida* (Portuguese); *laim* (Russian); *lima, lima ácida* (Spanish); *manao* (Thai); *chanh ta* (Vietnamese)

DESCRIPTION The fresh unripe, greenish fruits, about 5 cm (2 in.) in diameter, are mostly used but sometimes also ripe (yellow) ones. They are boiled in brine and then smoked and dried to produce black limes (*loomi, lumi, omani*), a popular spice in Middle Eastern cuisine.

THE PLANT Lime or West Indian lime (also known as Mexican lime or key lime) is a small, thorny tree with strongly aromatic leaves, pure white flowers and small fruits with several seeds.[1,2] A second species (or variety) is known as Tahiti lime, Persian lime or seedless lime (*C. latifolia*). It has slightly larger, seedless fruits.[1,2]

ORIGIN Lime originated somewhere in Southeast Asia. It is the main sour citrus fruit in tropical Asia (India and China) where lemons are mostly unknown. The plants were distributed to Africa and the New World and are now cultivated in the southern United States, West Indies, Mexico, Peru, Brazil, Egypt, Ivory Coast, India and South Africa.[1,2]

CULTIVATION Propagation is from seeds rather than grafting. The only other *Citrus* grown from seeds is bitter orange, *C. aurantium*.[1,2] The tree requires a tropical climate and is the most frost-tender of all *Citrus* species.

HARVESTING Fruits are hand-picked (using protective gloves) when fully mature but still unripe and pale green in colour.

CULINARY USES Limes, lime juice or lime peels are widely used in marmalade, jam, sorbet, chutney, pickles, salad dressings and desserts. It is particularly important in sauces, fish and meat dishes and the juice is sometimes used as a marinade for raw fish, using a "cooking" method known as *ceviche* (where the acidity of the lime juice denatures the proteins in the fish). The juice is popular as a drink and is used in punches and cocktails. The zest is used like lemon zest (see *C. limon*) but the latter lacks the refreshing lime flavour. West Indian lime leaves (dried or preferably fresh) are sometimes used in Asian cooking to flavour chicken soup, curries, sambals, and fish dishes but makrut lime is much more popular. Essential oil distilled from the fruits is used in carbonated drinks, confectionery, alcoholic beverages, desserts, jellies, puddings and meat products.[3] Black limes are used with pulses, seafood, meat and rice dishes, to which they add a sour, smoky and slightly bitter taste. Powdered black lime is an ingredient of Persian Gulf-style *baharat* spice mixtures.

FLAVOUR COMPOUNDS Essential oil distilled from lime fruits has (+)-limonene as main compound, with numerous minor constituents.[3,4] The fresh, floral-like aroma of the oil was found to be mainly due to geranial, neral and linalool, while the intense sweet note comes from 7-methoxycoumarin.[4]

NOTES Lime fruits are very acidic and low in sugar (0.8%) with 46 mg/100 g of vitamin C.[3]

(+)-limonene   7-methoxycoumarin

1. **Morton, J. 1987.** Mexican Lime. pp. 168–172. In: *Fruits of warm climates*. Julia F. Morton, Miami, Florida.
2. **Saunt, J. 2000.** *Citrus varieties of the world*. Sinclair International Ltd, Norwich.
3. **Harborne, J.B., Baxter, H. 2001.** *Chemical dictionary of economic plants*. Wiley, New York.
4. **Chisholm, M.G., Wilson, M.A., Gaskey, G.M. 2003.** Characterization of aroma volatiles in key lime essential oils (*Citrus aurantiifolia* Swingle). *Flavour and Fragrance Journal* 18: 106–115.

Lime leaves and fruits (*Citrus aurantiifolia*)

Seedless limes (*C. aurantiifolia*)

Loomi (smoked, dried limes) – *Citrus aurantiifolia*

Tahiti lime (*Citrus aurantiifolia*)

Tahiti lime flowers (*Citrus aurantiifolia*)

107

Citrus hystrix
lime leaf • makrut lime • papeda

Citrus hystrix DC. (Rutaceae); *makroet-lemmetjie*, *lemmetjie-blaar* (Afrikaans); *ba bi da* (Chinese); *limettier hérissé* (French); *Indische Zitrone* (German); *jeruk purut* (Indonesian); *limau purut* (Malay); *ma kruut* (Thai)

DESCRIPTION Lime leaves are easily recognized by the winged leaf stalk (petiole) that is the same size and shape as the leaf blade itself. Also distinctive are the scalloped margins and the fresh lemon aroma. The fruits are small, rounded to pear-shaped and usually have a bumpy surface.

THE PLANT A small, thorny tree with winged and leaf-like petioles, purple-tinged flowers (not pure white as in true limes) and rough, warty fruits.[1,2]

ORIGIN The tree has been cultivated for such a long time in Southeast Asia that its original distribution and origins are now obscure.[1] It is found over large parts of tropical Asia (Sri Lanka to Myanmar, Thailand, Malaysia, Indonesia and the Philippines).

CULTIVATION Trees are easily cultivated in tropical regions and they adapt quite well to temperate conditions provided there is no severe frost. They are ideal for large containers.

HARVESTING Leaves and green fruits are harvested and sold fresh on local markets. In recent years, dried lime leaves (or powdered leaf) have become readily available amongst the herbs and spices in most supermarkets. The leaves and fruits may also be stored frozen to retain the flavour.

CULINARY USES Fresh leaves and fruits are considered to be much better than the dried ones. Their aromatic and astringent flavour is strongly associated with the cuisines of Southeast Asia, which are becoming increasingly popular in Western countries. The fruit rind (zest) is commonly used in spicy Thai curries and other dishes, including the famous *tom yam* soup. The leaves give a recognizable aroma and flavour to Thai and Vietnamese chicken dishes. In Bali, Cambodia, Java, Laos, Malaysia, Myanmar, Thailand and Singapore, whole leaves are liberally added to soups and stews, as well as fish, seafood and chicken dishes (often alongside Indonesian bay leaf). Shredded leaves or grated rind can be mixed into curries, salads, sauces, soups and spice pastes. The fruits yield only small amounts of very acidic juice but are sometimes used in marmalade or as candied whole fruits.

FLAVOUR COMPOUNDS The characteristic aroma of makrut lime leaves is due to (–)-(*S*)-citronellal, which is the main compound in the essential oil (up to 80%).[3,4] Minor constituents include citronellol, nerol and limonene.[3,4] The fruit peel of makrut lime and common lime are chemically quite similar, with limonene and β-pinene as main compounds in both.[3,4]

NOTES The small, kumquat-like fruits of kalamansi or calamondin (*Citrus madurensis*, also treated as ×*Citrofortunella microcarpa*) are used in the Philippines to flavour dishes and desserts (and to make drinks, with liberal amounts of sugar and honey added to the sour juice).

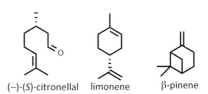

(–)-(*S*)-citronellal   limonene   β-pinene

1. Saunt, J. 2000. *Citrus varieties of the world*. Sinclair International Ltd, Norwich.
2. Burkill, I.H. 1966. *A dictionary of the economic products of the Malay Peninsula*, Vol. 1, pp. 567–568. Crown Agents for the Colonies, London.
3. Lawrence, B.M., Hogg, J.W., Terhune, S.J., Podimuang, V. 1971. Constituents of the leaf and peel oils of *Citrus hystrix* DC. *Phytochemistry* 10: 1404–1405.
4. Jantan, I., Ahmad, A.S., Ahmad, A.R., Ali, N.A.M., Ayop, N. 1996. Chemical composition of some *Citrus* oils from Malaysia. *Journal of Essential Oil Research* 8: 627–632.

Leaves and fruits of makrut lime (*Citrus hystrix*)

Fruits (*Citrus hystrix*)

Dried leaves (*Citrus hystrix*)

Leaves and thorns (*Citrus hystrix*)

Flowers (*Citrus hystrix*)

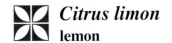

Citrus limon
lemon

Citrus ×limon (L.) Osb. [= *C. medica* × *C. ×aurantium*] (Rutaceae); *suurlemoen* (Afrikaans); *ning meng, yang ning meng* (Chinese); *citron* (French); *Zitrone* (German); *limone* (Italian); *remon* (Japanese); *limão cravo* (Portuguese); *limonero* (Spanish)

DESCRIPTION The fruits are ellipsoid in shape and turn bright yellow when they ripen. Most cultivars have several seeds and acidic juice but seedless and sweet forms have also been developed. The rough lemon (*C. jambhiri*) has rounded fruits with a rough surface. Lemon peel is most often used fresh but can also be dried, candied or pickled (in salt). The essential oil obtained from the peel is an important source of flavour.

THE PLANT Lemons differ from limes (*C. limon*) in the purplish-white flowers, usually larger fruits with a pronounced point or nipple, a thicker fruit rind and a somewhat sweeter taste.[1] Cultivars[2] include 'Lisbon', 'Eureka' (the common supermarket lemon) and the Italian 'Sorrento' (traditional limoncello lemons).

ORIGIN Lemon is believed to be of hybrid origin.[1,2] The Punjab region of Pakistan and India has been proposed as the region of origin,[2] from where lemons spread to China, Southeast Asia and Arabia many centuries ago, and to Europe and the New World only during the Middle Ages.

CULTIVATION Lemon trees are grown commercially in almost all warm regions of the world. It is often grafted onto the more robust rough lemon (*C. jambhiri*).[1,2] Lemons can survive short spells of subzero temperatures and are an ideal choice for even the smallest of herb gardens. It makes an attractive centre-piece when grown in a large ceramic pot.

HARVESTING Lemons are hand-picked and can be left on the tree for lengthy periods until needed, or stored for several months under controlled conditions.

CULINARY USES Lemons are commonly used in salad dressings, mayonnaises, marinades, sauces, rice dishes (e.g. risotto), vegetable dishes, meat dishes, and especially in fish and other seafood dishes.[3] Lemon juice, lemon slices or lemon peel are also an essential ingredient and garnish of many alcoholic and non-alcoholic beverages (lemonade, homemade lemon syrup, cocktails, punches, liqueurs, fruit juices and ice teas, as well as ice creams, sorbets, jams, marmalades and sweets. Lemon zest (the glandular outer peel or exocarp, also called the *flavedo*), obtained by scraping, peeling or grating the outer layer of the fruits, is used to flavour a wide variety of confectionery items,[3] including pies (e.g. lemon meringue pie), tarts, mousses, creams and condiments such as lemon pickle and lime chutney. A lemon twirl is the typical garnish of a Dry Martini. Candied lemon peel is a popular flavour and texture ingredient of biscuits and cakes. The juice is a valuable antioxidant that prevents the browning of fresh fruit and vegetables.[3]

FLAVOUR COMPOUNDS The sour taste is due to citric acid (5%) and the lemon aroma to (+)-limonene and citral (= geranial plus neral).[4] Lemons contain 0.58 % vitamin C (ascorbic acid).[4]

NOTES Vitamin C in the form of lemons ("limes") was once used to counteract scurvy on ships, hence the name "limey" for British seamen. Ascorbic acid means "no scurvy acid" (Latin *a* = no; *scorbutus* = scurvy).

limonene   geranial   neral

citric acid   ascorbic acid (vitamin C)

1. **Mabberley, D.J. 2008.** *Mabberley's plant-book* (3rd ed.). Cambridge University Press, Cambridge.
2. **Saunt, J. 2000.** *Citrus varieties of the world.* Sinclair International Ltd, Norwich.
3. **Larousse. 1999.** *The concise Larousse gastronomique.* Hamlyn, London.
4. **Harborne, J.B., Baxter, H. 2001.** *Chemical dictionary of economic plants.* Wiley, New York.

Lemons, the fruits of *Citrus limon*

Dried lemon peel

Seedless lemon

Flowers and buds (*Citrus limon*)

Rough lemon (*Citrus jambhiri*)

Lemon tree (*Citrus limon*)

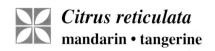

Citrus reticulata
mandarin • tangerine

Citrus reticulata Blanco (Rutaceae); *nartjie* (Afrikaans); *chu, ju, chieh* (Chinese); *mandarine* (French);
Mandarine (German); *santara, suntara* (Hindi); *mikan* (Japanese); *mandarino* (Italian); *mandarina* (Portuguese);
mandarino (Spanish); *som cheen* (Thai)

DESCRIPTION Ripe fruits are easily recognized by the loose peel and fruit segments that can easily be peeled by hand.

THE PLANT A small tree with dark green leaves, white fragrant flowers and small, broad fruit. Various types of mandarins[1] include the common mandarin or tangerine (*C. reticulata*, e.g. the popular 'Clementine'), the satsuma mandarin or *unshiu mikan* (*C. unshiu*), Mediterranean mandarin (*C. deliciosa*) and king mandarin (*C. nobilis*). Hybrids include tangors (tangerine × sweet orange) and tangelos (tangerine × pomelo; e.g. 'Minneola' and the large-fruited Jamaican 'Ugli'). The popular yuzu (*C. ×junos*), a hybrid between mandarin and possibly the ichang papeda (*C. cavaleriei*, formerly *C. ichangensis*).[2]

ORIGIN Mandarin probably originated in northeastern India or southwestern China. It has been grown at least since 1000 BC in China and spread from there to the rest of Asia (e.g. Japan, AD 1000) and Europe (early 1800s).[2]

CULTIVATION Mandarins are cold tolerant and are commercially cultivated in many countries.

HARVESTING Ripe fruits are hand-harvested during the very short fruiting season.

CULINARY USES Dry mandarin peel is a popular spice, used traditionally to flavour liqueurs, confectionery and sweet potatoes. Fresh fruits may be used in jams, marmalades, preserves, cooking, confectionery and cold drinks. The aromatic zest of yuzu (*C. ×junos*) is used in Japanese sauces (*ponzu, yuzukosh*), yuzu vinegar, yuzu syrup, yuzu tea, liquor (e.g. *yuzukomachi*), marmalade and as garnish with *chawanmushi* and miso soup. Sliced yuzu with honey and sugar are made into Korean *yujacha* ("yuza tea"). In Western countries, yuzu

is gaining popularity as flavourant of carbonated drinks, beers, ciders and sweets. Oranges or candied orange peel (*C. sinensis*) provide the flavour for many famous dishes (e.g. *sauce bigarade* with roast duck), desserts, confectionery (e.g. *orangine*) and marmalades. Bitter orange (*C. ×aurantium*) is used for marmalade and crystallized (candied) peel and the rind oil in liqueurs such as Cointreau, Curaçao and Grand Marnier. Bergamot oil (from *C. bergamia*) is the flavourant in Earl Grey tea. The fragrant fruits of Buddha's hand (*C. media* 'Fingered', a form of citron) are used in China to flavour sweets and tea.[2] *Chinotto* is an Italian cola-like soft drink flavoured with *C. myrtifolia* fruits.

FLAVOUR COMPOUNDS The peels of mandarin, sweet orange and grapefruit all have limonene, myrcene and α-sinensal as major volatile compounds. However, the distinctive flavour of true mandarin peel (but not clementine)[3] is ascribed to methyl-*N*-methylanthranilate.[4] Yuzu peel has yuzu lactone and 6-methyl-5-hepten-2-ol as major aroma constituents.[5]

NOTES The bitter taste of navel orange juice, grapefruit and bitter orange is caused by flavonoids (respectively limonin, naringin and neohesperidin).

limonene   myrcene   α-sinensal

methyl-*N*-methyl -anthranilate   yuzu lactone   6-methyl-5 -hepten-2-ol

1. Saunt, J. 2000. *Citrus varieties of the world.* Sinclair International Ltd, Norwich.
2. Mabberley, D.J. 2008. *Mabberley's plant-book* (3rd ed.). Cambridge University Press, Cambridge.
3. Harborne, J.B., Baxter, H. 2001. *Chemical dictionary of economic plants.* Wiley, New York.
4. Chisholm, M.G., Jell, J.A., Cass, D.M. 2003. Characterization of the major odorants found in the peel oil of *Citrus reticulata* Blanco cv. Clementine using gas chromatography-olfactometry. *Flavour and Fragrance Journal* 18: 275–281.
5. Song, H.S. et al. 2000. Quantitative determination and characteristic flavour of *Citrus junos* (yuzu) peel oil. *Flavour and Fragrance Journal* 15: 245–250.

Mandarins and mandarin peel (*Citrus reticulata*)

Mandarin (*Citrus reticulata*)

Zest and candied citrus peel

Sweet orange (*Citrus sinensis*)

Buddha's hand (*Citrus media* 'Fingered')

Myrtle-leaved orange (*Citrus myrtifolia*)

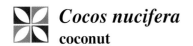

Cocos nucifera
coconut

Cocos nucifera L. (Arecaceae/Palmae); *kokosneut, klapper* (Afrikaans); *noix de coco* (French); *Kokosnuß* (German); *kelapa* (Indonesian); *kelapa* (Malay); *nuez de coco* (Spanish); *maphrao on* (Thai)

DESCRIPTION The large fruits (coconuts) have a smooth outer skin (epicarp), a fibrous middle layer (mesocarp) and a hard inner shell (endocarp) which encloses the nut. Each nut comprises a small embryo and a large, white, fleshy endosperm with a central cavity filled with a watery liquid (coconut water). The main culinary products are the dried and shredded or powdered oily endosperm (copra) and coconut milk (macerated fresh endosperm, forming a milky liquid).[1,2]

THE PLANT This palm tree reaches 30 m (ca. 100 ft), but only 10 m (ca. 33 ft) in dwarf cultivars and bears clusters of large fruits of up to 2 kg (4.4 lbs).

ORIGIN It is thought to have originated from the Malesian region (between Southeast Asia and the western Pacific).[1,2] Sea currents and indigenous people distributed it to the Pacific and Indian oceans some 3 000 years ago, and more recently Portuguese and Spanish explorers took it to all tropical regions. Coconut was once a major source of cooking oil and many by-products (such as coir for rope) but has reverted back to being a multi-purpose subsistence plant.[2]

CULTIVATION Coconut palms require humid, wet and warm conditions, with the mean daily temperature not dipping below 12 °C (53.6 °F), as well as full sun and sandy soil.[3] It tolerates saline conditions.

HARVESTING Coconuts are picked by workers climbing the trees (using notches cut into the stems) or by using knives mounted on long bamboo poles. Harvesting and processing is labour intensive. About 40 billion coconuts are produced each year,

mainly in the Philippines, Indonesia, India and Sri Lanka.[1–3]

CULINARY USES Dried coconut and coconut flour are used worldwide in confectionery (biscuits, cakes), while fresh endosperm is used in Indonesian and Polynesian vegetable, fish and meat dishes. Coconut milk adds a distinctive flavour to Indian curries, stews and rice. It is used as a marinade for meat and may be added to desserts and pastries. Edible oil (coconut butter), extracted from copra, is used to manufacture ghee, margarine, ice cream, imitation dairy products and chocolate. Sugary sap, tapped from flowering stalks, provides palm wine and palm vinegar after fermentation.

FLAVOUR COMPOUNDS The distinctive coconut flavour is caused by several δ-lactones of aliphatic hydroxy-carboxylic acids,[4] ranging in chain length from 8 to 14, such as R-(+)-δ-octalactone and R-(+)-δ-dodecalactone. The most important flavour compound is considered to be 5-decanolide (5-pentyloxan-2-one),[4] also found in wine and butter. It has a sweet, creamy, milk-like flavour.

NOTES The endosperm contains 65% saturated oil but no cholesterol – this is a common error in the literature.[3]

(R)-(+)-δ-octalactone   (R)-(+)-δ-dodecalactone

5-decanolide

1. **Mabberley, D.J. 2008.** *Mabberley's plant-book* (3rd ed.). Cambridge University Press, Cambridge.
2. **Harries, H.C. 1995.** Coconut. In: Smartt, J., Simmonds, N.W. (Eds), *Evolution of crop plants* (2nd ed.), pp. 389–494. Longman, London.
3. **Adkins, S.W., Foale, M., Samosir, Y.M.S. (Eds). 2006.** Coconut revival–new possibilities for the 'tree of life'. Proceedings of the International Coconut Forum held in Cairns, Australia, 22–24 November 2005. ACIAR Proceedings No. 125, 103 pp.
4. **Jirovetz, L., Buchbauer, G., Ngassoum, M.B. 2003.** Solid-phase-microextraction-headspace aroma compounds of coconut (*Cocos nucifera*) milk and meat from Cameroon. *Ernährung/Nutrition* 27: 300–303.

Coconut fruits and products (*Cocos nucifera*)

Coconut milk

Coconut palm (*Cocos nucifera*)

Green coconuts (*Cocos nucifera*)

Coffea arabica
coffee • Arabian coffee

Coffea arabica L. (Rubiaceae); *koffie* (Afrikaans); *ka fei* (Chinese); *café arabica* (French); *Kaffee, Arabicakaffee* (German); *caffè* (Italian); *arabika koohii* (Japanese); *café* (Portuguese); *cafeto* (Spanish); *kafae* (Thai)

DESCRIPTION Washed and dried coffee seeds ("green coffee beans") are pale yellowish grey but turn dark reddish-brown when roasted. Instant coffee is produced by spray-drying or freeze-drying. Decaffeinated coffee has become available since 1930.[1] Coffee is sometimes flavoured with vanilla or cinnamon.

THE PLANT An evergreen shrub or small tree with glossy leaves, fragrant white flowers and two-seeded fleshy fruits (drupes). *Coffea canephora* (robusta or Congo coffee) is used mainly for instant coffee and *C. liberica* (Liberian or Abeokuta coffee) is used to add a bitter flavour to blends.[1]

ORIGIN Highlands of southwestern Ethiopia and the adjacent Boma plateau of Sudan.[1] Initially, coffee was a masticatory or infusions of the leaves or fruit pulp were used as beverages. The practice of using the seeds for coffee appears to have started in Ethiopia not long before the 14th century AD. Between AD 1400 and 1700 Arabia (Yemen) maintained a monopoly but Dutch and Portuguese explorers took coffee seeds and seedlings to other tropical regions of the world, where plantations were established.[1,2] Brazil and Colombia are today the main producers. At more than $10 billion (in 2000), coffee is second only to petroleum in annual export value.[2]

CULTIVATION Plants are usually grown from seeds and thrive in tropical or warm temperate climates but can survive subzero temperatures. They do well in containers. Well-drained soil (pH 5 to 7), mulching, regular watering and occasional feeding are recommended.

HARVESTING Fruits are picked by hand when red and fully ripe. They are either directly dried or depulped in order to remove the two seeds.

CULINARY USES Coffee or coffee extract is an important flavouring used in a wide range of desserts, confectionery, ice cream, sweets, candies and liqueurs (e.g. Tia Maria).

FLAVOUR COMPOUNDS More than 800 volatile compounds are formed when coffee is roasted.[2,3] The main aroma compounds responsible for the familiar coffee fragrance are (E)-(+)-β-damascenone, furfurylmercaptan, 3-mercapto-3-methylbutylformate (or -acetate), 2,5-dimethyl-4-hydroxy-3[2H]-furanone and guaiacol.[2,3] Roasting does not affect the level of caffeine (about 150 mg per cup). During roasting (typically at 185 to 240 °C for about 12 minutes), residual water in each thick-walled cell in the coffee bean is converted to steam, which results in high pressure and complex chemical reactions (including caramelization), during which sugars combine with amino acids, peptides and proteins to create bittersweet glycosylamines and melanoidins (the dominant taste compounds in coffee).[2] The best way to appreciate the complexity of coffee aroma is espresso – strong, black, Italian-style coffee made by passing steam through firmly tamped down, fine to medium-ground coffee. A skilled barista (coffee bar technician) will use 30 seconds percolation time to produce about 30 ml of extract with the all-important crema on top.[2]

NOTES Another important source of flavour (and caffeine) is the cola nut (*Cola acuminata*) from West Africa.

(E)-(+)-β-damascenone   furfurylmercaptan   3-mercapto-3-methylbutylacetate

2,5-dimethyl-4-hydroxy-3[2H]furanone   guaiacol   caffeine

1. Wrigley, G. 1995. Coffee. In: Smartt, J., Simmonds, N.W. (Eds), *Evolution of crop plants* (2nd ed.), pp. 439–442. Longman, London.
2. Illy, E. 2002. The complexity of coffee. *Scientific American* 286: 86–91.
3. Buffo, R.A., Cardelli-Freire, C. 2004. Coffee flavour: an overview. *Flavour and Fragrance Journal* 19: 99–104.

116

Green and roasted coffee beans, the seeds of *Coffea arabica*

Fruits (*Coffea arabica*)

Roasting coffee beans

Flowers (*Coffea arabica*)

Dried cola nuts (*Cola acuminata*)

Coriandrum sativum
coriander • cilantro • Chinese parsley

Coriandrum sativum L. (Apiaceae); *koljander, koljanderblare* (Afrikaans); *hu sui, xiang sui, yan sui* (Chinese); *coriandre, persil arabe* (French); *Koriander, Chinesische Petersilie* (German); *dhania, dhaanya* (Hindi); *coriandro* (Italian); *ketumbar, daun ketumbar* (Malay); *coriandro, cilantro* (Spanish); *phak chee* (Thai)

DESCRIPTION Coriander fruits ("seeds") are either oblong and 3–5 mm (0.12–0.20 in.) in diameter (var. *vulgare*, cultivated in India and other tropical countries) or globose and 1.5–3 mm (0.059–0.12 in.) in diameter (var. *microcarpum*, grown in temperate regions; they have a higher volatile oil content of up to 1%). The strongly aromatic leaves, known as coriander leaf, cilantro (in North America) or Chinese parsley are only used fresh. "Cilantro" should not be confused with "culantro" or "Mexican coriander" (see *Eryngium foetidum*).

THE PLANT Annual herb with broad lower leaves and finely divided upper leaves. The small flowers occur in umbels and have the outer petals enlarged.

ORIGIN Eastern Mediterranean region and western Asia.[1] It is mentioned in the Bible and fruits were found in the tomb of Tutankhamen.[1]

CULTIVATION Coriander is very easily grown from seed and will thrive in full sun under a wide range of climatic conditions and soil types.

HARVESTING For use as fresh herb, young plants are gathered, roots and all. For use as spice, the ripe fruits are harvested (in the early morning, while wet with dew) from plants left to mature.

CULINARY USES Fresh leaves are an essential ingredient of many culinary traditions, including Indian, Chinese, Middle Eastern, Mexican, South American and Thai. They are a garnish and flavour ingredient (added near the end of the cooking time) in many dishes (soup, seafood, stew and curry) and spice pastes or sauces (salsa, guacamole, *zhoug*).

Fresh roots are used in Asian cuisines (e.g. Thai soups and curry pastes). Coriander fruits are a popular spice for meat dishes, sausages (e.g. South African *boerewors*), marinades, pickles, soups and vegetable dishes. Coriander is an essential component of Asian and Indian spice mixtures (e.g. *garam masala*, together with cumin) and curry powders. It is used in confectionery (bread, cakes, biscuits) and to flavour gin and liqueurs such as chartreuse and *Izarra*.[2]

FLAVOUR COMPOUNDS The distinctive smell of cilantro is loved by some but hated by others. This dislike is strongest amongst East Asians and Caucasians and appears to be genetically determined.[3] Various aldehydes are associated with the typical cilantro smell, especially (*E*)-2-decenal, which represents less than 10% of total aroma compounds,[4] as well as (*E*)-2-dodecenal. Coriander fruits are chemically quite different and have (+)-linalool as main compound (60–80%), but with α-pinene, geranyl acetate and myrcene also contributing to flavour variations.[5]

NOTES Decenal is found in the emissions of several species of stinkbug and other insects,[3] hence the "stinkbug-like" smell of cilantro.

(*E*)-2-decenal   (*E*)-2-dodecenal

(+)-linalool

1. **Zohary, D., Hopf, M. 2000.** *Domestication of plants in the Old World* (3rd ed.). Clarendon Press, Oxford.
2. **Larousse. 1999.** *The concise Larousse gastronomique*. Hamlyn, London.
3. **Mauer, K.L. 2011.** *Genetic determinants of cilantro preference.* MSc thesis, Department of Nutritional Sciences, University of Toronto.
4. **Eyres, G., Dufour, J.-P., Hallifax, G., Sotheeswaran, S., Marriott, P.J. 2005.** Identification of character-impact odorants in coriander and wild coriander leaves using gas chromatography-olfactometry (GC-O) and comprehensive two-dimensional gas chromatography–time-of-flight mass spectrometry (GC×GC-TOFMS). *Journal of Separation Science* 28: 1061–1074.
5. **Ramasamy, R., Maya, P., Keshava Bhat, K. 2007.** Aroma characterization of coriander (*Coriandrum sativum* L.) oil samples. *European Food Research and Technology* 225: 367–374.

Leaves (cilantro) and fruits (coriander; powdered, regular and Indian) of *Coriandrum sativum*

Fruits (*Coriandrum sativum*)

Plants (*Coriandrum sativum*)

Flowers and green fruits (*Coriandrum sativum*)

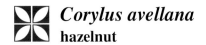

Corylus avellana
hazelnut

Corylus avellana L. (Betulaceae); *haselneut* (Afrikaans); *ou zhe* (Chinese); *hazelnoot* (Dutch); *noisette* (French); *Haselnuß* (German); *nocciola* (Italian); *avely* (Portugueae); *avellana* (Spanish)

DESCRIPTION Hazelnut (cobnuts) are spherical to oval (ca. 15 mm or just over ½ in. in diameter) with a smooth shell enclosed in a fibrous husk (involucre). The closely related filbert (*C. maxima*) has larger, more oblong nuts and longer involucres that cover most of the nut. It is sometimes regarded as a mere cultivar of *C. avellana*. American hazelnut (*C. americana*) and Turkish hazelnut (*C. colurna*) are other sources of hazelnuts.[1]

THE PLANT A deciduous shrub or small tree with rounded, markedly toothed leaves. Male flowers (slender, hanging catkins) and female flowers (in small erect groups) are borne on separate trees and appear before the leaves.

ORIGIN The common hazel (one of about 18 species)[1] is indigenous to Europe and western Asia, from the British Isles (where it is an important component of hedgerows) northwards and eastwards to the Caucasus and northwestern Iran. Hazelnuts have been used as human food all over Europe since prehistoric times[2] and the plant features prominently in folklore.[1]

CULTIVATION The nuts were apparently first cultivated by the Romans.[1] Plants can be propagated from seeds or vegetatively by layering, and are grown in temperate countries. Commercial production is now centred in Turkey (ca. 70% of global trade),[3] with smaller quantities originating from Spain, Italy and France. Modern cultivars, mostly based on the filbert (*C. maxima*) and hybrids with *C. avellana*, are grown in plantations. A small percentage of nuts are wild-harvested.

HARVESTING Nuts are picked by hand in autumn, often while still enclosed in the husks. The nuts are sold with shell and all or more often after being mechanically dehulled.

CULINARY USES Roasted hazelnuts are an important flavour ingredient in chocolates, nut truffles, sweetmeats and confectionery (biscuits, cakes, pastries), to which they also add a delicate crisp texture. Hazelnut flavour is commonly used in alcoholic beverages. Whole or grated hazelnuts are used to some extent in cooking, especially in chicken and fish dishes, as well as in stuffings and stews. Ground hazelnuts are added to butter and various types of spreads for flavour. Extracted oil may be used as salad dressing. Roasted and salted hazelnuts are popular dessert nuts in many parts of the world.

FLAVOUR COMPOUNDS The main chemical compounds responsible for the distinctive aroma of hazelnuts are filbertone (5-methyl-2-hepten-4-one) and 1,3-dimethoxybenzene (resorcinol dimethyl ether).[3] Roasting results in numerous additional flavour compounds such as pyrazines, ketones, aldehydes, furans and pyrones, all contributing to some extent to the complex aroma of roasted hazelnuts[4].

NOTES 1,3-dimethoxybenzene has also been found in port wine.[5]

filbertone   dimethoxybenzene

1. **Mabberley, D.J. 2008.** *Mabberley's plant-book* (3rd ed.). Cambridge University Press, Cambridge.
2. **Zohary, D., Hopf, M. 2000.** *Domestication of plants in the Old World* (3rd ed.). Clarendon Press, Oxford.
3. **Alasalvar C., Shahidi, F., Cadwallader, K.R. 2003.** Comparison of natural and roasted Turkish Tombul hazelnut (*Corylus avellana* L.) volatiles and flavor by DHA/GC/MS and descriptive sensory analysis. *Journal of Agricultural and Food Chemistry* 51: 5067–5072.
4. **Alasalvar C., Pelvan, E., Bahar, B., Korel, F., Ölmez, H. 2012.** Flavour of natural and roasted Turkish hazelnut varieties (*Corylus avellana* L.) by descriptive sensory analysis, electronic nose and chemometrics. *International Journal of Food Science & Technology* 47: 122–131.
5. **Rogerson, F.S.S., Azevedo, Z., Fortunato, N., de Freitas, V.A.P. 2002.** 1,3-Dimethoxybenzene, a newly identified component of port wine. *Journal of the Science of Food and Agriculture* 82: 1287–1292.

Hazelnuts, the seeds of *Corylus avellana*

Hazelnuts with shells (*Corylus avellana*)

Hazelnuts in their husks (*Corylus avellana*)

Products flavoured with hazelnuts

Leaves and young fruits (*Corylus avellana*)

Young fruits (*Corylus avellana*)

121

Crambe maritima
seakale • sea kale

Crambe maritima L. (Brassicaceae); *seekool* (Afrikaans); *chou marin* (French); *Meerkohl* (German)

DESCRIPTION The edible parts are the etiolated (blanched) leaf stalks, together with the unfolded young leaf blades and flower stalks (about 20 cm or 8 in. in length).

THE PLANT A leafy perennial herb with large, glaucous, collard-like leaves, robust flowering stalks, attractive white flowers and spherical, indehiscent, single-seeded fruits. Both *Crambe hispanica* and *C. hispanica* subsp. *abyssinica* (formerly known as *C. abyssinica*) are sources of industrial oil obtained from cold-pressed seeds.[1]

ORIGIN Seakale occurs naturally along the sea coasts of western Europe and part of western Asia (North Atlantic to the Black Sea).[1] It is a traditional food plant in Scotland and England. Local people used to heap up loose shingle around the root crowns in spring, thus bleaching the young shoots as they emerge. It is the only indigenous commercial vegetable and is produced mainly in Kent and Lincolnshire.[1] Cultivation started in the 18th century and reached a peak of popularity in the early 19th century. Seakale is commonly grown as an ornamental plant in botanical gardens in Europe and is often seen in herb gardens.

CULTIVATION Seakale is a halophyte and therefore tolerant of saline conditions. It is easily grown from seeds, root cuttings (taken in winter) or by division (in spring). A deep, rich sandy soil is ideal. In early spring, the plants are covered with special forcing pots to blanch them (by excluding sunlight). Blanching (forcing) is traditionally achieved by covering the crowns with sand or gravel but black polythene are nowadays often used.

HARVESTING The white or pale yellow (sometimes reddish) etiolated leaves or flowering stalks are cut off before they unfold. The product is similar to broccoli in appearance but should be used as soon as possible after harvesting.

CULINARY USES Seakale is not really a culinary herb but rather a vegetable, boiled or steamed and eaten like asparagus. It may be dipped in melted butter and eaten with salt and pepper or served with a béchamel sauce.[2,3] Alternatively, the vegetable may be enjoyed fresh as salad, served with vinaigrette or it may be sautéed with garlic.[2] The flavour is described as nut-like and slightly bitter.

FLAVOUR COMPOUNDS The bitter taste is ascribed to the presence of glucosinolates, at a level of 5.4 to 7.3 μmol per g fresh weight.[4] Epi-progoitrin has been identified as the main compound, representing more than 80% of total glucosinolates.[4] Enzymatic breakdown products include goitrin and other isothiocynates known to reduce the production of thyroid hormones. However, studies have shown that these compounds, when part of a normal and balanced diet, have no harmful effect, possibly due to the inactivation of the specific glucosinolate-degrading enzyme, myrosinase, during cooking.[5]

NOTES Seakale contains 0.18% vitamin C.[3]

epi-progoitrin   goitrin

1. **Mabberley, D.J. 2008.** *Mabberley's plant-book* (3rd ed.). Cambridge University Press, Cambridge.
2. **Larousse. 1999.** *The concise Larousse gastronomique*. Hamlyn, London.
3. **Harborne, J.B., Baxter, H. 2001.** *Chemical dictionary of economic plants*. Wiley, New York.
4. **Quinsac, A., Charrier, A., Ribaillier, D., Péron, J.-Y. 1994.** Glucosinolates in etiolated sprouts of sea-kale (*Crambe maritima* L.). *Journal of the Science of Food and Agriculture* 65: 201–207.
5. **McMillan, M., Spinks, E.A., Fenwick, G.R., 1986.** Preliminary observations on the effect of dietary brussel sprouts on thyroid function. *Human Toxicology* 5: 15–19.

Seakale – *Crambe maritima*

Seakale with rhubarb forcer

Fruiting plant (*Crambe maritima*)

Flowering plant (*Crambe maritima*)

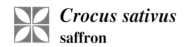

Crocus sativus
saffron

Crocus sativus L. (Iridaceae); *saffraan* (Afrikaans); *fan hong hua* (Chinese); *safran* (French); *Safran* (German); *zafferano* (Italian); *safuran* (Japanese); *açafrão* (Portuguese); *azafrán* (Spanish); *ya faran* (Thai)

DESCRIPTION Saffron threads are the bright red stigmas and styles obtained from fresh flowers of the saffron crocus. About 150 flowers are required to produce only 1 g (0.035 oz.) of dry saffron threads, so that saffron is the most expensive spice in the world.

THE PLANT A small bulbous plant with thread-like leaves and lilac-purple flowers.

ORIGIN Saffron is a sterile cultigen (autotriploid), developed in ancient Greece from native species, most likely *Crocus cartwrightianus*.[1] Saffron is believed to feature on Minoan pottery and frescoes from 1600 BC. It has played an important role in history[1,2] as a source of spice and orange-yellow dye. The name saffron is derived from *zafarán* or *za'fran*, the Arabian word for yellow.

CULTIVATION Saffron is propagated by division of the corms when dormant in summer. The plants require a Mediterranean climate with moderately cold, wet winters and hot, dry summers, full sun and a rich, damp but well-drained soil. It tolerates short periods of frost (−10 °C; 14 °F). Full flowering occurs in mid-autumn in a short, synchronized burst of one or two weeks. About 300 tons are produced each year[1] – mainly in Iran and Spain but also Turkey, Greece, Italy and Kashmir (considered the best quality).[1]

HARVESTING The flowers open at dawn and are picked in the early morning before they wilt. The styles and stigmas are removed by hand and then dried. Harvesting and processing is a labour-intensive activity that contributes to the high price of saffron. In modern food processing, safflower (false saffron, sometimes fraudulently sold as true saffron), turmeric (Indian saffron) or artificial colours and flavours are mostly used.[2]

CULINARY USES Saffron is famous as the essential ingredient of French bouillabaisse, Spanish paella and Italian risotto. It is an important spice in many traditional Mediterranean, Arabian and Indian rice and curry dishes, as well as semolina and milk puddings. As a reddish-yellow dye and flavourant, it is still much used in the production of butter, cheese, pastries, confectionery (e.g. saffron cakes in Cornwall,[1] saffron bread in Sweden), sauces and liqueurs (e.g. chartreuse).

FLAVOUR COMPOUNDS The compounds responsible for the unique colour, taste and aroma of saffron are all derived from carotenoids.[3] The colour comes from crocin (and crocetin), while the bitter taste is due to a glucoside, picrocrocin.[3] Several volatile compounds such as isophorone and especially safranal are responsible for the flavour and aroma,[3–5] often described as "hay-like", or "seabreeze-like". A minor compound, 2-hydroxy-4,4,6-trimethyl-2,5-cyclohexadien-1-one is now considered to be the most powerful aroma constituent of saffron, followed by safranal.[4,5]

NOTES Saffron is the richest known source of vitamin B_2. The maximum safe dose of saffron is 1 g (0.035 oz.) per day.

1. **Mabberley, D.J. 2008.** *Mabberley's plant-book* (3rd ed.). Cambridge University Press, Cambridge.
2. **Farrel, K.T. 1999.** *Spices, condiments and seasonings.* Aspen Publishers, Gaithersburg, USA.
3. **Cadwallader, K.R., Back, H.H., Cai, M. 1997.** *ACS Symposium Series, No. 166 (Spices),* American Chemical Society, Washington, pp. 66–79.
4. **Tarantilis, P. A., Polissiou, M., 1997.** Isolation and identification of the aroma constituents of saffron (*Crocus sativus*). *Journal of Agricultural and Food Chemistry* 45: 459–462.
5. **D'Auria, M., Mauriello, G., Rana, G.L. 2004.** Volatile organic compounds from saffron. *Flavour and Fragrance Journal* 19: 17–23.

Saffron, the styles and stigmas of *Crocus sativus*

Saffron (*Crocus sativus*)

Leaves and flowers (*Crocus sativus*)

Paella with saffron

125

Cuminum cyminum
cumin

Cuminum cyminum L. (Apiaceae); *komyn* (Afrikaans); *zi ran qin* (Chinese); *cumin* (French); *Mutterkümmel, Römischer Kümmel* (German); *jeera, zira* (Hindi); *jinten* (Indonesian); *comino* (Italian); *jintan putih* (Malay); *cominho* (Portuguese); *comino* (Spanish); *yee raa* (Thai)

DESCRIPTION The small dry fruits (often referred to as "seeds") are narrowly oblong, about 5 mm (less than ¼ in.) long and greenish or greyish-brown in colour. Cumin (*jeera*) should not be confused with *shah jeera* ("black cumin"), a similar and related spice from Pakistan and India with slightly larger and often somewhat curved fruits. To add to the confusion, the name "black cumin" is often used for the angular black seeds of *Nigella sativa* (Ranunculaceae), also known as fitches, black seed, *kalonji* (*kalanji*) or onion seed.[1]

THE PLANT Cumin is a small, annual herb with thin, non-woody stems, fennel-like leaves and small white or pink flowers arranged in umbels.

ORIGIN Based on historical and archaeological records, cumin is believed to be indigenous to the Mediterranean region and western Asia.[1,2] It is one of the oldest of spices, recorded from 5000 BC in Egypt, the Middle East (mentioned in the Bible)[2] and widely used in ancient Greece, Rome and medieval Europe to season soups, stews, fish and poultry. An old French culinary term *cominée* refers to dishes containing cumin.[3]

CULTIVATION Cumin is propagated by sowing the whole fruits ("seeds"). It has a very long history of cultivation in Egypt, the Middle East, Turkey, Greece and India, which later spread to Morocco, northern Europe, Russia, Central America, China and Japan and other areas.[2]

HARVESTING Mature fruits are hand-harvested just before they are shed.

CULINARY USES Cumin is responsible for the distinctive spicy flavour and slightly bitter taste of curry powder, of which it is an essential ingredient (along with chilli pepper and turmeric). It is also used in other spice mixtures and chutneys in Pakistan and India. The characteristic taste of a falafel is partly due to cumin. Cumin is an important ingredient of North African lamb stews and couscous as well as northern European soups, sauces, fish and meat dishes, including cold meats. It is widely used in bread-making (e.g. rye bread), cheese-making (e.g. Munster cheese) and confectionery, adding a spicy flavour to biscuits and pastries. Cumin is commonly used in pickling mixtures and is a traditional ingredient of German sauerkraut and kummel liqueur.[2]

FLAVOUR COMPOUNDS Cumin fruits contain an essential oil rich in cuminaldehyde (also known as 4-isopropylbenzaldehyde). Cuminaldehyde usually represents 40 to 65% of the total volatiles[2,4] and is accompanied by smaller amounts of *p*-mentha-1,3-diene-7-al, *p*-menth-3-ene-7-al and γ-terpinene.[4] The essential oil of *Bunium persicum* (black *zira*, *shah jeera*) contains γ-terpinene as main compound.[5]

NOTES Black *zira* (*kaala zeera*) or *shah jeera* (*Bunium persicum* fruits) is an important culinary spice used in roasted spice mixtures and meat dishes in Pakistan and northern India (Kashmir, Punjab and Bengal, including Moghul cookery with its Iranian influence), as well as Afghanistan, Bangladesh, Tajikistan and Iran. It is almost unknown outside of these regions.

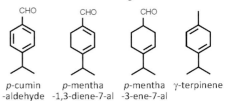

p-cumin   *p*-mentha   *p*-mentha γ-terpinene
-aldehyde -1,3-diene-7-al -3-ene-7-al

1. Mabberley, D.J. 2008. *Mabberley's plant-book* (3rd ed.). Cambridge University Press, Cambridge.
2. Farrel, K.T. 1999. *Spices, condiments and seasonings.* Aspen Publishers, Gaithersburg, USA.
3. Larousse. 1999. *The concise Larousse gastronomique.* Hamlyn, London.
4. Li, R., Jiang, Z.Y. 2004. Chemical composition of the essential oil of *Cuminum cyminum* L. from China. *Flavour and Fragrance Journal* 19: 311–314.
5. Azizi, M., Davareenejad, G., Bos, R., Woerdenbag, H.J., Kayser, O. 2009. Essential oil content and constituents of black zira (*Bunium persicum* [Boiss.] B. Fedtsch.) from Iran during field cultivation (domestication). *Journal of Essential Oil Research* 21: 78–82.

Cumin, the fruits of *Cuminum cyminum*

Cumin (*C. cyminum*)

Black cumin (*Bunium persicum*)

Flowering plant (*Cuminum cyminum*)

Flowers (*Cuminum cyminum*)

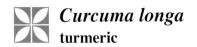

Curcuma longa
turmeric

Curcuma longa L. (= *C. domestica* Valeton) (Zingiberaceae); *borrie* (Afrikaans); *jiang huang, yu jin* (Chinese); *curcuma* (French); *Gelbwurz, Safranwurz, Kurkuma* (German); *haldi* (Hindi); *curcuma* (Italian); *taamerikku* (Japanese); *kunyit* (Malay); *curcuma, tumérico* (Spanish); *kha min* (Thai); *khuong hoàng, nghe* (Vietnamese)

DESCRIPTION The rhizomes are finger-like and bright orange inside when freshly cut but bright yellow when dried and orange-yellow when powdered. They are easily confused with the bright orange but more conical Indonesian mango ginger (*C. mangga*) or the similarly shaped but pale yellow mango ginger (*C. amada*).[1] Wild turmeric (*C. aromatica*) is also used as a source of dye.[2] The stoloniferous zedoary or *e-zhu* (*C. zedoaria*) is an alternative source of turmeric in China.

THE PLANT A leafy, stemless, ginger-like plant with large, oblong leaves and attractive yellow and white flowers borne in oblong clusters.

ORIGIN Turmeric is an ancient cultigen not found in the wild but is believed to be indigenous to southern Asia (possibly India). It has been used as a spice and dye since ancient times and is also important in religious rites and traditional customs. From India the plant is believed to have spread to Southeast Asia, the Far East and Polynesia (as far as Fiji) and much later to tropical America.[1]

CULTIVATION Turmeric is a sterile triploid and is propagated vegetatively by division of the branched rhizomes. It requires a tropical climate and high rainfall.

HARVESTING The rhizomes are dug up when mature, after which they may be boiled, dried and powdered to produce turmeric powder. India is the most important producer, with nearly 350 000 tons per year but smaller amounts are also produced in Bangladesh, Pakistan and various South and Southeast Asian countries.[1]

CULINARY USES Turmeric is best known as the spice that gives curry powder and mustard powder their bright orange or yellow colour. The fresh rhizomes, powdered rhizomes or extracts are important ingredients of Indian and Asian cooking. They are added to a wide range of dishes, not only for colour but also for flavour, including meat, fish, vegetables and rice dishes. A well-known example is Malaysian yellow rice (*nasi kunyit*), eaten on ceremonial or festive occasions and the origin of the *geelrys* ("yellow rice") of Cape cookery. Turmeric is used as a natural food dye in many food products, including mustard sauces, Worcestershire sauce, sandwich spread, butter, cheese, beverages, dairy products, drinks, confectionery, pickles and soup powders. Mango ginger has the flavour and aroma of raw mango and is used in curries and pickles.[1] Indonesian mango ginger is similar to mango ginger in flavour and uses.

FLAVOUR COMPOUNDS The bright yellow colour is due to pigments (curcuminoids) of which curcumin is the main compound.[3] The essential oil has a warm, spicy, earthy and slightly bitter flavour and contains mainly sesquiterpenes, of which *ar*-turmerone and turmerone are the major compounds.[4]

NOTES Curcumin has a wide range of medicinal properties.[3]

curcumin

ar-turmerone   turmerone

1. **Nayar, N.M., Ravindran, P.N. 1995.** Herb spices. In: Smartt, J., Simmonds, N.W. (Eds), *Evolution of crop plants* (2nd ed.), pp. 491–494. Longman, London.
2. **Mabberley, D.J. 2008.** *Mabberley's plant-book* (3rd ed.). Cambridge University Press, Cambridge.
3. **Esatbeyoglu, T., Huebbe, P., Ernst, I.M.A., Chin, D., Wagner, A.E., Rimbach, G. 2012.** Curcumin—from molecule to biological function. *Angewandte Chemie, International Edition*, 51: 5308–5332.
4. **Gopalan, B., Goto, M., Kodama, A., Hirose, T. 2000.** Supercritical carbon dioxide extraction of turmeric (*Curcuma longa*). *Journal of Agricultural and Food Chemistry* 48: 2189.

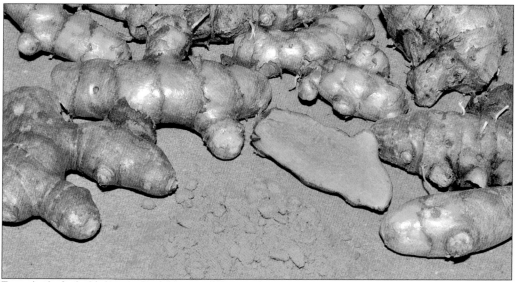

Turmeric, the fresh, dried or powdered rhizomes of *Curcuma longa*

Dried turmeric (*Curcuma longa*)

Indonesian mango ginger (left) and mango ginger (right)

Turmeric plant (*Curcuma longa*)

Turmeric flowers (*C. longa*)

Zedoary plant (*C. zedoaria*)

129

Cymbopogon citratus
lemongrass

Cymbopogon citratus (DC.) Stapf (Poaceae); *sitroengras* (Afrikaans); *xiang mao coa* (Chinese); *verveine de Indes* (French); *Lemongras, Zitronengras* (German); *sereh* (Indonesian); *citronella* (Italian); *remon gurasu* (Japanese); *serai* (Malay); *sontol* (Spanish); *takrai* (Thai); *xa* (Vietnamese)

DESCRIPTION The finger-sized fleshy leaf bases and young stems are white to pale green and often slightly purplish, especially when viewed in longitudinal section.

THE PLANT Lemongrass is a robust perennial tuft with thick stems and broad, bluish-green, aromatic leaves. It is a cultigen that rarely forms flowers.[1] This aromatic grass is often confused with similar-looking grasses that are commercial sources of essential oil.[2] These are most accurately identified by their main chemical constituents (see NOTES). Malabar lemongrass (*C. flexuosus*) is the source of East Indian lemongrass oil. It is sometimes used in Asian dishes in the same way as lemongrass. Ginger grass, palmarosa or rusha (*C. martinii*) has essential oil that is very similar to geranium oil. Two others are Ceylon citronella grass (*C. nardus*) and Java citronella grass (*C. winterianus*).

ORIGIN Southern India and Sri Lanka, as well as Malaysia[1] have been proposed as likely candidates, but the exact geographical origin of lemongrass has remained unknown. It is mainly grown and used in Bali, Cambodia, Indonesia, Laos, Malaysia, Singapore, Sri Lanka, Thailand and Vietnam.

CULTIVATION Lemongrass is propagated by division of the clumps. The crop requires regular watering and a warm or tropical climate, as it is frost-tender.

HARVESTING Young stems are broken off by hand and the leaves trimmed and discarded.

CULINARY USES Lemongrass is an essential component of Asian cooking. The flavour is subtle and never dominates. It is particularly well known as a flavour ingredient of "lemongrass soup" and the famous *tom yum*. There are many variants of this "spicy and sour" soup in Thailand and Laos. The inner fleshy part of the stem may be finely sliced or pounded to a paste and eaten raw.[3] When used as a culinary herb, the whole stem is often bruised before adding it to the dish. It can be removed later, after the flavour has been released. Lemongrass is used in a wide variety of dishes, including curries, stir-fries, soups and stews, particularly those based on fish, seafood and poultry. It is an important ingredient of spice mixtures, spicing liquors and sauces. Stalks are sometimes used as skewers for prawns or seafood satay.[3] Lemongrass may be added to tea for flavour, or it can be mixed with lemon verbena and lemon balm to make lemon tea.

FLAVOUR COMPOUNDS Lemongrass is a traditional source of natural citral, which is a mixture of geranial and neral (ca. 40% of each).[4] The subtle lemon flavour is ascribed to these two main ingredients.

NOTES *Cymbopogon flexuosus* oil is similar to lemongrass oil (dominated by citral), while so-called citronella oil, obtained from *C. nardus* and *C. winterianus*, has mainly citronellal, with smaller amounts of geraniol and citronellol. Palmarosa oil, distilled from *C. martinii*, yields almost pure geraniol (and is a substitute for oil of roses).

geranial   neral   citronellal   citronellol

1. **Burkill, I.H. 1966.** *A dictionary of the economic products of the Malay Peninsula*, Vol. 1, pp. 724–728. Crown Agents for the Colonies, London.
2. **Mabberley, D.J. 2008.** *Mabberley's plant-book* (3rd ed.). Cambridge University Press, Cambridge.
3. **Hutton, W. 1997.** *Tropical herbs and spices*. Periplus Editions, Singapore.
4. **Pino, J.A., Rosado, A. 2000.** Chemical composition of the essential oil of *Cymbopogon citratus* (DC.) Stapf. from Cuba. *Journal of Essential Oil Research* 12: 301–302.

Lemongrass, the stems and leaf bases of *Cymbopogon citratus*

Lemongrass (*Cymbopogon citratus*)

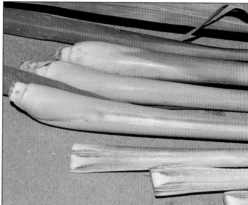

Lemongrass (*Cymbopogon citratus*)

Lemongrass plant (*Cymbopogon citratus*)

Elettaria cardamomum
cardamom

Elettaria cardamomum (L.) Maton (Zingiberaceae); *kardamom* (Afrikaans); *kardemumma* (Finnish); *cardamome* (French); *Cardamomen* (German); *ilaayacii* (Hindi); *cardamomo* (Italian); *karadamomo* (Japanese); *cardamomo* (Portuguese, Spanish); *kardemumma* (Swedish); *aila cheddi* (Tamil); *luk grawan* (Thai); *kakule* (Turkish)

DESCRIPTION Small, near-ripe fruits (three-valved capsules) containing numerous dark brown seeds. They are green in colour when simply air-dried but may be pure white as a result of bleaching with steam or sulphur before drying.[1] There are several other spices known as cardamom.[1,2] The large brown fruits of black cardamom, derived from *Amomum subulatum* (Nepal cardamom), is popular in India and Pakistan for flavouring savoury dishes. The brown colour results from the partial roasting and smoking of the fruits when they are dried over open fires. Even larger are the fruits of *A. costatum* (Chinese black cardamom), which is important in Sichuan and Vietnamese cuisine.

THE PLANT A tall perennial herb with erect stems bearing broad leaves and horizontal flowering stems bearing clusters of white and pink flowers at ground level.[1]

ORIGIN Cardamom is indigenous to the forests of India and Sri Lanka, where it has been wild-harvested for centuries.[1,2] Along with black pepper and dried ginger, this "queen of spices" formed the bulk of the oriental spice trade since the third century BC. In the modern era it has become an important crop, with Guatemala as the main producer (5 000 tons per annum), followed by India (4 000 tons).[2] Minor producers include Tanzania, Sri Lanka and Papua New Guinea.[2]

CULTIVATION Cultivated cardamom does not differ from the wild type. Plants can be propagated from seeds or by division. They require a tropical climate and thrive at higher latitudes in the shade of forest trees. Plants do not respond well to fertilization (it lowers the quality) but irrigation during the dry season improves the yield.[2]

HARVESTING Near-ripe fruits are usually hand-picked individually but the whole cluster is sometimes harvested, thus sacrificing the immature fruits.

CULINARY USES The delicious sweetish and pungent taste of cardamom has found its way into many different culinary traditions. In its native India, cardamom forms an important component of curries and curry powders. It is also widely used in rice, vegetable and meat dishes, as well as sweet desserts. The seeds are traditionally used to flavour Arabian coffee and black Turkish tea.[3] In Europe and America, cardamom is well known as an essential ingredient of gingerbread and sweet pastries. Scandinavians (especially Swedes and Finns) are particularly fond of cardamom and large amounts are used in confectionery, desserts, stewed fruits, mulled wines, meat dishes and sausages.[3] Ground cardamom is added to hamburger patties and meatloaf – one teaspoon per kg (2.2 lb.), while roughly bruised seeds are used in sausages as well as in breads, buns and brioches.[3]

FLAVOUR COMPOUNDS The essential oil contains 1,8-cineole (eucalyptol) and α-terpinyl acetate as major compounds, but the typical aroma of cardamom is ascribed to trace amounts of unsaturated aliphatic aldehydes such as (*E*)-4-decenal.[4]

NOTES For Ethiopian cardamom (korarima), see *Aframomum corrorima*.

1,8-cineole   α-terpinyl acetate   (*E*)-4-decenal

1. **Burkill, I.H. 1966.** *A dictionary of the economic products of the Malay Peninsula*, Vol. 1, pp. 910–915. Crown Agents for the Colonies, London.
2. **Nayar, N.M., Ravindran, P.N. 1995.** Herb spices. In: Smartt, J., Simmonds, N.W. (Eds), *Evolution of crop plants* (2nd ed.), pp. 491–494. Longman, London.
3. **Swahn, J.O., 1999.** *The lore of spices*. Senate Publishing, London.
4. **Noleau, I., Toulemonde, B., Richard, H. 1987.** Volatile constituents of cardamom (*Elettaria cardamomum*) cultivated in Costa Rica. *Flavour and Fragrance Journal* 2: 123–127.

Cardamom, the fruits and seeds of *Elettaria cardamomum* (with black cardamom – *Amomum* species)

Cardamom seeds (with Ethiopian cardamom)

Fruits and seeds (*E. cardamomum*)

Plant with flowers (*Elettaria cardamomum*)

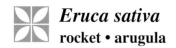

Eruca sativa
rocket • arugula

Eruca sativa Mill. [= *E. vesicaria* (L.) Cav. subsp. *sativa* (Mill.) Thell.] (Brassicaceae); *roket* (Afrikaans); *zi ma cai* (Chinese); *roquette* (French); *Rauke* (German); *rucola* (Italian); *roketsuto* (Japanese); *rúgula* (Portuguese); *arúgula* (Spanish)

DESCRIPTION Fresh leaves of rocket are dark green above and paler beneath, deeply lobed and have a pungent aroma. The leaves of wild rocket (*Diplotaxis tenuifolia*) are much smaller and narrower but have a similar aroma. *Eruca sativa* is known as rocket or garden rocket in Britain, Australia, Canada and New Zealand, but as arugula, eruca or rocket salad in America.

THE PLANT A weedy annual herb with erect stems, broadly lobed leaves and white or cream-coloured flowers. Rocket is often confused with wild rocket (*Diplotaxis tenuifolia*), which has narrower leaves and bright yellow flowers.[1]

ORIGIN Rocket is indigenous to Mediterranean Europe, while wild rocket has a wider distribution in Europe. Both species are important traditional salad plants in southern Europe and have become popular all over the world in recent years.

CULTIVATION Both species are amongst the easiest of herbs to grow but require warm conditions and full sun. They are fast-growing, cool-season crops that thrive even in poor soil and tend to become weedy.[1] Sowing is usually done at three-weekly intervals.[1]

HARVESTING Fresh leaves or young seedlings are harvested by hand in the early morning. In recent years, germinated seeds (sprouts) have also become popular.

CULINARY USES The pungent, mustard-like aroma of rocket (also commonly referred to as arugula or rucola) has become a characteristic feature of French and Italian restaurants where liberal quantities are used as garnishes (for soups, pizzas and other dishes) and as traditional salad ingredients (e.g. the *mesclun* or *mescladisse* of southern France and the *misticanza* of Italy). It is often combined with chervil, corn salad, dandelion, endive, lettuce, purslane and various other herbs, resulting in a somewhat bitter salad that is usually served with vinaigrette and olive oil. Anchovies, walnuts, mozzarella cheese or sundried tomatoes may be added for extra taste. In Spain, rocket leaves are served as a separate garnish for a traditional dish known as *gazpachos*. *Rucolino* is a digestive alcoholic drink prepared from macerated rocket leaves (a speciality of the Italian island of Ischia).

FLAVOUR COMPOUNDS The typical aroma of rocket is due to a complex mixture of methylthioalkyl isothiocyanates and -nitriles, as well as (Z)-3-hexen-1-ol and -esters. The main compounds are usually 4-methylthiobutyl isothiocyanate and 5-methylthiopentanonitrile,[2,3] together with 5-methylthiopentyl isothiocyanate, (Z)-3-hexen-1-ol, (Z)-3-hexen-1-yl butanoate and (Z)-3-hexen-1-yl 2-methylbutanoate.[2] 4-Mercaptobutyl glucosinolate is usually the main compound in *E. sativa* and invariably so in *D. tenuifolia*.[4]

NOTES Rocket is a commercial oilseed in India, Pakistan and Iran (similar to rapeseed or canola). The seed oil, known as rocket oil or jamba oil, is used as a cooking oil, salad oil and lubricant.

4-methylthiobutyl isothiocyanate

5-methylthio-pentanonitrile

1. **Morales, M., Janick, J. 2002.** Arugula: A promising speciality leaf vegetable. In: Janick, J., Whipkey, A. (Eds), *Trends in new crops and new uses*, pp. 418–423. ASHS Press, Alexandria, VA.

2. **Jirovetz, L., Smith, D., Buchbauer, G. 2002.** Aroma compound analysis of *Eruca sativa* (Brassicaceae) SPME headspace leaf samples using GC, GC-MS, and olfactometry. *Journal of Agricultural and Food Chemistry* 50: 4643–4646.

3. **Miyazawa, M., Maehara, T., Kurose, K. 2002.** Composition of the essential oil from the leaves of *Eruca sativa*. *Flavour and Fragrance Journal* 17:187–190.

4. **Bennett, R.N., Carvalho, R., Mellon, F.A., Eagles, J., Rosa, E.A. 2007.** Identification and quantification of glucosinolates in sprouts derived from seeds of wild *Erica sativa* L. (salad rocket) and *Diplotaxus tenuifolia* L. (wild rocket) from diverse geographical locations. *Journal of Agricultural and Food Chemistry* 55: 67–74.

Rocket, the leaves of *Eruca sativa* (left) and wild rocket, the leaves of *Diplotaxis tenuifolia* (right)

Flowers and fruits (*Eruca sativa*)

Flowering plants (*Eruca sativa*)

Wild rocket plant (*Diplotaxus tenuifolia*)

Eryngium foetidum
culantro • eryngo • sawtooth coriander

Eryngium foetidum L. (Apiaceae); *Meksikaanse koljander* (Afrikaans); *ci yán sui, yang yuan sui* (Chinese); *chardon étoile fétide, panicaut fétide, herbe puante, coriandre mexicain* (French); *Langer Koriander, Mexicanischer Koriander* (German); *pereniaru korianda* (Japanese); *ketumbar java* (Malay); *cilantro extranjero* (Mexican Spanish); *culantro, racao, recao* (Spanish); *phak chee farang* (Thai); *uzun kişniş* (Turkish); *ngò gai* (Vietnamese)

DESCRIPTION The stiff and leathery leaves are markedly toothed along their margins and have a strong, pungent smell. The herb is usually referred to as culantro in English-speaking Caribbean countries but has many English names, including long coriander, wild or Mexican coriander, fitweed, stinkweed, saw-leaf herb or sawtooth coriander.[1] Other regional names include *chadron benee* (Dominica), *coulante* (Haiti), *recao* (Puerto Rico) or *shado beni* and *bhandhania* (Trinidad and Tobago).[1] It should not be confused with cilantro (*Coriandrum sativum*).

THE PLANT A biennial herb with oblong leaves borne in basal rosettes. The small white flowers are arranged in dense oblong heads, surrounded by whorls of leaf-like bracts.[1,2]

ORIGIN Tropical America (especially the Caribbean Islands) but widely cultivated and naturalized in parts of Africa and the United States.[1] It spread from China to all parts of Asia.[3]

CULTIVATION Culantro is easily propagated by sowing the small, dry, two-seeded fruits. The plants grow in partial shade and require rich soil and regular watering.[1]

HARVESTING Young shoots are harvested for use as a fresh herb (but it retains the distinctive flavour when dried). Culantro is well known throughout Latin America and the Far East but is rather poorly known in the United States, Canada and Europe despite large volumes imported (mainly from Puerto Rico and Trinidad) to meet the demand from ethnic populations.[1] In China, the herb is grown and harvested in Guangdong, Guangxi, Hainan Island and Yunnan.[2]

CULINARY USES Culantro is used as a spice and seasoning in a wide range of vegetable and meat dishes, chutneys, preserves, sauces and snack foods.[1] In Asian cuisine (especially in Thailand, Malaysia and Singapore) it is commonly used with cilantro (or replaces cilantro) in curries, noodle dishes and soups.[1] It is often used to mask the smell of beef and other strong-smelling food items. For this reason it is added to Thai raw beef salad (*larp*), offal soup, and *tom yam* soup made with beef rather than seafood.[4] Culantro is popular in Latin America and especially in Puerto Rico, where it is an essential ingredient of the local version of salsa, a spicy sauce made from chillies, garlic, onion, tomatoes and lemon juice and usually eaten with tortilla chips. It is also an ingredient of *sofrito* or *recaito*, a Puerto Rican spice mixture used in a wide range of food items and sauces, including soups, stews and rice dishes.[1]

FLAVOUR COMPOUNDS The leaves contain an essential oil rich in aliphatic aldehydes, of which (*E*)-2-dodecenal is the main compound (nearly 60%), accompanied by small amounts of 2,3,6-trimethylbenzaldehyde, dodecanal and (*E*)-2-tridecenal.[5]

NOTES The similarity with cilantro (and stink bugs) is due to 2-dodecenal (see *Coriandrum sativum*).

(*E*)-2-dodecenal

dodecanal

(*E*)-2-tridecenal

2,3,6-trimethyl-benzaldehyde

1. **Ramcharan, C. 1999.** Culantro: A much utilized, little understood herb. In: Janick, J. (Ed.), *Perspectives on new crops and new uses*, pp. 506–509. ASHS Press, Alexandria, VA.
2. **Hu, S.-Y. 2005.** *Food plants of China*. The Chinese University Press, Hong Kong.
3. **Burkill, I.H. 1966.** *A dictionary of the economic products of the Malay Peninsula*, Vol. 1, p. 944. Crown Agents for the Colonies, London.
4. **Hutton, W. 1997.** *Tropical herbs and spices*. Periplus Editions, Singapore.
5. **Thakuri, B.C., Chanotiva, C.S., Padalia, R.C., Mathela, C.S., 2006.** Leaf essential oil of *Eryngium foetidum* L. from far western Nepal. *Journal of Essential Oil Bearing Plants* 9: 251–256.

Culantro, long coriander or eryngo, the leaves of *Eryngium foetidum*

Culantro (*Eryngium foetidum*)

Plant (*Eryngium foetidum*)   Flowers (*Eryngium foetidum*)

137

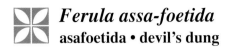

Ferula assa-foetida
asafoetida • devil's dung

Ferula assa-foetida L. (Apiaceae); *duiwelsdrek* (Afrikaans); *a wei* (Chinese); *férule persique, ase fétide* (French); *hing* (Hindi); *Stinkasant, Teufelsdreck* (German); *assafetida* (Italian); *agi* (Japanese); *anghuzeh* (Persian); *hingu* (Sanskrit); *asafetida* (Spanish); *dyvelsträck* (Swedish)

DESCRIPTION The oleoresin gum known as asafoetida or *hing* is a waxy, golden brown resinous material with a very strong, garlic-like, sulphurous odour (hence the name devil's dung; *Teufelsdreck* in German). The pure product is too strong to be used directly as a culinary spice and is therefore sold as a powder, diluted with gum arabic and flour, so that small quantities can be sprinkled on food. The product is obtained from various *Ferula* species, including *F. assa-foetida*, *F. foetida* (nowadays perhaps the most important commercial source) and *F. narthex*.[1,2] The name *assa-foetida* comes from *aza* (Persian for gum) and *foetida* (Latin for foul-smelling) but there is doubt about the correct application of the botanical name.[3]

THE PLANT *Ferula* species used for gum production are robust perennial herbs with compound leaves and large umbels of yellow flowers.

ORIGIN *Ferula assa-foetida* occurs in western Iran, *F. foetida* from eastern Iran to Pakistan and India and *F. narthex* in Afghanistan.[1] These are the best known sources of culinary asafoetida but there are several other plants of the Apiaceae (Umbelliferae) that are sources of resinous gums used in perfumes, incenses and traditional medicine.[2]

CULTIVATION *Ferula* species are mostly wild-harvested from native or naturalized plants.

HARVESTING Gum is obtained from mature plants just before they flower. The top part of the plant is cut off and the collar above the root is exposed by removing the soil. The milky latex is collected once it has solidified. A fresh cut is then made and

the gum again collected. The process is repeated for up to three months until there is no longer any exudate to collect.

CULINARY USES Asafoetida is an important spice in Middle Eastern and South Asian cuisines and is widely used (especially in India) to flavour meat dishes, stews, gravies, sauces, mushrooms and pickles. When used sparingly, the offensive smell is lost during cooking, leaving a pleasant, garlic-like aroma. The spice is not popular in Western cuisine but is (or was) allegedly an essential ingredient of Worcestershire sauce.

FLAVOUR COMPOUNDS The gum or oleoresin is a complex mixture of compounds, including ferulic acid esters, polysaccharide gums based on glucose, galactose and galacturonic acids, as well as terpenoids and coumarins.[4,5] The odour is ascribed to various sulphides and sulfanes; *sec*-butyl-propenyl disulphide (both *E* and *Z* isomers) are usually the main sulphur compounds.[4,5]

NOTES A famous North African (Libyan) condiment known as *silphium* or *laserpitium* was used in ancient Roman cuisine and Greek traditional medicine but is thought to have been over-exploited to the point of extinction because of its rarity and high value.[2] *Silphium* seems to have been replaced by asafoetida.

sec-butyl-(*E*)-propenyl-disulphide

sec-butyl-(*Z*)-propenyl-disulphide

1. **Mabberley, D.J. 2008.** *Mabberley's plant-book* (3rd ed.). Cambridge University Press, Cambridge.
2. **Langenheim, J.H. 2003.** *Plant resins*, pp. 412–417. Timber Press, Portland.
3. **Chamberlain, D.F. 1977.** The identity of *Ferula assa-foetida* L. *Notes from the Royal Botanic Garden, Edinburgh* 35: 229–233.
4. **Rajanikanth, B., Ravindranath, B., Shankaranarayana, M.L. 1984.** Volatile polysulphides of asafoetida. *Phytochemistry* 23: 899–900.
5. **Degenhardt, A. et al. 2012.** Novel insights into the flavour chemistry of asafetida. In: *Recent advances in the analysis of food and flavors*, Chapter 12, pp. 167–175. American Chemical Society.

Asafoetida, the dried gum from stems of *Ferula assa-foetida* and other *Ferula* species

Plant (*Ferula foetida*)

Asafoetida

Flowers (*Ferula foetida*)

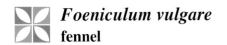
Foeniculum vulgare Mill. (Apiaceae); *vinkel* (Afrikaans); *hui xiang* (Chinese); *fenouil* (French); *Fenchel* (German); *sanuf, saunf* (Hindi); *finocchio* (Italian); *fenneru* (Japanese); *funcho* (Portuguese); *hinojo* (Spanish); *phak chi* (Thai)

DESCRIPTION All parts of the plant are edible – the compound leaves with their slender segments, the leaf stalks, white swollen stem bases and the small, dry fruits. They are greenish or yellowish brown, about 10 mm (⅜ in.) long, with five ribs on each of the two fruit halves (mericarps). A single resin canal is situated under each furrow. The flavour is similar to that of anise and liquorice, with a spicy, fresh aroma and a sweet anise-like taste. Fennel is sometimes wrongly called anise in the United States.

THE PLANT A perennial herb (up to 1.5 m or 5 ft) with compound, feathery leaves and small yellow flowers arranged in umbels. Sweet fennel, var. *dulce*, is commonly grown as a culinary herb and spice. The foliage is usually bright green but the decorative bronze fennel has purplish brown leaves. Florence fennel or *finocchio*, var. *azoricum*, is a smaller plant with white bulbous leaf bases that are eaten as a vegetable.[1]

ORIGIN Asia[1] and the Mediterranean region (now global, and often an invasive weed). It was well known to the ancient Greeks, Egyptians and Romans.

CULTIVATION Fennel is frost-hardy, versatile and easily grown from seeds sown in spring. It does best in full sun and deep, well-drained soil.

HARVESTING The fruits are harvested by hand or mechanically just before they ripen. They are used as a spice or may be steam-distilled to obtain fennel oil.

CULINARY USES Fennel fruits are a commercial spice that is widely used in cooking and baking by almost all culinary traditions of the world. They are an important ingredient of curry powders, spice mixtures, pickles, herbal teas, soups, chicken dishes, fish, seafood, processed meats, sausages and breads. Whole or powdered fruits are used in tomato dishes, pastas, borsch and sauerkraut, as well as vegetables (cabbage, celery, cucumbers, onions and potatoes).[2] The leaves are commonly added to fish and seafood dishes, both as garnish and for flavour (in the same way as dill). They give a fresh, herbaceous, aromatic, anise-like flavour to soups, stews and meat dishes. Florence fennel, raw and sliced, adds flavour to salads and to a variety of cooked vegetable and meat dishes. Powdered fruits or the anise-flavoured essential oil is widely used in beverages, teas, baked goods, ice creams, liqueurs and seasonings for meat products (e.g. pepperoni and other types of sausages).[2]

FLAVOUR COMPOUNDS The main aroma impact compounds in fennel fruits are *trans*-anethole and fenchone.[3] Two types of fennel oil are distinguished:[4] bitter fennel oil has 30 to 75% anethole, 12 to 33% fenchone and the level of α-pinene exceeds that of limonene; sweet fennel oil has 80 to 90% anethole, 1 to 10% fenchone and has more limonene than α-pinene. Anethole is also the main compound in anise (*Pimpinella anisum*).

NOTES Fennel is traditionally used in carminative and digestive medicines.

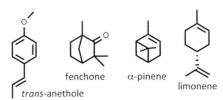

fenchone   α-pinene   limonene   *trans*-anethole

1. **Mabberley, D.J. 2008.** *Mabberley's plant-book* (3rd ed.). Cambridge University Press, Cambridge.
2. **Farrel, K.T. 1999.** Spices, condiments and seasonings. Aspen Publishers, Gaithersburg, USA.
3. **Zeller, A., Rychlik, M. 2006.** Character impact odorants of fennel fruit and fennel tea. *Journal of Agricultural and Food Chemistry* 54: 3686–3692.
4. **Harborne, J.B., Baxter, H. 2001.** *Chemical dictionary of economic plants.* Wiley, New York.

Fennel, the leaves and fruits of *Foeniculum vulgare*

Fennel fruits (*Foeniculum vulgare*)

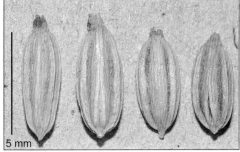
Fennel fruits (*Foeniculum vulgare*)

Florence fennel (*Foeniculum vulgare*)

Flowers (*Foeniculum vulgare*)

141

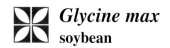

Glycine max
soybean

Glycine max (L.) Merr. (Fabaceae); *sojaboon* (Afrikaans); *da hou, huang dou* (Chinese); *fève de soja, haricot soja* (French); *Sojabohne* (German); *kedelai* (Indonesian); *soia* (Italian); *daizu* (Japanese); *kacang soya* (Malay); *frijol soya* (Spanish); *thua lueang* (Thai)

DESCRIPTION Soybeans are fermented to make a wide range of condiments (sauces and pastes) that are characteristic of the cuisines of East and Southeast Asia. Soy sauce or *shōyu* (Japanese) is a salty, earthy, brownish liquid used for seasoning food during cooking or as table condiment and dipping sauce.[1,2] It is produced by controlled fermentation of paste made from boiled soybeans (often with roasted and crushed wheat or barley added).[1,2] *Aspergillus* moulds (*A. oryzae, A. sojae*) break down the soy and wheat proteins into free amino acids and the starch into simple sugars. Lactic acid bacteria ferment the sugars into lactic acid, while yeasts turn the sugars into alcohol. Ageing and secondary fermentation result in the complex flavour. After fermentation, salt and water are added. A quicker method to make soy sauce and other HVPs (hydrolysed vegetable proteins) is through acid hydrolysis.[1]

THE PLANT An annual with trifoliate leaves, minute flowers and hairy pods, each with up to four rounded seeds.

ORIGIN Wild soya (*G. soja*) is indigenous to East Asia and was domesticated in China around 1100 BC.[3] Soy sauce started as a variant of fermented fish sauce but the two products eventually became separate entities. The making of soy sauce spread from China to Asia and later from Japan to Europe.[1,2]

CULTIVATION Soybeans are grown commercially on a large scale as an annual pulse crop.

HARVESTING Ripe pods and seeds are harvested mechanically.

CULINARY USES Soy-based products are a distinctive feature of Asian cuisines. These include soy sauce, soy paste (miso), soy curd (tofu or "soy cheese"), soy "milk" and fermented soy cakes (tempeh).[1–3] There are numerous types of soy sauce (*jeong yau*) in China, e.g. light soy sauce (*jeong tsing*), an opaque, salty, light brown sauce used mainly for seasoning, dark soy sauce (*lou chau*), made by prolonged ageing and the addition of caramel colour and/or molasses, and thick soy sauce (*jeong yau gou*), a dark-coloured dipping and finishing sauce thickened with starch, sugar and spices. In Japan, there are five basic types of *shōyu*, which typically have a sweet and sherry-like flavour (*koikuchi* is the most common). In Indonesia, the term *kecap* and *ketjap* is used for all fermented sauces (the origin of the English word "ketchup"). These include *kecap asin* (salty, light soy sauce) and *kecap manis* (thick, sweet soy sauce with palm sugar or molasses added). Soy sauce is known as *joseon ganjang* in Korea, *toyo* in the Philippines, *dóuyóu* in Malaysia and Singapore and *xi dau* in Vietnam. Korean *joseon ganjang* and Japanese *tamari* are dark, rich liquid by-products formed when fermented soybean pastes (Korean *doenjang* and Japanese *miso*) are made.

FLAVOUR COMPOUNDS Soy sauce adds an umami (savoury) taste to food because of the presence of glutamic acid and other glutamates. Glutamic acid is an abundant naturally occurring amino acid, used as food additive in the form of the sodium salt (monosodium glutamate or MSG).

NOTES Soy sauce was the original secret ingredient of Worcestershire sauce,[2,3] later replaced by HVP.

glutamic acid   monosodium glutamate

1. **Tanaka, N. 2000.** Shōyu: the flavour of Japan. *The Japan Foundation Newsletter* 27: 1–7.
2. **Shurtleff, W., Aoyagi, A. 2012.** *History of soy sauce (160 CE to 2012): extensively annotated bibliography and source book*. SoyInfo Center, Lafayette.
3. **Mabberley, D.J. 2008.** *Mabberley's plant-book* (3rd ed.). Cambridge University Press, Cambridge.

Soy products, with fruits and seeds of *Glycine max*

Seeds and fruits (*Glycine max*)

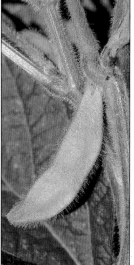

Soybean plant (*Glycine max*)   Flower, fruit (*G. max*)   Ripe fruits (*G. max*)

Glycyrrhiza glabra
liquorice • licorice

Glycyrrhiza glabra L. (Fabaceae); *soethout* (Afrikaans); *yang gan cao* (Chinese); *zoethout* (Dutch); *réglisse* (French); *Echtes Süßholz, Lakritze* (German); *glykoriza* (Greek); *liquirizia* (Italian); *you kanzou* (Japanese); *alcaçuz* (Portuguese); *orozuz, regaliz* (Spanish); *lakrits* (Swedish); *meyan kökü* (Turkish)

DESCRIPTION Dried rhizomes ("liquorice root") are woody pieces of underground stem, bright yellow inside and covered with fissured brown bark. "Licorice" is the spelling used in the United States and Canada. Chinese liquorice or *gan cao* ("sweet herb", *G. uralensis*) is a close relative used in the same way.[1]

THE PLANT A suffrutex (up to 1 m or ca. 3 ft) with branched and spreading rhizomes and erect above-ground stems bearing pale purple flowers and small, few-seeded pods.

ORIGIN Mediterranean region to central Asia.[1] Uses date back to the time of the ancient Assyrians and Romans.[1,2] Spanish monks at Pontefract in Yorkshire maintained a monopoly during the 16th century[1,2] and the town became the centre of the liquorice industry. The famous coin-shaped sweets known as Pontefract cakes (also Pomfret or Pomfrey cakes) were manufactured here during ca. 1660 to 1960.[1] Today, liquorice is grown commercially in many countries, including the Mediterranean region (North Africa, France, Italy, Spain), Russia, Turkey, India and China.[1,3]

CULTIVATION Liquorice can be propagated from seeds but plants are more often grown from pieces of rhizome. Plants grow best in warm, temperate regions and require full sun and deep soil.

HARVESTING Rhizomes are harvested in the autumn, three to five years after planting. The whole plant is dug up and the rhizomes cleaned and trimmed. They may also be boiled in water to create black liquorice syrup that is concentrated and dried by boiling and/or spray-drying.

CULINARY USES Liquorice juice is mostly used as a flavourant in confectionery (e.g. gingerbread), sweets, candies, desserts, stewed fruit and drinks such as liquorice water, mulled wine, root beer, dark beer (stout) and liqueurs.[2,3] Liquorice not only reduces the bitter taste of beer and ale but also increases the foaming. Liquorice water is a refreshing cold drink resembling coconut milk that is made by infusing sliced rhizomes in water and adding lemon juice or orange peel for extra flavour.[3] Liquorice is used to a limited extent as a culinary spice in China, adding flavour and a hint of sweetness to savoury foods, sauces and stews (especially pork). Liquorice is an ingredient of Chinese functional food mixtures known as *bupins*.[4] Rhizome pieces may be chewed as a sweet snack. Various types of hard or pliable liquorice sweets are made by mixing the juice (concentrate) with gum arabic, starch or icing sugar and flavouring agents.[2] Such sweet or salty black liquorice sweets (known as "drop") are particularly popular in the Netherlands.

FLAVOUR COMPOUNDS Liquorice extracts contain 5–10% glycyrrhizin, a triterpenoid glycoside (saponin) that is 50 times sweeter than sucrose, with a strong aftertaste typical of liquorice.[1,3]

NOTES Liquorice is a useful sweetener for diabetics but repeated use of large amounts may be harmful. Large quantities are used in the pharmaceutical industry and as flavouring agent for tobacco.[3]

glycyrrhizin

1. **Mabberley, D.J. 2008.** *Mabberley's plant-book* (3rd ed.). Cambridge University Press, Cambridge.
2. **Larousse. 1999.** *The concise Larousse gastronomique.* Hamlyn, London.
3. **Farrel, K.T. 1999.** *Spices, condiments and seasonings.* Aspen Publishers, Gaithersburg, USA.
4. **Hu, S.-Y. 2005.** *Food plants of China.* The Chinese University Press, Hong Kong.

Liquorice or licorice, the dried rhizomes of *Glycyrrhiza glabra*

Leaf and flowers (*Glycyrrhiza glabra*)

Anise drinks often contain liquorice extract

Liquorice sweets

Hibiscus sabdariffa
hibiscus • roselle • Jamaica sorrel • karkade

Hibiscus sabdariffa L. (Malvaceae); *hibiskus* (Afrikaans); *mei gui qie, shan qie zi* (Chinese); *karkadé, roselle, oseille de Guinée* (French); *Hibiscus, Roselle, Sabdariffa-Eibisch* (German); *karcadè* (Italian); *roozera* (Japanese); *rosela, vinagreira* (Portuguese); *rosa de Jamaica* (Spanish)

DESCRIPTION The sepals or calyces ("hibiscus flowers") are bright red when fresh but turn dark purple to almost black when dried. They have a refreshing, sweet-sour taste.

THE PLANT An erect annual herb (usually 2 m or ca. 6 ft in height) with lobed leaves and white or yellow flowers with dark centres. The fleshy red sepals (calyx) are surrounded by an outer row of bracts (epicalyx).

ORIGIN The plant is indigenous to Africa (Angola) and may have been domesticated in the Sudan about 6 000 years ago.[1] It was introduced to Asia and the Americas in the 17th century[1] and now grows in almost all warm regions of the world. Sudan and Egypt are the main producers, followed by Thailand and China.[1] Hibiscus is grown for local consumption in many parts of Africa, Asia (India, Indonesia, the Philippines) and Central America (Mexico, Brazil, West Indies).[1,2] The leaves are a popular vegetable, especially in Burma (Myanmar), and the bast fibres are an important product in parts of India and Thailand.[1,2]

CULTIVATION Seeds are mostly directly sown into the fields[1,2] but plants can also be grown from transplanted seedlings or even cuttings.[2] A tropical or subtropical climate with a long, warm season is required for the fruits to ripen. The plants are frost-tender and require deep, well-drained soil.

HARVESTING The fruits with their persistent calyces are snapped off by hand in the early morning and dried to a moisture content of 12%. Exports (mainly to Germany and the USA) exceed 15 000 tons per year.[1]

CULINARY USES Hibiscus is sometimes used as a condiment to flavour meat and fish dishes but its main use is as a natural food colourant and sweet-sour flavourant in herbal teas and alcoholic and non-alcoholic beverages. Tea and various cold drinks (*karkadé, da bilenni, agua de Flor de Jamaica* or "sorrel") are very popular in African and Latin American countries.[1,2] The natural pectin content makes it ideal for jellies, jams, chutneys, sauces and preserves.[2] Young leaves are important as culinary herb, e.g. in Burmese *chin baung kyaw* curry and Senegalese *thiéboudieune* (a fish and rice dish).[2]

FLAVOUR COMPOUNDS The sour taste is due to organic acids, including hibiscus acid (hydroxycitric acid) as well as ascorbic, citric, malic and tartaric acids.[3] The most intense aroma compounds were determined to be 1-octen-3-one and nonanal.[4] In fresh hibiscus, the floral and citrus aromas are due to linalool and octanal.[4] The intense red colour comes from anthocyanins (0.15%), of which the 3-sambubiosides of cyanidin and delphinidin are the main compounds.[3] Hibiscus is also rich in mucilage polysaccharides (15%).

NOTES Hibiscus tea is a gentle expectorant, laxative and diuretic.

hibiscus acid (hydroxycitric acid); 1-octen-3-one; nonanal; octanal; linalool

1. **McClintock, N.C., El Tahir, I.M. 2004.** *Hibiscus sabdariffa* L. In: Grubben, G.J.H., Denton, O.A. (Eds), Plant resources of tropical Africa 2. Vegetables, pp. 321–326. PROTA Foundation, Wageningen.
2. **Morton, J. 1987.** Roselle. In: *Fruits of warm climates*, pp. 281–286. Julia F. Morton, Miami.
3. **Ramírez-Rodriques, M.M., Plaza, M.L., Azero, A., Balaban, M.O., Marshall, M.R. 2011.** Physicochemical and phytochemical properties of cold and hot water extraction from *Hibiscus sabdariffa*. *Journal of Food Science* 76: C428–435.
4. **Ramírez-Rodriques, M.M., Balaban, M.O., Marshall, M.R., Rouseff, R.L. 2011.** Hot and cold water infusion aroma profiles of *Hibiscus sabdariffa*: fresh compared with dried. *Journal of Food Science* 76: C212–217.

Hibiscus, roselle, karkade or Jamaica sorrel, the dried calyces of *Hibiscus sabdariffa*

Roselle (*Hibiscus sabdariffa*)

Leaves and calyces (*Hibiscus sabdariffa*)

Flower (*Hibiscus sabdariffa*)

147

Humulus lupulus
hop • hops

Humulus lupulus L. (Cannabaceae); *hops* (Afrikaans); *houblon* (French); *Hopfen* (German); *luppolo* (Italian); *lúpulo* (Spanish)

DESCRIPTION Hop cones are fresh or dried female flower clusters, comprising overlapping, leaf-like bracts and bracteoles that hide the small flowers, with minute lupulin glands scattered on the surface. Brewers often use commercial hop extract in the form of pellets.

THE PLANT A perennial climber of up to 10 m (ca. 33 ft) that dies back to ground level in winter. The wind-pollinated male and female flowers occur on separate plants but only female plants are grown because seedless cones are mostly preferred.

ORIGIN Northern temperate zones of Asia, Europe and North America. The first record of wild hops being used for beer-brewing dates back to 3 000 years ago and comes from Finland.[1] Cultivation started in Bavaria in the 8th century, from where it spread to the rest of the world.[1]

CULTIVATION Female plants are propagated from cuttings. The annual stems ("bines") are supported by poles. Day length is critical for both vegetative growth and flowering. Modern cultivars, partly developed from North American cluster hops, produce high yields of good quality soft resin (up to 15%) and are resistant to fungal diseases.[1]

HARVESTING Hop cones are collected in autumn and kiln-dried at low temperatures to a moisture content below 10%.

CULINARY USES Hops is mainly used as a flavourant to add aroma and bitterness to beer and ale. Commercial lager beers usually have a low hop content, while pilsners and certain ales have a more pronounced hop flavour. According to the German purity law ("Reinheitsgebot") of 1516, only water, malt, hops and yeast are allowed in beer-brewing. Hops once had a preservative (bacteriostatic) function but this is no longer important because of the hygiene conditions of modern processing.

Shoots with male flowers (*jets de houblon*) are prepared like asparagus – Belgian and French dishes that include hop shoots are referred to as *à l'anversoise*. Hop shoots in cream are traditionally served with poached eggs.[2]

FLAVOUR COMPOUNDS The flavour compounds are produced as excretions of the lupulin glands that dry to spherical granules known as lupulin (up to 20% of the total dry weight of the cones).[1] The bitterness of beer is due to the *iso-α*-acids (humulone and related compounds),[3] while the major flavour compounds are linalool, geraniol, myrcene, ethyl-2-methylbutanoate, ethyl-3-methylbutanoate and ethyl-2-methylpropanoate. The most powerful aroma-actives occur in minute quantities and include sulphur-containing thiols (4MMP, 3MHA and 3MH, also important in wine), as well as β-damascenone and humulene (and various derivatives such as epoxides which are produced during the brewing process).[4]

NOTES The bitter compounds in beer stimulate appetite and gastric secretions, thereby improving digestion.

ethyl 2-methylbutanoate

ethyl 3-methyl -butanoate

ethyl 2-methyl -propanoate

3-mercapto-1-hexanol

β-damascenone

humulene

humulone

mercapto-4-methyl -pentan-2-one

3-mercapto -hexyl acetate

1. Neve, R.A. 1995. Hops. In: Smartt, J., Simmonds, N.W. (Eds), *Evolution of crop plants* (2nd ed.), pp. 33–35. Longman, London.
2. Larousse. 1999. *The concise Larousse gastronomique*. Hamlyn, London.
3. Hughes, P. 2000. The significance of *iso-α*-acids for beer quality. *Journal of the Institute of Brewing* 106: 271–276.
4. Kishimoto, T., Wanikawa, A., Kono, K., Shibata, K. 2006. Comparison of the odor-active compounds in unhopped beer and beers hopped with different hop varieties. *Journal of Agricultural and Food Chemistry* 54: 8855–886

Dried hop cones

Hop pellets

Hops, the female inflorescences of *Humulus lupulus*

Hopped ale and beer

Hop plant (*Humulus lupulus*)

Hop cones (*Humulus lupulus*)

149

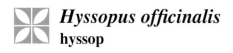

Hyssopus officinalis
hyssop

Hyssopus officinalis L. (Lamiaceae); *hisop* (Afrikaans); *hysope* (French); *Ysop* (German); *issopo* (Italian); *hisopo* (Spanish)

DESCRIPTION Hyssop herb comprises the fresh or dried stems and leaves, with or without flowers. Hyssop is also used as an extract or as essential oil distilled from above-ground parts.

THE PLANT An erect perennial shrublet up to 0.6 m (ca. 2 ft) high, somewhat woody at the base, with thin, square stems, small, oblong, opposite leaves and small, bright blue (rarely pale blue, violet, purple or white) flowers.

ORIGIN Hyssop is indigenous to southern Europe.[1] The historical records of hyssop (derived from *ezov* or *esob* in Hebrew), including several biblical references, actually refer to the Middle Eastern *Origanum syriacum*, a relative of marjoram and oregano.[1] Hyssop is well known as a culinary herb and ornamental garden plant, commonly seen in herb gardens. Commercial production occurs mainly in France, Hungary and the Netherlands.

CULTIVATION Hyssop is easily propagated from seeds or from cuttings. It tolerates drought and poor soil but grows best in full sun and well-drained, slightly acid soil. Plants should be strongly pruned to encourage new growth and are best replaced after three to five years.

HARVESTING Leafy stems are hand-harvested twice a year. The herb is used fresh or dried (or as concentrated extract or steam-distilled hyssop oil).

CULINARY USES Hyssop has been described as a unique and versatile culinary herb when used in small amounts. The taste is intensely minty, spicy and floral, with a pleasant bitterness. Fresh young leaves and flowers are used in salads, soups, sauces and meat dishes.[2] Infusions of the dried herb can replace sage in seasoning stocks, soups, rich sauces (e.g. for pasta and gnocchi), oily fish dishes, meat dishes, vegetables, stuffings, processed pork and other charcuterie products.[2] It can be baked into pita bread and used in cakes, custards, puddings, jams, candies and ice cream. Hyssop is well known as an ingredient of herbal liqueurs such as absinthe, Benedictine and chartreuse.[2] Classical absinthe recipes called for 500 to 1 000 g (1.1 to 2.2 lbs) of the herb (or 6 g / 0.2 oz of essence) per 100 litres / 219 gallon.[3]

FLAVOUR COMPOUNDS The essential oil usually contains pinocamphone, isopinocamphone and pinocarvone as main compounds[4] but some chemical variants may have large amounts of β-pinene, camphor, 1,8-cineole, linalool or limonene. Pinocamphone is similar to thujone in its neurotoxic activity but the levels of both these compounds in historic and modern absinthe were found to be within safe limits.[3]

NOTES Hyssop should not be confused with anise hyssop (*Agastache foeniculum*), a North American culinary herb that is used in much the same way as hyssop. It has a sweet, aniseed flavour. Other similar-looking herbs that are sometimes used for food flavouring include catnip (*Nepeta cataria* and related species) and calamint (*Calamintha nepeta*). All these species belong to the Lamiaceae family and have become popular as ornamental plants.

pinocamphone   isopino   pinocarvone   β-pinene
                -camphone

1. Fleisher, A., Fleisher, Z. 1988. The identification of biblical hyssop and origin of the traditional use of oregano-group herbs in the Mediterranean region. *Economic Botany* 42: 232–241.
2. Larousse. 1999. *The concise Larousse gastronomique*. Hamlyn, London.
3. Lachenmeier, D.M., Nathan-Maister, D., Breaux,T.A., Sohnius, E.-M., Schoeberl, K., Kuballa, T. 2008. Chemical composition of vintage preban absinthe with special reference to thujone, fenchone, pinocamphone, methanol, copper, and antimony concentrations. *Journal of Agricultural and Food Chemistry* 56: 3073–781.
4. Kerrola, K., Galambosi, B., Kallio, H. 1994. Volatile components of hyssop (*Hyssopus officinalis* L.). *Journal of Agricultural and Food Chemistry* 42: 776–781.

Hyssop, the leaves and flowers of *Hyssopus officinalis*

Flowers (*Hyssopus officinalis*)

Anise hyssop (*Agastache foeniculum*)

Blue catnip (*Nepeta ×faassenii*)

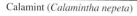

Calamint (*Calamintha nepeta*)

Catnip (*Nepeta cataria*)

151

Illicium verum
star anise • Chinese anise

Illicium verum Hook.f. (Illiciaceae); *steranys* (Afrikaans); *ba jiao hui xian* (Chinese); *badiane de Chine*, *anis étoilé* (French); *Echter Steranis, Badian* (German); *anice stellato* (Italian); *bunga lawang* (Indonesian, Malay); *anis estallado* (Spanish); *dok chan* (Thai); *hôi* (Vietnamese)

DESCRIPTION The attractive fruits have eight separate follicles and a distinctive sweet anise smell and taste, mainly located in the fruit wall and not the seeds. The similar-looking but poisonous fruits of Japanese anise or *shikimi* (*Illicium anisatum*) are less regular in shape but are best identified by their bitter and balsamic (not sweet anise-like) taste.[1]

THE PLANT An evergreen tree (up to 18 m or 60 ft), with solitary, yellow, magnolia-like flowers.[2]

ORIGIN North Vietnam and southern China.[3]

CULTIVATION The plant is no longer found in the wild but is cultivated (and used as a popular spice) in China, Japan, India, Malaysia, the Philippines, Thailand and Vietnam. It is grown from seeds or cuttings and requires a tropical climate.

HARVESTING The fruits are harvested by hand when ripe and then dried.

CULINARY USES Star anise is the most important spice in Chinese cuisine and associated cooking traditions of Asia[2] and is widely used in curries and chutneys. It is an essential ingredient of the main Chinese spice mixtures, including "five-spices powder" (*wu xiang fen*), "spicy liquid" (*lu shui*) and "major spices" (*da liao*).[2] Chinese five-spices powder is a mixture of star anise, fennel, cassia, clove and Sichuan pepper in a ratio of 5:5:4:3:1. This mixture has many uses.[2] Powdered and mixed with syrup or honey, it is applied to slow-cooked duck, spare ribs or suckling pig. *Lu shui* is a special spicy liquid or stock that is used repeatedly for boiling pork, lamb, beef, chicken and duck. The spices used in South China are star anise with cassia, clove, fennel, liquorice, kaempferia, black cardamom, Sichuan pepper and often tangerine peel; in North China, typically star anise with ginger, galangal, fennel, fenugreek, clove, cassia and black pepper.[2] The spices are enclosed in a gauze bag and boiled in a mixture of water, soy sauce, gin, rock sugar, onion and garlic for 30 minutes, after which the parboiled meat (and salt) is added. The liquid is permanently kept to its original volume by adding water, soy sauce and gin, and the flavour is maintained by adding more spices on a weekly basis.[2] *Da liao* is the main spice mixture used in everyday Chinese cooking to prepare meat dishes. It varies from place to place but star anise is invariably a main ingredient.[2] Star anise has become popular all over the world for flavouring confectionery, biscuits, pastries and sweet desserts, especially poached fruits and fruit mousses. Star anise oil has largely replaced anise (*Pimpinella anisum*) in the production of anise liqueurs (anisette, sambuca) and anise brandies (arak, ouzo, pastis, Pernod, raki), as well as anise-flavoured sweets.

FLAVOUR COMPOUNDS *Trans*-anethole is the main flavour ingredient (as in real anise and fennel), representing 80% or more of the volatile compounds in star anise oil.[3]

NOTES *Illicium* species are a source of shikimic acid, used to produce modern antiviral drugs.

trans-anethole   shikimic acid

1. **Burkill, I.H. 1966.** *A dictionary of the economic products of the Malay Peninsula*, Vol. 1, p. 944. Crown Agents for the Colonies, London.
2. **Hu, S.-Y. 2005.** *Food plants of China*. The Chinese University Press, Hong Kong.
3. **Farrel, K.T. 1999.** *Spices, condiments and seasonings*. Aspen Publishers, Gaithersburg, USA.

Star anise, the fruits of *Illicium verum*

Leaves and flower (*Illicium verum*)

Flower (*Illicium verum*)

Leaves of *Illicium anisatum*

153

Juglans regia
common walnut • English walnut • Persian walnut

Juglans regia L. (Juglandaceae); *okkerneut* (Afrikaans); *hu tao ren* (Chinese); *noix* (French); *Walnuß* (German); *noce* (Italian); *nogal* (Spanish)

DESCRIPTION Walnut seeds are enclosed in a smooth brown shell that, in turn, is surrounded by a fleshy husk that peels away at maturity. The nut (seed) itself is shaped like two halves of a human brain. Walnuts and walnut oil have a delicious and distinctive taste and are commonly used as a food flavourant.

THE PLANT A large deciduous tree (up to 30 m or 100 ft) with 5–11 leaflets per leaf.[1] North American black walnut (*J. nigra*) has 11–13 toothed leaflets per leaf and is mainly grown for timber because the nuts, although very tasty, are difficult to extract.

ORIGIN Central Asia (Balkans to China).[1] Extensive forests occur in Kyrgyzstan. The tree was introduced to southern Europe and the Middle East by the ancient Persians, Greeks and Romans. From here it spread to the rest of Europe, North America and other parts of the world. Exports exceed 2 million tons per year, with China and the United States (California) as the main producers.

CULTIVATION Commercial cultivars, selected for their large fruits and thin shells, are propagated by grafting. They require deep, rich soil and a temperate climate with cold winters.

HARVESTING The nuts are harvested by hand or mechanically.

CULINARY USES Walnuts and walnut flour are important flavour ingredients of breads, biscuits, cakes, sweets, chocolates, pastries and ice cream. The seed oil is a popular salad oil, especially in France. Chopped or crushed walnuts add flavour to salads (e.g. Waldorf salad) and many dishes (e.g. meat, poultry, fish, stuffings, pastas and pâtés). Walnut paste is a special feature of Georgian cuisine, used in savoury dishes and appetizers.[2] It is made from finely ground walnuts, garlic, dried coriander and other ingredients. Walnut is used to flavour wines and liqueurs[3] (e.g. *brou de noix*, a French liqueur made from the green husks).

FLAVOUR COMPOUNDS The aroma compounds have been identified as lipid-derived volatiles[4] and include hexanal, pentanal, 1-hexanol, 1-pentanol and 1-penten-3-ol. The nut contains gallic acid and several other minor acids. The brown colour of walnut husks (sometimes used as a natural food colourant) is due to juglone and 1,4-naphthoquinone.[3] Walnut oil is rich in polyunsaturated fatty acids, especially linolenic and linoleic acids.

NOTES Seed kernels of Chinese black canarium (*Canarium pimela*) are used in southern China in the same way as walnuts in northern China – they are lightly fried and mixed with chicken meat (formerly a popular ingredient of Cantonese dishes).[5] The fresh or salted fruit flesh of several *Canarium* species is used in pickles and herbal teas.[5] A paste made from cooked candlenuts (*Aleurites moluccana*) is an important flavour ingredient of Indonesian and Malaysian curries.

hexanal pentanal 1-hexanol

1-pentanol

1-penten-3-ol juglone gallic acid

1. **Mabberley, D.J. 2008.** *Mabberley's plant-book* (3rd ed.). Cambridge University Press, Cambridge.
2. **Margvelashvili, J. 1991.** *The classic cuisine of Soviet Georgia.* Prentice Hall Press, New York.
3. **Stampar, F., Solar, A., Hudina, M., Veberic, R., Colaric, M. 2006.** Traditional walnut liqueur – cocktail of phenolics. *Food Chemistry* 95: 627–631.
4. **Elmore, J.S., Nisyrios, I., Mottram, D.S. 2005.** Analysis of the headspace aroma compounds of walnuts (*Juglans regia* L.). *Flavour and Fragrance Journal* 20: 501–506.
5. **Hu, S.-Y. 2005.** *Food plants of China.* The Chinese University Press, Hong Kong.

Common walnuts, the seeds of *Juglans regia*

Common walnut (*Juglans regia*)

Black walnut (*Juglans nigra*)

Chinese black canarium (*Canarium pimela*)

Candlenuts (*Aleurites moluccana*)

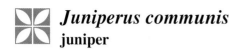 *Juniperus communis*
juniper

Juniperus communis L. (Cupressaceae); *jenewerbessie* (Afrikaans); *kataja* (Finnish); *genièvre* (French); *Wacholderbeeren* (German); *ginepro* (Italian); *junipa* (Japanese); *junipeo* (Korean); *enebro, junipero* (Spanish); *enbär* (Swedish)

Description The female cones ("juniper berries") are spherical, fleshy, berry-like, about 6 mm (¼ in.) in diameter and dark purple-blue to black when ripe. They have a woody, resinous and turpentine-like aroma and a warm, pine-like taste.

The plant An evergreen shrub or small tree of about 6 m (ca. 20 ft) in height with crowded, small, flat, needle-shaped leaves. Male and female cones occur on separate plants. The cones take 18 months to ripen.

Origin Northern temperate zone (Asia, Europe and North America).[1] Juniper berries have been used since the Middle Ages or earlier as a spice and adulterant of pepper. About 200 tons are harvested each year from wild plants in central Europe.[1]

Cultivation Juniper is mostly wild-harvested but there are many garden cultivars grown as ornamental plants. Plants are usually grown from cuttings and adapt to a wide range of environmental conditions.

Harvesting The ripe cones ("berries") are wild-harvested by hand and air-dried (usually from September to October).

Culinary uses Juniper is used as a spice in European cuisine (especially in northern Europe and Scandinavia), mainly to flavour meat dishes.[2] Fresh or slightly crushed dry cones add a fresh, aromatic, spicy, peppery, resinous and slightly bitter taste to meat (wild birds, hare, rabbit, venison, reindeer, wild boar and pork)[2] but also cabbage and sauerkraut, as well as sausages, marinades and sauces. The French terms *à la liégeoise* or *à l'ardennaise* refer to meat dishes (mainly venison or game birds) that are flavoured

with juniper or flamed with gin (or juniper and alcohol).[2] Juniper oil is used in confectionery, meat products, jellies and puddings.[3] Gin, developed in the Netherlands in the 17th century, is flavoured with immature green juniper cones. The name is derived from *genever* or *jenever*, the Dutch word for juniper. Juniper cones, leaves or essential oil are also used in aromatic brandies (e.g. *brinjevac*, *genièvre*, *péquet* and *schiedam*), wines, bitter herbal liqueurs and Scandinavian beers (e.g. *sahti* – Finnish rye beer). The fresh cones with their high sugar content of ca. 25% may be fermented and distilled to produce schnapps (e.g. *Steinhäger*). In Germany, juniper cones are sometimes added to *Latwerge*, an unsweetened, jam-like condiment usually made from cooked plums. Juniper wood chips are traditionally used to smoke salmon and cured meats.

Flavour compounds The ripe cones contain 0.5 to 2% essential oil.[3] The pine-like or turpentine-like flavour of juniper cones and juniper oil is due to a complex mixture of volatile compounds that usually include α-pinene, sabinene, terpinen-4-ol, limonene and myrcene as main constituents.[3,4]

Notes Infusions of juniper berries or juniper oil are traditional diuretic and antiseptic medicines. The oil is used in aromatherapy.

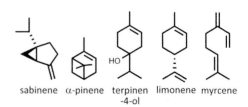

sabinene   α-pinene   terpinen-4-ol   limonene   myrcene

1. Mabberley, D.J. 2008. *Mabberley's plant-book* (3rd ed.). Cambridge University Press, Cambridge.
2. Larousse. 1999. *The concise Larousse gastronomique.* Hamlyn, London.
3. Harborne, J.B., Baxter, H. 2001. *Chemical dictionary of economic plants.* Wiley, New York.
4. Farrel, K.T. 1999. *Spices, condiments and seasonings.* Aspen Publishers, Gaithersburg, USA.

Juniper berries, the dried cones of *Juniperus communis*

Foliage and immature cones (*Juniperus communis*)

Trees (*Juniperus communis*)

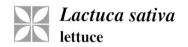

Lactuca sativa
lettuce

Lactuca sativa L. (Asteraceae); *blaarslaai* (Afrikaans); *sheng cai* (Chinese); *laitue* (French); *Gartensalat* (German); *lattuga* (Italian); *selada* (Malay); *alface* (Portuguese); *lechuga* (Spanish)

DESCRIPTION The most popular types include leaf lettuce (looseleaf, cutting and bunching types), romaine or cos lettuce (long upright heads) and crisphead lettuce ("iceberg" type, popular in the United States). Less well known are butterhead, summercrisp, stem and oilseed lettuces. Lettuce leaves have a very high water content (ca. 95%) and are almost invariably eaten fresh.

THE PLANT An annual with soft leaves that exude milky latex from the thick leaf midribs.

ORIGIN Mediterranean region and the Near East. Lettuce is a cultigen derived from wild lettuce (*L. serriola*).[1] It was initially grown for seeds in Egypt (since 4500 BC), and only later as a leaf crop.[1] Lettuce is the most popular of all salad leaves, with annual production estimated to exceed 23 million tons worldwide.

CULTIVATION Lettuce is an annual crop, propagated from seeds. It is best grown in full sun, in well-drained, nitrogen-rich soil under cool conditions. Heat causes bolting (and a bitter taste).

HARVESTING Heads of lettuce (or loose leaves) are picked in the early morning and rapidly transported to the market.

CULINARY USES Lettuce is usually classified as a vegetable but it can also be regarded as a culinary herb because it is almost invariably eaten fresh and adds subtle flavours to the salad in which it is usually the main ingredient. Lettuce leaves are added (either tossed or composed) to green salads and mixed salads of which there are countless variations all over the world.[2] Caesar, Greek, Niçoise and Waldorf salads are famous examples. A Caesar salad is made from romaine (cos) lettuce with croutons, Parmesan cheese, lemon juice, raw or coddled egg, Worcestershire sauce, garlic and black pepper. Its creation is attributed to the Italian American restaurateur Caesar Cardini. A typical Greek salad has tomato wedges, cucumber, green bell pepper, onion, feta cheese and kalamata olives, is seasoned with salt, black pepper and dried oregano and dressed with olive oil. The term "Greek salad" is also used in the United States, Australia, South Africa and the United Kingdom to refer to a lettuce-based salad with the aforementioned ingredients, and dressed with oil and vinegar. A Niçoise salad (named after the city of Nice) is made from romaine lettuce, Niçoise olives and anchovies, served with Dijon vinaigrette. A Waldorf salad is traditionally made from fresh apple, celery and walnuts, dressed with mayonnaise and served on a bed of lettuce. It was first created in the 1890s at the Waldorf Hotel in New York City. Lettuce is a popular garnish (e.g. on hamburgers) and may be braised, stuffed, puréed, cooked or served as a chiffonnade (chopped, softened in butter, with stock, milk or cream added).[2]

FLAVOUR COMPOUNDS The main aroma compound of fresh lettuce leaves is *cis*-3-hexen-1-ol, a compound responsible for the distinctive smell of freshly mown lawn grass.[3] The bitter taste of lettuce (and endives) is due to lactucopicrin and related minor sesquiterpene lactones.[4]

NOTES A few examples of other salad herbs are also shown opposite.

cis-3-hexen-1-ol   lactucopicrin

1. Zohary, D., Hopf, M. 2012. *Domestication of plants in the Old World*. Clarendon Press, Oxford.
2. Larousse. 1999. *The concise Larousse gastronomique*. Hamlyn, London.
3. Arey, J., Winer, A.M., Atkinson, R., Aschman, S.M., Morrison, C.L. 1991. The emission of (Z)-3-hexen-1-ol, hexenylacetate and other oxygenated hydrocarbons from agricultural plant species. *Atmospheric Environment* 5/6: 1063–1076.
4. Seo, M.W., Yang, D.S., Kays, S.J., Lee, G.P., Park, K.W. 2009. Sesquiterpene lactones and bitterness in Korean leaf lettuce cultivars. *HortScience* 24: 246–249.

Lettuce, the leaves of *Lactuca sativa*

Lettuce (*L. sativa* 'Madrileña')

Lettuce seedlings (*L. sativa*)

Lettuce (*L. sativa* 'Hoya de Roble')

Prickly lettuce (*L. serriola*)

Endive (*Cichorium endivia*)

Corn salad (*Valerianella locusta*)

159

Laurus nobilis
bay • bay laurel • sweet bay • true laurel

Laurus nobilis L. (Lauraceae); *lourierblaar* (Afrikaans); *laurier* (Dutch); *laurier* (French); *Lorbeer* (German); *laur, alloro* (Italian); *laurel* (Spanish)

DESCRIPTION Laurel leaves (bay leaves) are dark green above and paler below, with a sweet, spicy aroma and a bitter taste. The fresh or dried leaves are used (or essential oil distilled from fresh leaves).

THE PLANT An evergreen shrub or small tree, rarely up to 20 m (66 ft) high, with male and female flowers on separate plants and purple-black, olive-like fruits. Many unrelated plants are also called "bay" or "laurel" due to superficial similarities with true laurel.

ORIGIN Mediterranean region. In ancient Greece, laurel was used to form a crown or wreath of honour for poets and heroes. Laurel became a symbol of victory in Roman times. During the Renaissance, persons who received a doctorate were decorated with a berried branch of laurel, hence the terms "baccalaureus" or "bachelor" of art/science (from the Latin *bacca*, a berry and *laureus*, of the laurel). The expressions "poet laureate", "Nobel laureate" and "resting on your laurels" have the same origin. Laurel is widely cultivated as an ornamental tree and popular culinary spice.

CULTIVATION Trees are usually propagated from cuttings and less often from seeds. It prefers a warm Mediterranean climate but grows surprisingly well under a wide range of environmental conditions and is resistant to drought and frost. It is an attractive and rewarding plant to grow as a feature in the herb garden or in a container on the patio.

HARVESTING Leaves are picked by hand and air-dried. They are usually sold whole and can be kept for up to one year under favourable storage conditions. Commercial production is centred in the Mediterranean region.

CULINARY USES Bay leaves or bay leaf seasoning is one of the most popular of all food flavourings. The leaves are a classic component of a *bouquet garni* (a small bundle of herbs cooked with a dish but removed before it is served). A traditional use of whole bay leaf is in *bobotie* (a South African meat dish of Malay and Dutch origin). Fresh or dried leaves are used sparingly in stews and in fish, chicken and meat dishes. They add a delicious flavour to milk puddings. The ground spice can be used in marinades, pasta sauces, pâtés, pickles, soups, stocks and vegetables. Bay leaf is an essential component of mixed pickling spices and has commercial applications in processed meat products (corned beef, chicken loaf, mortadella and sausages). The essential oil is used in non-alcoholic beverages.

FLAVOUR COMPOUNDS The leaves and leaf oil contain eucalyptol (1,8-cineole) as main compound, with smaller amounts of α-terpinyl acetate, linalool, methyleugenol and other aromatic constituents adding to the complexity of the aroma.

NOTES Filé powder, made from the leaves of the North American *Sassafras albidum* (Lauraceae), is an essential flavour ingredient in Cajun and Creole cooking (e.g. in *filé gumbo*). It should be free of safrole, a potential carcinogen and banned food substance. Salam leaf (*Syzygium polyanthum*), also known as Indonesian bay leaf or *daun salam*, is used in Indonesian and Malay cooking.

1,8-cineole   α-terpinyl   linalool   methyleugenol
                   acetate

1. **Mabberley, D.J. 2008.** *Mabberley's plant-book* (3rd ed.). Cambridge University Press, Cambridge.
2. **Farrel, K.T. 1999.** *Spices, condiments and seasonings.* Aspen Publishers, Gaithersburg, USA.
3. **Fiorini, C., Fourasté, I., David, B., Bessière, J.M. 1997.** Composition of the flower, leaf and stem essential oils from *Laurus nobilis. Flavour and Fragrance Journal* 12: 91–93.
4. **Harborne, J.B., Baxter, H. 2001.** *Chemical dictionary of economic plants.* Wiley, New York.

Bay leaf or sweet bay, the leaves of *Laurus nobilis*

Female flowers (*Laurus nobilis*)

Male flowers (*Laurus nobilis*)

Fruits (*Laurus nobilis*)

Salam leaf (*Syzygium polyanthum*)

Sassafras (*Sassafras albidum*)

161

Lavandula angustifolia
common lavender • true lavender • English lavender

Lavandula angustifolia Mill. (= *L. officinalis* Chaix, *L. vera* DC.) (Lamiaceae); *laventel* (Afrikaans); *lavande* (French); *Echter Lavendel* (German); *lavanda* (Italian); *lavanda* (Spanish)

DESCRIPTION Lavender leaves are typically silvery-hairy, with a strong floral fragrance. The leaves, flowers, flower buds and essential oil are all used as flavourants.

THE PLANT A much-branched aromatic shrub (1 m or ca. 3 ft high and wide) with attractive purplish blue flower spikes. The classification of species and cultivars is complicated and common names are not precisely applied.[1] English or Old English lavender usually refers to *L. angustifolia* (even though it is not indigenous to England); French lavender usually refers to *L. dentata* (or *L. stoechas*); Spanish lavender often applies to *L. stoechas* (less often to *L. dentata* or sometimes *L. lanata*); Dutch lavender or lavandin is used for *L. ×intermedia*, a group of hybrids between *L. angustifolia* and *L. latifolia*.

ORIGIN True lavender is indigenous to the western Mediterranean region.[1,2] Various species have been used since ancient times for their aromatic and medicinal properties. The ancient Greeks called it *nardus*. It was one of the holy herbs of Solomon's temple and is mentioned in the Bible as *nard*. Lavender has only recently become a culinary herb and is not mentioned in any of the classical books on southern French cuisine. Since the 1970s, it has been added to commercial herb mixtures sold in the United States as "Herbes de Provence". "Herbes de Provence" were not traditionally standardized (or sold as such) but comprised various combinations of rosemary, thyme, basil, savory and other herbs collected in the countryside.

CULTIVATION Lavender is easily propagated from cuttings, seeds or by division. It is resistant to drought and frost, and grows best in full sun and well-drained, sandy soil. Regular pruning is beneficial. Lavender has been called "queen of the herb garden" and is a popular and attractive choice for home gardens.

HARVESTING Flowering tops are harvested by hand (or mechanically, for large-scale extraction of the essential oil).

CULINARY USES Lavender herb (or commercial mixtures called "Herbes de Provence") is used in grills (fish and meat) and in vegetable stews. Flower buds or leaves are used to make tea blends, tisanes, lavender sugar, lavender syrup, lavender scones and marshmallows. Lavender is sometimes combined with cheeses made from sheep or goat's milk. Flowers are used in confectionery, often as a garnish on cakes, and go particularly well with chocolate. Lavender oil is added to non-alcoholic beverages and liqueurs. The herb is one of several botanical flavourants of vermouths (aromatized wines),[3] used in popular cocktails such as the Martini, Manhattan and Negroni. Lavender honey is a Provençal speciality.

FLAVOUR COMPOUNDS Lavender oil contains linalool and linalyl acetate as main constituents and many minor ones (e.g. 1-octen-3-yl acetate, lavandulol, lavandulyl acetate, 1,8-cineole, camphor, limonene and α-pinene)[4] contributing to the complex herbal-floral fragrance.

NOTES Lavender oils are mainly used in perfumery and in soaps and detergents. Flowers and fruits are included in aromatic potpourris and sachets.

linalyl acetate   linalool   lavandulol   lavandulyl acetate   1-octen-3-yl acetate

1. **Upson, T., Andrews, S. 2004.** *The genus* Lavandula. Timber Press, Portland.
2. **Mabberley, D.J. 2008.** *Mabberley's plant-book* (3rd ed.). Cambridge University Press, Cambridge.
3. **Tonutti, I., Liddle, P. 2010.** Aromatic plants in alcoholic beverages. A review. *Flavour and Fragrance Journal* 25: 341–350.
4. **Harborne, J.B., Baxter, H. 2001.** *Chemical dictionary of economic plants*. Wiley, New York.

Leaves of *Lavandula angustifolia*, *L. dentata* and *L. ×intermedia*

Lavandin (*Lavandula ×intermedia*)

Spanish lavender (*L. stoechas*)

Lavender or English lavender (*L. angustifolia*)

163

Lepidium sativum
cress • garden cress • pepper grass

Lepidium sativum L. (Brassicaceae); *kers, tuinkers* (Afrikaans); *jia du xing cai* (Chinese); *cresson alénois* (French); *Gartenkresse* (German); *alim* (Indonesian); *crescione* (Italian); *agrião, mastruco* (Portuguese); *berro de huerta, berro alenois, mastuerzo* (Spanish)

DESCRIPTION Fresh leaves, young seedlings or sprouts are used. They have a peppery taste. Sprouts and seedlings are easily recognized by the deeply three-lobed cotyledons (seedling leaves).

THE PLANT An erect annual of up to 0.6 m (2 ft) high with compound leaves and minute white flowers.[1] Seeds are borne in capsules of about 6 mm (¼ in.) long. Several similar and related plants are known as cress.[2] Best known of all is watercress, *Nasturtium officinale*. Land cress or winter cress, *Barbarea verna* (*Winterkresse* in German, *barbarée* in French) has lobed leaves and yellow flowers. Nasturtium (*Tropaeolum majus*) is sometimes called Indian cress. Spoon cress or spoonwort, *Cochlearia officinalis* (*Löffelkresse* in German, *cochléaire* in French) was once an important source of vitamin C (to prevent scurvy) but has since declined in popularity after *Citrus* fruits became more readily available.[2]

ORIGIN Garden cress is thought to be indigenous to western Asia[2] (but perhaps originally Ethiopia or Iran).[1] It was used by the ancient Egyptians as a food source and became well known in parts of Europe (including Britain, France, Italy and Germany), where it is still a minor crop.

CULTIVATION Garden cress is propagated from seeds and grown as a spicy salad herb, often in kitchen gardens. It is a short-lived plant that matures in three to four months. Sprouts or seedlings are mostly used. Commercial growers germinate the seeds in special containers and sell these, container and all, as an early spring salad. *Lepidium* is the traditional component of so-called "mustard and cress", a mixture of mustard and cress seedlings that are used as salad herb. The cress takes longer to germinate and is sown four days before the mustard.

HARVESTING For culinary use, cress leaves are harvested before flowering. Sprouts are gathered a few days after germination. Young plants should be carefully cleaned by removing the thicker stems and yellowing leaves, taking care not to soak them in water. For seed production, plants are left to mature, after which they are cut, dried and processed to collect the seeds.

CULINARY USES Cress is most commonly used in the form of sprouts or young seedlings, as a tasty and healthy garnish for salads and sandwiches. Young plants may be used as a garnish in grilled (broiled) dishes[3] and are added to soups. They have a slightly peppery and somewhat bitter taste, and a slightly pungent odour. Cress purée was once a popular food item in France, where several types of cress are still used to some extent (watercress being the most popular).[3]

FLAVOUR COMPOUNDS Like other members of the mustard family, cress has sulphur-containing compounds (glucosinolates). The major compound in *Lepidium* leaves and seeds is glucotropaeolin, also known as benzyl glucosinolate.[4]

NOTES Cress seeds are an important commercial source of cooking oil in India; the seeds yield 25% of a yellowish-brown, semi-drying oil.[1]

glucotropaeolin
(benzyl glucosinolate)

1. **Jansen, P.C.M. 2004.** *Lepidium sativum* L. In: Grubben, G.J.H., Denton, O.A. (Eds), *Plant resources of tropical Africa 2. Vegetables*, pp. 365–367. PROTA Foundation, Wageningen.
2. **Mabberley, D.J. 2008.** *Mabberley's plant-book* (3rd ed.). Cambridge University Press, Cambridge.
3. **Larousse. 1999.** *The concise Larousse gastronomique*. Hamlyn, London.
4. **Williams, D.J., Critchley, C., Pun, S., Chaliha, M., O'Hara, T.J. 2009.** Differing mechanisms of simply nitrile formation on glucosinolate degradation in *Lepidium sativum* and *Nasturtium officinale* seeds. *Phytochemistry* 70: 1401–1409.

Cress or garden cress, the seedlings

Cress seedlings (*Lepidium sativum*)

Cress seeds (*Lepidium sativum*)

3 mm

Flowering plant (*Lepidium sativum*)

Flowers and fruits (*Lepidium sativum*)

Mustard and cress

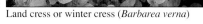

Land cress or winter cress (*Barbarea verna*)

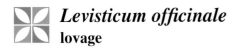

Levisticum officinale
lovage

Levisticum officinale Koch (= *Ligusticum levisticum* L.) (Apiaceae); *lavas, Maggikruid* (Afrikaans); *ou dang gui* (Chinese); *livèche* (French); *Liebstöckel, Maggikraut* (German); *levistico* (Italian); *ligústico* (Spanish)

DESCRIPTION The leaves are compound, with broad, wedge-shaped leaflets that are toothed along the upper margins. They have a spicy, slightly bitter taste.

THE PLANT A sparse, erect perennial herb, up to 2 m (ca. 6 ft) high, with small yellow flowers arranged in flat-topped double umbels. Lovage is superficially similar to angelica (*Angelica archangelica*) but the latter is a more robust plant with leaflets toothed along their lower margins.

ORIGIN Eastern Mediterranean region.[1] Lovage is said to represent the characteristic flavour of Roman cuisine (along with *silphion* and *garum*). It is now found throughout Europe and North America, not only as a cultivated herb and crop plant but also as a naturalized weed.[1]

CULTIVATION Plants are propagated from seeds or by division of the rootstocks. It tolerates partial shade and low temperatures and grows well in slightly alkaline, rich, well-drained soil.

HARVESTING For home use, healthy young leaves are picked as required. Extracts and essential oils are produced on a commercial scale from leaves and roots, harvested in autumn.

CULINARY USES Fresh leaves are still popular as a culinary herb in parts of Europe (mainly Britain, Germany and Romania), used in much the same way as celery. They are used in salads, soups, sauces, sour pickles, vinegars, new potatoes, stews and meat dishes. Blanched stems may be added to salads and can be candied (like angelica). Fruits are sometimes used as a spice. Above-ground parts are distilled to produce lovage leaf oil. The roots are eaten as a vegetable but are more often used in the form of extracts, essential oils or powdered dry root to flavour a wide range of food items, including alcoholic herbal tonics, liqueurs, confectionery, jellies, puddings, frozen dairy products, meat products, as well as savoury and sweet sauces.[2]

FLAVOUR COMPOUNDS Lovage is chemically complex. The leaf essential oil has α-terpenyl acetate as main compound (about 30%), accompanied by several alkylphthalides such as ligustilide and sedanolide. These compounds, which also give celery its distinctive flavour, represent the major portion (ca. 70%) of lovage root oil.[3] Of special interest is sotolon (sotolone), the major aroma-impact compound in lovage leaves and roots,[4] with a very low detection threshold. Sotolon has a spicy aroma at high concentrations but is caramel-like (maple syrup-like) at low concentrations. The German name for lovage is *Maggikraut*, named after Maggi sauce, a German spice sauce similar in appearance and taste to soy sauce and Worcestershire sauce. Maggi sauce contains neither lovage nor sotolon. However, it does contain homofuraneol, the 5-ethyl homologue of sotolon, to which the similarity in flavour can be ascribed.[4]

NOTES Lovage is used in traditional medicine as a carminative and diuretic.

sotolon

α-terpinyl acetate

homofuraneol

(*Z*)-ligustilide   sedanolide

1. **Mabberley, D.J. 2008.** *Mabberley's plant-book* (3rd ed.). Cambridge University Press, Cambridge.

2. **Harborne, J.B., Baxter, H. 2001.** *Chemical dictionary of economic plants.* Wiley, New York.

3. **Raal, A., Arak, E., Orav, A., Kalais, T., Muurisepp, M. 2008.** Composition of the essential oil of *Ligusticum officinalis* W.D.J. Koch from some European countries. *The Journal of Essential Oil Research* 20: 318–322.

4. **Blank, I., Schieberle, P. 1993.** Analysis of the seasoning-like flavour substance of a commercial lovage extract (*Levisticum officinale* Koch). *The Flavour and Fragrance Journal* 8: 191–195.

Leaves and fruits of lovage, *Levisticum officinale*

Plant (*Levisticum officinale*)

Leaves and flowers (*Levisticum officinale*)

167

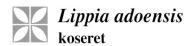

Lippia adoensis
koseret

Lippia adoensis Hochst. ex Walp. (Lamiaceae); *koseret* (Amharic)

DESCRIPTION Koseret comprises the coarsely chopped and dried twigs (four-angled in transverse section), together with the leaves (with prominent veins, a rough surface texture and a strong lemony and spicy aroma).

THE PLANT A much-branched woody shrub 1 to 3 m (ca. 3 to 10 ft) high, with leaves arranged in whorls of three at each node. The tiny purple flowers are borne in dense oblong heads. About three but up to 12 of these stalked flower heads occur at each node.[1] The plant superficially resembles *Aloysia citrodora* (previously known as *Lippia citriodora*), which is also commonly cultivated in Ethiopia. The latter has loose panicles of flowers (not compact heads) and a strong lemon smell.

ORIGIN The species is known only from Ethiopia and Eritrea, where the wild variety is found in disturbed areas and along forest margins. The cultivated variety is morphologically closely similar but has a markedly different aroma.

CULTIVATION Plants are easily propagated from cuttings. They require a warm climate, are tolerant of drought but do best in deep, well-drained soil. The shrubs are regularly pruned to promote new growth.

HARVESTING Young stems and leaves are harvested by hand, cut into about 25 mm (1 in.) sections and allowed to air-dry.

CULINARY USES Koseret is an important traditional Ethiopian culinary herb, commonly sold on the local markets as a coarsely chopped, spice herb. It is used specifically to add a strong, lemony flavour to butter. The spiced butter is an essential ingredient of *kifto*, a traditional minced beef dish.[2,3] The flavoured butter is melted over low heat, after which minced beef and chilli powder spice (*mitmita*) are added.[4] The ingredients are thoroughly mixed and the *kifto* is then ready to be served raw (as a type of tartare). It can also be cooked over low heat, stirring frequently.[4]

FLAVOUR COMPOUNDS The essential oil of the cultivated variety has linalool as main compound, with small amounts of sesquiterpenes, including germacrene D, α-copaene, β-cadinene and α- and β-caryophyllene.[5] The wild variety of the species differs in lacking linalool and having limonene as main compound, together with perillaldehyde, piperitone and citral A, none of which are found in the cultivated variety.[5]

NOTES Fresh leaves of the wild form of koseret are used to wash wooden kitchen utensils and to give them a fresh smell.[3] Ethiopian *Lippia abyssinica* (*aricharaku*) is similarly utilized, while *ged-hamer* (*L. carviodora*) is a traditional herbal tea and flavouring of the Somali people.[4] The American *L. alba* (known as bushy lippia or *hierba nigra*) is used to flavour Mexican mole sauces.

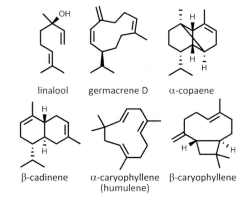

linalool   germacrene D   α-copaene

β-cadinene   α-caryophyllene (humulene)   β-caryophyllene

1. Demissew, S. 2006. Verbenaceae. In: Hedberg, I., Kelbessa, E., Edwards, S., Demissew, S., Persson, E. (Eds), *Flora of Ethiopia and Eritrea*, Vol. 5, pp. 499–514. National Herbarium, Addis Ababa and University of Uppsala, Uppsala.
2. Jansen, P.C.M. 1981. *Spices, condiments and medicinal plants in Ethiopia, their taxonomy and agricultural significance.* Agricultural Research Reports 906. Centre for Agricultural Publishing and Documentation, Wageningen.
3. Demissew, S. 1993. A description of some essential oil bearing plants in Ethiopia and their indigenous uses. *Journal of Essential Oil Research* 5: 465–479.
4. Bekele, G. 1992. *Ethiopian & foreign cook guide book.* Team Work, Addis Ababa.
5. Abegaz, B., Asfaw, N., Lwande, W. 1993. Constituents of the essential oils of wild and cultivated *Lippia adoensis* Hochst. ex Walp. *Journal of Essential Oil Research* 5: 487–491.

Koseret, the dried leaves of *Lippia adoensis*

Leaves and flowers (*Lippia adoensis*)

Koseret on the market

Plant (*Lippia adoensis*)

169

Mangifera indica
mango • amchur • amchoor

Mangifera indica L. (Anacardiaceae); *mango, veselperske* (Afrikaans); *mang guo* (Chinese); *mangue* (French); *Mango* (German); *aam* (Hindi); *mango* (Italian); *mango, mangou* (Japanese); *mangga* (Malay); *manga* (Portuguese); *mango* (Spanish); *mamuang* (Thai)

DESCRIPTION Green (unripe) fruits are sliced and dried to produce *a(a)mchur* (*amchoor*): pale whitish brown chunks or pale greyish brown powder with a sour, astringent taste and a slightly resinous, fruity flavour.

THE PLANT A tropical tree (up to 30 m / 100 ft) that can become very old and produce up to an astounding 29 000 fruits in a single season.[1] Fruits of modern varieties are less fibrous and resinous than older types.

ORIGIN Mango is an ancient cultigen and its exact place of origin is uncertain.[1,2] It is now believed to have originated in northeastern India and adjoining parts of Bangladesh and Myanmar, and not in Southeast Asia, where the greatest diversity is currently found.[2] Mango was introduced to Africa by about AD 1000 and much later to tropical America.[1,2] It has become one of the most important commercial tropical fruits.

CULTIVATION Mangoes are polyembryonic, with up to five embryos in each seed.[1,2] One of the embryos is usually a hybrid, while the others are vegetatively identical to the parent tree. For this reason, mangoes are often grown from seeds. For the home garden, grafted trees of dwarf varieties are more practical. Mangoes require a tropical or subtropical climate, deep fertile soil, and wet summers with relatively dry winters.

HARVESTING Fruits are usually picked when mature but still green because they are easily ripened under controlled conditions. The fruits are washed to remove the sap which leaks from the cut stems.

CULINARY USES *Amchur* is an acidic condiment and spice that is widely used as an essential ingredient of many South Asian dishes, curries, marinades and sauces. Unripe fruits are the basis of Indian and African chutneys, pickles and hot spicy sauces. Ripe fruits or mango juice are added to fruit salads, cold desserts, purées, ice creams, yogurts and soufflés.

FLAVOUR COMPOUNDS The flavour and aroma of mangoes are chemically complicated and not associated with any single chemical compound.[3,4] The resinous and lemony odour and taste are due to δ-3-carene, limonene, terpinolene and other monoterpenoids,[3] while ethyl 2-methylpropanoate and methyl benzoate are amongst the most important aroma-active compounds, responsible for the fruity fragrance of ripe mangoes.[4]

NOTES Although not classified as herbs or spices, many fruits and fruit juices (apricots, cherries, figs, plums, tomatoes) are essential in adding flavour and acidity to dishes, confectionery, beverages, pickles and chutneys. Cherries are used for pickles, sweet and sour dishes and as condiment to go with duck and game. Chinese plum or Japanese apricot (*Prunus mume*) is used in juices, alcoholic drinks, pickles and preserves (Chinese *suanmeizi*, Japanese *umeboshi*). The flavour and colour of tomatoes (*Lycopersicon esculentum*) are essential in tomato sauce or ketchup, and in many pasta sauces, purées, salads, soups and soufflés. The red carotenoid pigment (lycopene) also adds health benefits to dishes in which cooked or processed tomato is used.

δ-3-carene

limonene   terpinolene
(δ-terpinene)

ethyl 2-methyl propanoate

methyl benzoate

1. Litz, R.E. (Ed.) 2009. *The mango: Botany, production and uses* (2nd ed.). CABI, Oxfordshire.
2. Morton, J. 1987. Mango. pp. 221–239. In: *Fruits of warm climates*. Julia F. Morton, Miami, Florida.
3. Pino J.A., Mesa, J., Muños, Y., Martí, M.P., Marbot, R. 2005. Volatile components from mango (*Mangifera indica*) cultivars. *Journal of Agricultural and Food Chemistry* 53: 2213–2223.
4. Pino, J.A., Mesa, J. 2006. Contribution of volatile compounds to mango (*Mangifera indica*) aroma. *Flavour and Fragrance Journal* 21: 207–213.

Dried green mango, *aamchur*, *aamchoor* (*Mangifera indica*)

Green mangoes (*Mangifera indica*)

Ripe mangoes (*Mangifera indica*)

Sweet cherry (*Prunus avium*)

Mume leaves (*Prunus mume*)

Tomatoes (*Lycopersicon esculentum*)

171

Melissa officinalis
lemon balm • balm • sweet balm

Melissa officinalis L. (Lamiaceae); *sitroenkruid* (Afrikaans); *citronnelle, mélisse* (French); *Zitronenmelisse* (German); *melissa, cedronella* (Italian); *melissa* (Spanish)

DESCRIPTION The fresh or dried leaves have a wrinkled appearance, with prominent veins, toothed margins and a characteristic lemon-like smell.

THE PLANT A perennial herb of 0.6 m (2 ft) high, bearing opposite leaves and small, two-lipped, white flowers. The plant is also known as balm or balm mint but should not be confused with bee balm (*Monarda didyma*), a popular and attractive North American garden plant that was once used in traditional medicine ("Oswego tea")[1] and to flavour venison.[2]

ORIGIN Lemon balm occurs naturally in the eastern Mediterranean region and Asia Minor. The herb is recorded in classical Greek and Roman literature. In recent times it has become a popular medicinal and culinary herb and is cultivated on a commercial scale in most parts of Europe. It has become invasive in the United States.

CULTIVATION Lemon balm is a popular garden herb in most temperate parts of the world. Propagation is very easy, from seeds, cuttings or by division of mature plants. Plants will survive even severe frost and do best in full sun or partial shade in moist, fertile and well-drained soil. Lemon balm tends to spread rapidly, so that occasional pruning may be beneficial. Cultivars differ in the colour of the leaves ('Aurea', 'Variegata'), in the composition of essential oil ('Lemonella', 'Lime') and in growth habit ('Quedlinburger', 'Quedlinburger Niederliegende').

HARVESTING Fresh leaves are picked for culinary use as required. They can also be successfully dried for later use.

CULINARY USES The aromatic leaves with their distinctive lemon aroma are used as ingredient of green salads, fruit salads, desserts, ice cream, milk puddings, soups, stuffings and sauces (e.g. lemon balm pesto). The flavour goes well with a wide range of dishes, including eggs, omelettes, vegetables, stews, chicken, pork and fish. Lemon balm, often in combination with other herbs such as peppermint and spearmint, adds flavour and colour to herbal teas (both hot and iced), tisanes, fruit drinks, punches, claret cups, homemade lemon liqueurs and cocktails. The sweet-scented flowers are distilled to make *Eau de carmes*[3] and various other lemon drinks and liqueurs such as Benedictine and chartreuse.

FLAVOUR COMPOUNDS The lemon flavour is due to the presence of citronellal and citral (a mixture of citral A and B, alternatively known as geranial and neral, respectively). Smaller amounts of citronellol, geraniol and nerol are also present in the essential oil.[3]

NOTES The name *Melissa* is the Greek word for "bee" and the plant has the reputation, since ancient times, that it attracts honeybees.[4] The pheromone of honeybees (synthesized in the Nasonov gland of the bee) includes geraniol as major compound, together with smaller quantities of citral and nerolic acid.[4] The combination of these compounds is included in artificial bee attractants.

citronellal   citral A (geranial)   citral B (neral)

citronellol   geraniol   nerol   nerolic acid

1. **Mabberley, D.J. 2008.** *Mabberley's plant-book* (3rd ed.). Cambridge University Press, Cambridge.
2. **Tilford, G.L. 1997.** *Edible and medicinal plants of the west.* Mountain Press, Missoula.
3. **Larousse. 1999.** *The concise Larousse gastronomique.* Hamlyn, London.
4. **Harborne, J.B., Baxter, H. 2001.** *Chemical dictionary of economic plants.* Wiley, New York.
5. **Burgett, M. 1980.** The use of lemon balm (*Melissa officinalis*) for attracting honeybee swarms. *Bee World* 61: 44–46.

Lemon balm (*Melissa officinalis*)

Lemon balm plants (*Melissa officinalis*)

Flowers (*Melissa officinalis*)

Bergamot, bee balm (*Monarda didyma*)

173

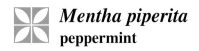

Mentha piperita
peppermint

Mentha ×piperita L. (Lamiaceae); *peperment* (Afrikaans); *la bo he* (Chinese); *menthe poivrée* (French); *Pfefferminze* (German); *gamathi pudinah* (Hindi); *pepaa minto* (Japanese); *menta piperina* (Italian); *hortela-pimenta* (Portuguese); *menta piperita* (Spanish); *bac hà* (Vietnamese)

DESCRIPTION The leaves are dark green with a reddish purple tinge, and have a characteristic peppermint smell.

THE PLANT An aromatic perennial herb with spreading stems and lilac-pink flowers borne in elongated clusters on the stem tips. A form with larger leaves, rounded flower clusters and a distinct lemon smell is known as bergamot mint, orange mint or eau de cologne mint (*M. ×piperita* var. *citrata*). Peppermint is similar to field mint (corn mint, Japanese mint, *pudinah* – *M. arvensis*) but this important mint of Asian cuisines has axillary clusters of flowers (not terminal spikes as in peppermint and spearmint).

ORIGIN Peppermint is a cultigen and hybrid between water mint (*M. aquatica*) and spearmint (*M. spicata*) that originated in England in the 17th century.[1,2] Corn mint is indigenous to Europe and Asia.[1,2] The distinct difference between these two mints is clearly shown in the photographs.

CULTIVATION Plants are propagated with the greatest of ease from stem cuttings or by division and can be planted in a sunken box or pot to contain excessive spread. Peppermint tolerates shade and cold winters but corn mint is adapted to warm and tropical conditions. Peppermint is grown on a large commercial scale the Midwest of the United States for peppermint oil, while Japanese peppermint oil is produced in Japan from a selection known as corn mint (*M. arvensis* var. *piperescens*).[1,2] Peppermint with its menthol flavour has become the basis of a large industry.

HARVESTING Fresh leaves are harvested at any time for culinary purposes. To produce menthol-rich essential oil, flowering stems (of both peppermint and corn mint) are harvested mechanically and extracted by steam distillation.[2,3]

CULINARY USES Peppermint herb and oil are used to flavour chocolates, puddings, ice creams, jellies, confectionery and beverages (e.g. crème de menthe).[2,3] Peppermint leaves are used to some extent in cooking (soups, stews and vegetables) but spearmint (or field mint) is usually preferred. Bundles of field mint are a common sight at oriental fresh produce markets. Field mint is an important ingredient in Chinese, Indian, Japanese, Malaysian and Thai cuisines. Fresh leaves are used as herbal tea and ingredient and garnish in salads. It sometimes replaces the laksa plant (*Persicaria odorata*) as main ingredient in *laksa* (Malaysian noodle soup).

FLAVOUR COMPOUNDS The distinctive flavour of peppermint and field mint is due to menthol, which causes a cooling sensation.[1,4] Peppermint yields 1–3% essential oil rich in menthol (30–50%) and menthone (15–33%), with some menthyl acetate and menthofuran.[4] Corn mint contains 75–90% menthol and is an industrial source of this flavourant. The oil is cooled to near freezing temperature, causing the menthol to crystallize.[4] The residue (known as dementholised corn mint oil) still contains 55% menthol, with menthone and menthylacetate.[4] (–)-Menthol is the only one of four isomers that can bind to the proteins of cold-sensitive receptors in mammalian mucous membranes, causing the unusual cooling effect.

NOTES Peppermint oil is mostly used to flavour toothpaste, oral hygiene products, chewing gum, sweets, lozenges and cough syrups.[3,4]

(–)-menthol (–)-menthone (–)-menthyl acetate (+)-mentho-furan

1. **Lawrence, B.M. 2007.** *Mint: The genus* Mentha. CRC Press, Boca Raton.
2. **Mabberley, D.J. 2008.** *Mabberley's plant-book* (3rd ed.). Cambridge University Press, Cambridge.
3. **Farrel, K.T. 1999.** *Spices, condiments and seasonings.* Aspen Publishers, Gaithersburg, USA.
4. **Harborne, J.B., Baxter, H. 2001.** *Chemical dictionary of economic plants.* Wiley, New York.

Peppermint (*Mentha* ×*piperita*) (top left), with eau-de-cologne mint (*M.* ×*piperita* var. *citrata*) (bottom left), pineapple mint (*M. suaveolens* 'Variegata') (top right) and ginger mint (*M.* ×*gracilis*) (bottom right)

Peppermint (*Mentha* ×*piperita*)

Peppermint flowers (*Mentha* ×*piperita*)

Field mint or Japanese mint (*Mentha arvensis*)

Eau de cologne mint (*Mentha* ×*piperita* var. *citrata*)

175

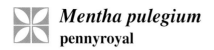

Mentha pulegium
pennyroyal

Mentha pulegium L. (Lamiaceae); *poleiment* (Afrikaans); *chun e bo he* (Chinese); *polei* (Dutch); *menthe pouliot, herbe aux puces* (French); *Flohkraut, Poleiminze* (German); *menta puleggia* (Italian); *penirooyaru minto* (Japanese); *menta poleo* (Spanish); *poleijmynta* (Swedish); *cây bac hà hăng* (Vietnamese)

DESCRIPTION The leaves are very small, with minutely toothed margins and a characteristic sweet smell resembling that of peppermint.

THE PLANT A robust perennial herb with ground-hugging stems and clusters of tiny pale violet flowers. The leaves of Corsican mint (*M. requienii*) are much smaller, scarcely exceeding 6 mm (¼ in.) in length, with smooth margins.

ORIGIN Europe and western Asia. It has become invasive in parts of Australia and the United States.[1] Corsican mint is western Mediterranean in origin and is naturalized in Britain.[1]

CULTIVATION Both species are easily propagated from cuttings or by division. They are hardy but will do best in moist, partly shady places. These plants are ideal along paths and stepping stones in the herb garden, because they rapidly form attractive groundcovers that emit a sweet, minty smell when bruised underfoot. They tend to become invasive in the garden and can be contained by using sunken containers, slate tiles or stones.

HARVESTING The leaves are simply picked when required. Pennyroyal oil (poisonous!) is distilled from flowering tops.

CULINARY USES Pennyroyal was an important culinary herb (and flavourant of wine and sailors' drinking water)[1] in ancient Rome and Greece, and during the Middle Ages. Today it is seldom used but can be sparingly combined with other herbs and spices to enrich the flavour of soups, stews, stuffings, vegetables and puddings. Corsican mint can also be used in small amounts in mint sauces (to complement lamb), as well as soups, salads, vegetable dishes, desserts, canned fruits and cold drinks. Both these mints are added to herbal teas and tisanes in small amounts.

FLAVOUR COMPOUNDS The main compound in pennyroyal is usually (+)-pulegone (about 80%)[2] but it appears to be quite variable, depending on provenance and method of extraction and analysis. Portuguese oil, for example, had menthone, pulegone and neo-menthol as the main constituents[3] while Turkish oil had isomenthone, pulegone, piperitenone and piperitone.[4] Headspace analysis gave mostly isomenthone, with some pulegone and small amounts of menthol.[4] The essential oil of *M. requienii* appears to be quite similar, yielding pulegone (78%) and isomenthone (18%), with small amounts of limonene, isopulegone and menthone.[5]

NOTES Pennyroyal has been used in traditional medicine as folk remedy and abortifacient,[1] as well as a repellent for insects and snakes. The oil is used in soaps and hygiene products, as well as in aromatherapy but great care should be taken, as it is very poisonous. Pennyroyal should not be confused with American pennyroyal or American false pennyroyal (*Hedeoma pulegioides*, also of the family Lamiaceae), which produces an oil of similar composition (American pennyroyal oil).[2]

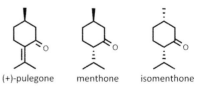

(+)-pulegone   menthone   isomenthone

1. **Mabberley, D.J. 2008.** *Mabberley's plant-book* (3rd ed.). Cambridge University Press, Cambridge.
2. **Harborne, J.B., Baxter, H. 2001.** *Chemical dictionary of economic plants*. Wiley, New York.
3. **Teixeira B., Marques, A., Ramos, C. et al. 2012.** European pennyroyal (*Mentha pulegium*) from Portugal: Chemical composition of essential oil and antioxidant and antimicrobial properties of extracts and essential oil. *Industrial Crops and Products* 36: 81–87.
4. **Yasa, H., Onar, H.C., Yusufoglu, A.S. 2012.** Chemical composition of the essential oil of *Mentha pulegium* L. from Bodrum, Turkey. *Journal of Essential Oil Bearing Plants* 15: 1040–1043.
5. **Chessa, M., Sias, A., Piana, A., Mangano, G.S., Petretto, G.L., Masia, M.D., Tirillini, B., Pintore, G. 2012.** Chemical composition and antibacterial activity of the essential oil from *Mentha requienii* Bentham. *Natural Products Research* 27: 93–99.

Leaves and flowers of pennyroyal (*Mentha pulegium*)

Leaves and flowers (*Mentha pulegium*)

Flowers (*Mentha pulegium*)

Corsican mint (*Mentha requienii*)

177

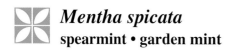

Mentha spicata
spearmint • garden mint

DESCRIPTION Leaves are broad, hairless and bright green, with toothed margins and prominent veins giving a wrinkled appearance. The leaves emit a strong spearmint smell even when lightly touched.

THE PLANT A hardy perennial herb with creeping rhizomes and erect flowering stems of up to 0.6 m (2 ft) high. White flowers are borne in terminal spikes. A form known as Bowles's mint is a backcross with a *M. ×villosa* hybrid.[1] Ginger mint, also known as Scotch spearmint (*M. ×gracilis*, previously known as *M. cardiaca*) has large, often variegated leaves and is chemically similar to spearmint, of which it is a hybrid (with *M. arvensis*).[1] Spearmint is similar to horse mint or longleaf mint (*M. longifolia*) but the leaves are much broader, and also to apple mint or woolly mint (*M. suaveolens*) but it lacks hairs. A decorative form with variegated leaves is known as pineapple mint (*M. suaveolens* 'Variegata'). Round-leaf mint (*M. ×rotundifolia*) is probably merely a form of *M. suaveolens*. Water mint (*M. aquatica*) has rounded (not spike-like) inflorescences.

ORIGIN Spearmint is a European garden hybrid between *M. suaveolens* and *M. longifolia*.[1] It became popular because of its distinctive aroma and is today an important commercial crop in the essential oil industry.

CULTIVATION Plants are easily grown from cuttings and flourish in rich, moist soil. It is one of the most commonly grown culinary herbs. Commercial cultivation is centred in the United States (Washington State).[2]

HARVESTING Vigorous young leafy shoots are hand-picked and rapidly delivered to fresh produce markets. For spearmint oil production, mechanical methods are used.[2]

CULINARY USES Mint is best known as the essential component of mint sauce or mint jelly, the traditional accompaniment for roast leg of lamb in England. It is included in beverages, sherbets, soups, stews, sauces, vinegars, veal, fish, beef, yogurt, green peas, spinach, carrots, eggplant and boiled potatoes.[2] Mint tea is a popular drink made from fresh or dried mint leaves, sweetened with honey and served hot or cold with a slice of lemon. Mint julep is a famous southern USA (Kentucky) cocktail made with spearmint leaves, bourbon, sugar and crushed ice.

FLAVOUR COMPOUNDS Spearmint oil is dominated by (–)-carvone (more than 50%),[3] which gives it the typical (non-cooling) spearmint flavour. Also present are limonene, phellandrene and dihydrocarvone.[3] It is interesting to note than (+)-carvone, the major flavourant of caraway (*Carum carvi*), has a quite different smell. The level of carvone appears to be higher from spring to autumn, with limonene exceeding carvone in winter.[4] Commercial Scotch spearmint oil has carvone and limonene as main compounds.[5]

NOTES Spearmint oil is used mainly in chewing gum, sweets, confectionery, ice cream, cordials, toothpaste, oral hygiene products and tobacco.

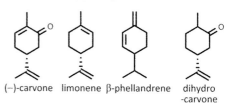

(–)-carvone   limonene   β-phellandrene   dihydro-carvone

1. **Mabberley, D.J. 2008.** *Mabberley's plant-book* (3rd ed.). Cambridge University Press, Cambridge.
2. **Farrel, K.T. 1999.** *Spices, condiments and seasonings.* Aspen Publishers, Gaithersburg, USA.
3. **Harborne, J.B., Baxter, H. 2001.** *Chemical dictionary of economic plants.* Wiley, New York.
4. **Benyoussef, E.-H., Yahiaoui, N., Khelfaoui, A., Aid, F. 2005.** Water distillation kinetic study of spearmint essential oil and of its major components. *Flavour and Fragrance Journal* 20: 30–33.
5. **Kothari, S.K., Singh, U.B. 1995.** The effect of row spacing and nitrogen fertilization on Scotch spearmint (*Mentha gracilis* Sole). *Journal of Essential Oil Research* 7: 287–297.

Mint or spearmint, the leaves of *Mentha spicata*

Spearmint (*Mentha spicata*)

Water mint (*Mentha aquatica*)

Ginger mint (*Mentha ×gracilis*)

Round-leaf mint (*Mentha ×rotundifolia*)

Horse mint (*Mentha longifolia*)

Apple mint (*Mentha suaveolens*)

Pineapple mint (*Mentha suaveolens* 'Variegata')

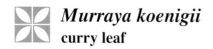

Murraya koenigii
curry leaf

Murraya koenigii (L.) Spreng. (Rutaceae); *kerrieblare* (Afrikaans); *ma jiao ye, ka li cai, duo ye jiu li xiang* (Chinese); *feuilles de curry* (French); *Curryblätter* (German); *karipatta, mitha neem* (Hindi); *duan kari* (Indonesian, Malay); *hojas de curry* (Spanish); *karivapilai, karuveppilei* (Tamil); *bai karee, hom khaek* (Thai); *lá cà ri* (Vietnamese)

DESCRIPTION Pinnately compound leaves, with 8–12 pairs of lance-shaped, minutely gland-dotted and highly aromatic leaflets.

THE PLANT A small evergreen tree of 3–6 m (ca. 10–20 ft) in height. Small white flowers are borne in terminal clusters, followed by edible, single-seeded, fleshy fruits that turn red and then black when they ripen.[1] Curry leaf should not be confused with the unrelated European curry plant (*Helichrysum italicum*), that is rarely also used as a culinary herb.[2]

ORIGIN Sri Lanka and India, where it is an important part of the culinary (and medicinal) traditions. It has been introduced to many other parts of the world by Indian immigrants over the last few centuries. It is commonly found in many kitchen gardens, supplying fresh leaves for daily culinary use, especially to flavour curry.

CULTIVATION Trees are most easily propagated from root suckers that typically develop around the trees but can also be grown from seeds or root cuttings. As an understorey shrub[1] it is best grown in partial shade, with a few hours of morning sun and shade in the afternoon. Mature plants can withstand light frost but in cold areas it is better to grow the plant in a large container that can be taken indoors in winter.

HARVESTING Fresh leaves are mostly sold on local markets or the dried individual leaflets (with leaf rachis removed) are packed as a spice. The dried or powdered leaflets are considered inferior because they lose much of their flavour upon drying.[2]

CULINARY USES Curry leaf is an indispensable ingredient of Indian dishes. It has also become popular in Thailand, Malaysia[3] and Indonesia (for fish curries and other dishes), as well as in South Africa and Réunion.[2] Curry leaf is often roasted (toasted) in a little oil to release the flavours.[2] Garam masala and *sambar podi* are two examples of "curry powders"[2] that typically contain cumin, pepper and/or chilli pepper as main ingredients, but usually also fenugreek, turmeric, brown mustard, cardamom, cloves, coriander, ginger and tamarind. Curry leaves are essential for making *sambar*, a traditional pigeon pea and lentil stew. *Sambar* powder is made from pan-roasted spices (coriander, chilli, fenugreek, pepper, brown mustard and others), combined with cilantro and curry leaves.

FLAVOUR COMPOUNDS The flavour is due to essential oil that varies from region to region. β-Caryophyllene, β-phellandrene, α-pinene and β-pinene are usually amongst the major compounds,[4] as are linalool (floral and spicy odour, released after enzymatic hydrolysis)[5] and octyl acetate (fruity odour, apparently lost during steam distillation).[5]

NOTES "Curry" as we know it today is a British concept based on the Tamil word *kari* (meaning "sauce") and is now loosely applied to a wide range of spicy dishes.[2]

β-caryophyllene   β-phellandrene   linalool

α-pinene   β-pinene   octyl acetate

1. **Parmar, C., Kaushal, M.K. 1982.** *Murraya koenigii.* pp. 45–48. In: *Wild Fruits.* Kalyani Publishers, New Delhi.
2. **Gernot Katzer's Spice Pages** (http://gernot-katzers-spice-pages.com).
3. **Hutton, W. 1997.** *Tropical herbs and spices.* Periplus Editions, Singapore.
4. **Raina, V.K., Lal, R.K., Tripathi, S., Khan, M., Syamasundar, K.V., Srivastava, S.K. 2002.** Essential oil composition of genetically diverse stocks of *Murraya koenigii* from India. *Flavour and Fragrance Journal* 17: 144–146.
5. **Padmakumari, K.P. 2008.** Free and glycosidically bound volatiles in curry leaves (*Murraya koenigii*) (L.) Spreng. *Journal of Essential Oil Research* 20: 479–481.

Curry leaf, the fresh or dried leaves of *Murraya koenigii*

Flowers (*Murraya koenigii*)

Fruits (*Murraya koenigii*)

Branch with fruits (*Murraya koenigii*)

181

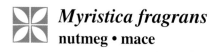

Myristica fragrans
nutmeg • mace

Myristica fragrans L. (Myristicaceae); *neut, muskaatneut, foelie* (Afrikaans); *rou dou kou, yu guo hua* (Chinese); *noix de muscade, fleur de muscade* (French); *Muskatnuß, Muskatblüte* (German); *buah pala, sekar pala* (Indonesian); *meesu, mirisutika* (Japanese); *noce moscata, mace* (Italian); *buah pala, sekar pala* (Malay); *nuez moscada, macis* (Spanish); *chan thet, dok chan thet* (Thai)

DESCRIPTION The ripe, dried and shelled seeds (nutmeg) are about 25 mm (ca. 1 in.) long and 16 mm (1 in.) in diameter, greyish in colour, with a wrinkled surface. They have a spicy flavour and aroma. The hard seeds are easily grated or broken and reveal dark brown oil canals where most of the essential oil is concentrated. The aril around the seed (mace) is net-like and bright red when fresh but pale orange-brown when dried.

THE PLANT An evergreen tree with simple, leathery leaves and small white flowers. Female trees bear smooth, peach-like fruits that split open at maturity. Adulterants and alternatives include Bombay nutmeg (*M. malabarica* and *M. beddomii*), Brazilian nutmeg (*Cryptocarya moschata*) and Madagascar nutmeg (*Ravensara aromatica*).[1]

ORIGIN Nutmeg is indigenous to Indonesia.[1] The Banda Islands, a group of ten small volcanic islands in the Indonesian province of Maluku, were once the world centre of a monopoly, enforced by successive periods of Portuguese, Dutch, French and British colonial invasions.[1] Nutmeg has an interesting history as hallucinogenic drug in Asia and as a spice in Europe, where it became popular around 1100 AD.[1] Commercial plantations were established, mainly in Indonesia and Grenada but also in several other tropical regions.[1] The annual production has been estimated at 10 000 tons.[1]

CULTIVATION There is no reliable method of distinguishing between male and female seedlings, so that propagation nowadays relies mostly on grafting. The plant requires wet, tropical conditions and is rarely grown in gardens.

HARVESTING Fruits are hand-picked when ripe and slowly dried, after which the outer shell (fruit wall) is removed, leaving the seed and aril to be sorted and graded.[2]

CULINARY USES Nutmeg has an exceptionally wide range of culinary uses as a spice.[2,3] It is used in ground or grated form (using a nutmeg grater) in sweet and spicy dishes, confectionery, desserts, custards, meat dishes, processed meats, sausages, vegetable dishes and especially potatoes, omelettes and soufflés. It is used in Indonesian soups and sweets, Japanese curry powders, Scottish haggis, Dutch potato and vegetable dishes and French béchamel sauce, mince meat, onion soup and snails.[3] Grated nutmeg is also used in fortified wines, mulled wines and cider, eggnog, liqueurs and other alcoholic drinks (e.g. sprinkled on Barbados rum punch). Mace has a more delicate flavour. It is included in spice mixtures and may be used in the same way as nutmeg in sauces and in chicken, fish and pork dishes.[2,3]

FLAVOUR COMPOUNDS Nutmeg essential oil (5–15% yield) has sabinene, α-pinene, β-pinene and terpinen-4-ol as the main components.[4] The distinctive flavour and aroma, however, are ascribed to myristicin (5–8%) and elemicin (2%).[4] Mace oil is similar to nutmeg oil in composition but has a higher proportion of myristicin.[4]

NOTES Myristicin is a psychoactive substance (monoamine oxidase inhibitor) and toxic at high doses; nutmeg and mace should be used sparingly.

myristicin elemicin

α-pinene (+) sabinene terpinen-4-ol β-pinene

1. **Flach, M., Willink, M.T. 1989.** *Myristica fragrans* Houtt., pp. 192–196. In: Westphal, E., Jansen, P.C.M. (Eds), *Plant resources of South East Asia: a selection.* Pudoc, Wageningen.
2. **Farrel, K.T. 1999.** *Spices, condiments and seasonings.* Aspen Publishers, Gaithersburg, USA.
3. **Larousse. 1999.** *The concise Larousse gastronomique.* Hamlyn, London.
4. **Harborne, J.B., Baxter, H. 2001.** *Chemical dictionary of economic plants.* Wiley, New York.

Nutmeg and mace (*Myristica fragrans*)

Nutmeg fruits (*Myristica fragrans*)

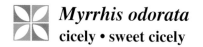

Myrrhis odorata
cicely • sweet cicely

Myrrhis odorata (L.) Scop. (Apiaceae); *mirrekerwel* (Afrikaans); *cerfeuil odorant, cerfeuil musqué* (French); *Süßdolde, Myrrhenkerbel* (German); *finocchio dei boschi, cerfoglio di spagna* (Italian); *suito-shiseri* (Japanese); *sisilli* (Korean); *Spansk körvel* (Swedish)

DESCRIPTION The leaves are pinnately compound, dentate and pale green. The small dry fruits are oblong, slender, up to 25 mm (1 in.) long, narrowly beaked and dark glossy brown when ripe. Both the leaves and fruits have a strong fragrance and aroma reminiscent of anise and liquorice.

THE PLANT A robust, aromatic, herbaceous perennial of up to 1.5 m (5 ft) in height. It has erect, grooved, hairy stems and bears small white flowers in compound umbels. As suggested by common names such as Spanish chervil, anise chervil and sweet chervil, the plant is superficially very similar to chervil (*Anthriscus cerefolium*) but differs in the large, fern-like leaves, the dark brown fruits and the sweet aroma. Sweet cicely is also known as garden myrrh. The term "myrrh" refers to aromatic resins, the most famous of which is collected as an exudate from the bark of East African myrrh trees (*Commiphora myrrha*) that grows in Kenya, Ethiopia and Somalia. Another type of myrrh is obtained by boiling the branches of the eastern Mediterranean myrrh shrub (*Cistus creticus*) in water and skimming off the aromatic resin. This plant is believed to be the biblical "myrrh".[1]

ORIGIN The plant is indigenous to central and southern Europe.[1] It has a long history of use as potherb and for strewing on medieval church floors.[1] This attractive and fragrant herb is now naturalized in most parts of Europe, where it is occasionally wild-harvested but also commonly grown in herb gardens.

CULTIVATION Plants are propagated from seeds or by dividing the rootstock of mature specimens. They are cold-tolerant and grow well in Scandinavia and other cold regions.

HARVESTING Leaves and stems are used fresh, while young fruits are picked when still green, before they become fibrous. Both leaves and fruits can be successfully dried for later use.

CULINARY USES Leaves, stems and fruits are used as a culinary herb to give a delicious, sweet, aniseed or liquorice taste to vegetable dishes, salads, soups, omelettes, stewed fruits, drinks and liqueurs. The fruits can be used to give a sweet spicy taste to meat dishes. Due to the natural sweetness of the herb it is used in rhubarb and stewed fruit dishes to reduce the acidity.[2] The roots were once eaten as a vegetable.[2]

FLAVOUR COMPOUNDS The leaves and fruits contain an essential oil rich in (*E*)-anethole as dominant compound, accompanied by small amounts of methyl eugenol and (*E*)-nerolidol.[3,4]

NOTES The sweet taste of culinary herbs and spices such as basil, fennel and star anise is not due to sugars but to the presence of high concentrations of one or both of two common phenylpropanoids, (*E*)-anethole (also refered to as *trans*-anethole) and estragole.[5]

(*E*)-anethole

methyl eugenol

(*E*)-nerolidol

1. **Mabberley, D.J. 2008.** *Mabberley's plant-book* (3rd ed.). Cambridge University Press, Cambridge.
2. **Kiple, K.F., Ornelas, K.C. (Eds), 2000.** *The Cambridge world history of food*. Cambridge University Press, Cambridge.
3. **Uusitalo, J.S., Jalonen, J.E., Aflatuni, A., Luoma, S.L. 1999.** Essential leaf oil composition of *Myrrhis odorata* (L.) Scop. grown in Finland. *Journal of Essential Oil Research* 11: 423–425.
4. **Dobravalskytè, D., Venskutonis, P.R., Zebib, B., Merah, O., Talou, T. 2013.** Essential oil composition of *Myrrhis odorata* (L.) Scop. leaves grown in Lithuania and France. *Journal of Essential Oil Research* 25: 44–48.
5. **Hussain, R.A., Poveda, L.J., Pezzuto, J.M., Soejarto, D.D., Kinghorn, A.D. 1990.** Sweetening agents of plant origin: Phenylpropanoid constituents of seven sweet-tasting plants. *Economic Botany* 44: 174–182.

Cicely, sweet cicely (*Myrrhis odorata*)

Flowers and young fruits (*Myrrhis odorata*)

Ripe fruits (*Myrrhis odorata*)

Plant (*Myrrhis odorata*)

Flowers (*Myrrhis odorata*)

Myrtus communis
myrtle

Myrtus communis L. (Myrtaceae); *mirt* (Afrikaans); *xiang tao mu* (Chinese); *myrte* (French); *Echte Myrte* (German); *mirto* (Italian); *maatoru* (Japanese); *mirto* (Portuguese); *mirto* (Spanish); *myrten* (Swedish)

DESCRIPTION The leaves are dark glossy green, with a sweet aroma and a bitter taste. The edible berries are dark purple to black when ripe. They resemble small guavas (note the persistent calyx) and have a sweet but somewhat resinous taste.

THE PLANT The true myrtle is an attractive evergreen shrub of 2–3 m (ca. 6–10 ft) high, with small white flowers bearing numerous stamens. The original (wild) form of the species has relatively large, dark green leaves (as shown here) but ornamental cultivars may have very small or variegated leaves.

ORIGIN The plant has been cultivated since ancient times, so that its exact origin is not known.[1] Plants are now found from the Mediterranean area to southern Asia. Myrtle is associated with Roman and Greek mythology, rituals and ceremonies.[1]

CULTIVATION Plants are easily propagated from cuttings. They adapt to almost any climate and are remarkably resistant to frost. Commercial cultivation is mainly aimed at the distillation of essential oil from the leaves (*eau d'Ange*), which is used in perfumery.[1] Myrtle is often grown as a focal point in the herb garden and makes an attractive clipped hedge.

HARVESTING Fresh leaves and ripe berries are hand-picked when required. The berries can be dried and stored for future use as spice.

CULINARY USES The Romans used myrtle leaves to flavour stews and wines and the plant is still locally important in the cuisines of Corsica and Sardinia.[2] On these islands they are added to bouillabaisse and dishes of pork and wild boar.[2] Myrtle fruits or leaf extracts are used to make a liqueur known as *mirto*.[2] Leaves and berries can be used in stuffings, or a roast or game bird wrapped in leaves for five to ten minutes before serving. Dried berries can be ground in a pepper mill to add a spicy flavour to stews and meat dishes. Leafy twigs or dried leaves can be added to the fire while roasting meat. The essential oil is used to flavour meat sauces and seasonings.[3]

FLAVOUR COMPOUNDS The aroma and flavour is due to the presence of essential oil (myrtle oil)[3] rich in limonene, 1,8-cineole, α-pinene and linalool.[4] Myrtenyl acetate and myrtenol occur only in some chemotypes.[5]

NOTES Since ancient times, myrtle has been regarded as sacred to Venus and as a symbol of beauty, love and purity. Kate Middleton, Diana Spencer and Queen Elizabeth all followed royal tradition by having myrtle in their bridal bouquets. The tradition was started by Princess Victoria in 1858 when she married the future Frederick III of Germany. A sprig was taken from the famous myrtle planted by her mother, Queen Victoria, at Osborne House on the Isle of Wight.

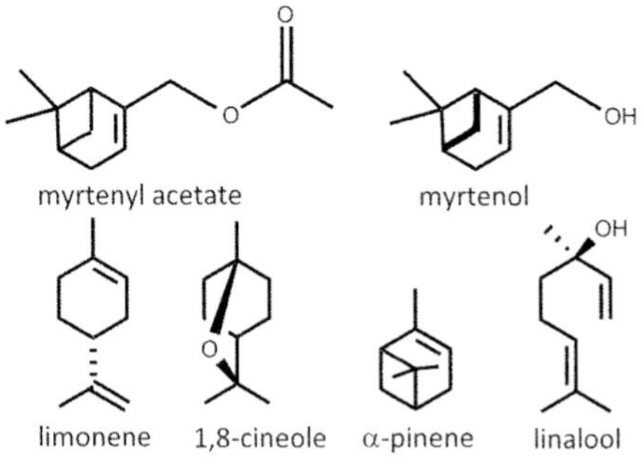

myrtenyl acetate   myrtenol

limonene   1,8-cineole   α-pinene   linalool

1. **Mabberley, D.J. 2008.** *Mabberley's plant-book* (3rd ed.). Cambridge University Press, Cambridge.

2. **Larousse. 1999.** *The concise Larousse gastronomique*. Hamlyn, London.

3. **Harborne, J.B., Baxter, H. 2001.** *Chemical dictionary of economic plants*. Wiley, New York.

4. **Tuberoso, C.I., Barra, A., Angioni, A., Sarritzu, E., Pirisi, F.M. 2006.** Chemical composition of volatiles in Sardinian myrtle (*Myrtus communis* L.) alcoholic extracts and essential oils. *Journal of Agricultural and Food Chemistry* 54: 1420–1426.

5. **Pereira, P.C., Cebola, M.-J., Bernardo-Gil, M.G. 2009.** Evolution of the yields and composition of essential oil from Portuguese myrtle (*Myrtus communis* L.) through the vegetative cycle. *Molecules* 14: 3094–3105.

Flower and fruit of myrtle (*Myrtus communis*)

Flower and young fruits (*Myrtus communis*)

Nasturtium officinale
watercress

Nasturtium officinale R. Br. [= *Rorippa nasturtium-aquaticum* (L.) Hayek] (Brassicaceae); *bronkors* (Afrikaans); *xi yang cai, dou ban cai* (Chinese); *cresson de fontaine* (French); *Brunnenkresse* (German); *crescione acquatico* (Italian); *mizu garashi* (Japanese); *selada air* (Malay); *agrião* (Portuguese); *berro* (Spanish)

DESCRIPTION Fresh leafy stems are succulent, dark green in colour and have a pungent, peppery taste and distinctive grassy aroma.

THE PLANT An aquatic perennial herb with spreading stems bearing irregularly compound leaves and white flowers. Seeds are borne in oblong capsules and are variously arranged in one or two rows, depending on the species. The generic name for the plant and its close relatives has recently reverted back from *Rorippa* to *Nasturtium*.[1] Brown watercress, also called winter watercress, is believed to be a hybrid (backcross) between cultivated *N. officinale* (seeds in two rows) and *N. microphyllum* (seeds in one row).[2] The leaves typically turn purplish brown in winter. The name "cress" is also used for several related culinary herbs, including garden cress (*Lepidium sativum*) and land cress (*Barbarea verna*).[1]

ORIGIN Watercress is indigenous to Europe and has become a cosmopolitan weed of freshwater habitats.[1] It has probably been foraged since ancient times but the recorded history as a health food goes back to classical Persia, Greece and Rome. Commercial cultivation only started in the 19th century. Watercress is grown in the United Kingdom in Hertfordshire, Hampshire, Wiltshire and Dorset. Alresford near Winchester, the centre of an annual Wintercress Festival, is considered to be the watercress capital. In the United States, the title of "watercress capital of the world" first went to Huntsville, Alabama but is nowadays associated with Oviedo, Florida.

CULTIVATION On a commercial scale, plants are grown in fresh water (ideally slightly alkaline) in large inundated beds.

HARVESTING The leafy stems are cut by hand, washed (but never left to soak in water) and packed. The picked stems have a limited shelf life of about two days when kept under refrigeration.

CULINARY USES Watercress has become very fashionable in recent years. The leaves are pulled of the stems when eaten raw in salads and sandwiches or used as a garnish. It makes a very tasty soup, often combined with leeks and potatoes. Cooked watercress is prepared in the same way as spinach.[3] Watercress purée or watercress dip is made from pounded leaves added to mashed potato or split peas, with variable quantities of the stems included for extra flavour.[3] *À la cressonnière* is the term used in France for dishes that contain watercress.[3] In Southeast Asia, it is cooked in soup and not eaten raw.

FLAVOUR COMPOUNDS The pungent, peppery taste is due to the presence of gluconasturtiin, a mustard oil glycoside. By crushing or chewing the leaves, the glycoside is mixed with the enzyme myrosinase, resulting in the formation of two hydrolysis products. These are hydrocinnamonitrile and phenethyl isothiocyanate, the two main flavour compounds.

NOTES Wild-harvested watercress is best avoided as it can transmit parasites.

gluconasturtiin → hydrocinnamonitrile / phenethyl isothiocyanate

1. **Mabberley, D.J. 2008.** *Mabberley's plant-book* (3rd ed.). Cambridge University Press, Cambridge.
2. **Kiple, K.F., Ornelas, K.C. (Eds). 2000.** *The Cambridge world history of food.* Cambridge University Press, Cambridge.
3. **Larousse. 1999.** *The concise Larousse gastronomique.* Hamlyn, London.

Watercress, the leaves of *Nasturtium officinale*

Flowers (*Nasturtium officinale*)

Leaves (*Nasturtium officinale*)

Plants (*Nasturtium officinale*)

Plants (*Nasturtium officinale*)

Nelumbo nucifera
lotus

Nelumbo nucifera Gaertn. (Nymphaeaceae); *lotus* (Afrikaans); *he hua, lian* (Chinese); *nelumbo, lotus sacré* (French); *Lotosblume* (German); *kanwal* (Hindi); *nelumbo* (Italian); *hasu, renkon* (Japanese); *bunga telpok, teratai* (Malay); *kamala* (Sanskrit); *loto sagrado* (Spanish); *bua luang* (Thai)

DESCRIPTION Lotus leaves in the form that they are sold are approximately 45 cm (ca. 18 in.) long and 25 cm (10 in.) wide. Fresh leaves have a sweet and slightly bitter taste, while dry leaves have a fresh, earthy taste described as tea-like. The small nuts ("seeds") have an almond-like flavour, while stamens (removed from the flowers) have a floral fragrance.

THE PLANT An aquatic plant with fleshy rhizomes and large, attractive, red, pink or white flowers. The one-seeded fruits (nuts) are borne in a distinctive flat-topped compound fruit.

ORIGIN Southern and eastern Asia to tropical Australia. Lotus is considered sacred in India, Tibet and China and features prominently in oriental religion and mythology. It has been cultivated since ancient times and is an important and versatile commercial food plant and flavourant.

CULTIVATION Plants are propagated from rhizome sections and grow easily in shallow water in tropical regions.

HARVESTING The leaves, flowers, stamens, nuts ("seeds") and rhizomes are hand-harvested as required. All parts are used fresh but can also be dried for later use.

CULINARY USES Lotus leaf (*he ye*) is used as a culinary herb or spice. Dried leaves are soaked in water for about an hour before use. In China, food is commonly wrapped in lotus leaves as a means of flavouring the dish (similar to the use of pandan leaf in Thailand). Well-known examples include *he ye zheng ji* ("chicken wrapped in lotus leaf") or Chinese tamale, a special pastry wrapped in lotus leaves.[2] Another method of flavouring dishes is to use the leaves as a lining for bamboo steamers when steaming meat or fish. The dried leaves can be enjoyed as tea or used to flavour tea. The flowers (petals) are an ingredient of traditional dishes, such as *he hua ji* ("chicken with lotus flowers").[2] Dried stamens are used to make a fragrant tea (*lian hua cha* in Chinese, *chè sen* in Vietnamese). The nuts (*lian zi*) are either eaten raw as nuts (after removal of the bitter embryos) or they may be dried and popped, like popcorn. They are also used to add an almond-like taste to soups and stews. Lotus seed paste, made from soft-boiled "seeds", is a starchy base for rice flour pudding and confectionery items such as mooncakes.[2] The rhizomes are boiled or roasted and eaten as a vegetable.[2] They may also be salted and dried or fried like potato crisps.

FLAVOUR COMPOUNDS The volatile flavour compounds of lotus leaves, stems and petals are phthalides, of which the main compounds are tetradecane isobutylidene phthalide (stems) and *n*-butylidene dihydrophthalide (petals).[3] The leaves contain tetradecane. The stamens have aldehydes and esters as flavour compounds, mainly undecanal and neryl acetate.[3]

NOTES The lotus effect refers to the very high water repellence of lotus leaves, caused by the peculiar micro- and nanoscopic architecture of the leaf surface. The principle has been commercialised in self-cleaning paints and coatings.

tetradecane   undecanal

isobutylidene phthalide   *n*-butylidene dihydrophthalide   neryl acetate

1. **Mabberley, D.J. 2008.** *Mabberley's plant-book* (3rd ed.). Cambridge University Press, Cambridge.
2. **Hu, S.-Y. 2005.** *Food plants of China.* The Chinese University Press, Hong Kong.
3. **Choi, H.-S. 2011.** Headspace flavor composition of pink-flowered lotus (*Nelumbo nucifera* Gaertner). *Analytical Chemistry Letters* 1: 194–201.

Lotus leaf, the dried leaves of *Nelumbo nucifera*

Plant (*Nelumbo nucifera*)

Nuts or "seeds" (*Nelumbo nucifera*)

Rhizomes (Chinese arrowroot, *Nelumbo nucifera*)

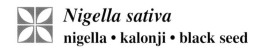

Nigella sativa
nigella • kalonji • black seed

Nigella sativa L. (Ranunculaceae); *nigella* (Afrikaans); *kamun aswad, shuniz* (Arabic); *pei hei zhong cao* (Chinese); *nigelle, poivrette, cumin noire* (French); *Schwarzkümmel* (German); *kalonji, kala jeera* (Hindi); *cuminella* (Italian); *cyah-daneh* (Persian); *nigela* (Portuguese); *niguiella, ajenuz común* (Spanish); *çörek otu* (Turkish)

DESCRIPTION The ripe seeds are typically black in colour (hence the name *Nigella*, a diminutive form of the word "black"), strongly angular in shape, with a rough surface, a thyme-like aroma and a slightly bitter taste. The seeds are sometimes referred to as "onion seeds" or "black cumin" (*kala jeera*) but these names are confusing. Black cumin more accurately refers to *Bunium persicum* (See NOTES under *Cuminum cyminum*). The names nigella and kalonji are less ambiguous.

THE PLANT The plant (fennel flower)[1] is an erect annual herb of about 0.3 m (ca. 1 ft) high with compound leaves and attractive white or pale blue flowers. The segmented fruit capsules and persistent styles are distinctive.

ORIGIN Mediterranean region and western Asia. It has been domesticated in ancient times in the Middle East and seeds were found in Tutankhamen's tomb.[1,2] The seeds are still very important as condiment, general tonic and traditional medicine in Arabia, Egypt, India and Pakistan and are sometimes referred to in Arabic as *Habbatul barakah* ("seed of blessing"). It was once an important spice in southern and central Europe ("fitches" in English[1] or *quatre-épices* or *toute-épices* in French).[3]

CULTIVATION Seeds are easy to germinate. Plants are tolerant of cold and some drought but require full sun to flourish.

HARVESTING The ripe seed capsules are harvested before they split open. They are then dried and the seeds extracted mechanically or by hand.

CULINARY USES The aromatic and pungent seeds are an important spice and are traditionally sprinkled over cakes and breads, such as Indian *naan* bread (especially Peshawari *naan*), Turkish *çörek* buns and Bosnian *somun* pastries. It is also used to flavour spicy dishes. The Bengali and East Indian spice mixture called *panch phoran* ("five spices") is a mixture of the seeds of nigella, black mustard, cumin, fennel and fenugreek in equal parts. The black mustard may be replaced with radhuni (*Bunium roxburghianum*) or celery seed. The mixture is never ground but always used as whole seeds (often tempered in a little oil or ghee) in curries, vegetable dishes and pickles.

FLAVOUR COMPOUNDS The aroma of the seeds resembles thyme and oregano due to the presence of thymol and thymoquinone in the essential oil.[4] The oil appears to be highly variable and may sometimes have *p*-cymene, carvacrol and α-thujene as major compounds.[4] Thymoquinone apparently gets converted to thymol. The presence of an aromatic alkaloid, nigelline (= damascenine) has also been reported.[5]

NOTES *Nigella damascena* (love-in-a-mist) is a very attractive blue-flowered species and popular garden plant. It is called *chernushka* in Russian and the seeds are used to flavour breads and cheeses in Russia, Poland, Turkey and the Balkan region.

thymol   thymoquinone   *p*-cymene   carvacrol

1. **Mabberley, D.J. 2008.** *Mabberley's plant-book* (3rd ed.). Cambridge University Press, Cambridge.
2. **Zohary, D., Hopf, M. 2000.** *Domestication of plants in the Old World* (3rd ed.). Clarendon Press, Oxford.
3. **Larousse. 1999.** *The concise Larousse gastronomique*. Hamlyn, London.
4. **Benkaci-Ali, F., Baaliouamer, A., Meklati, B.Y., Chemat, F. 2007.** Chemical composition of seed essential oils from Algerian *Nigella sativa* extracted by microwave and hydrodistillation. *Flavour and Fragrance Journal* 22: 148–153.
5. **Franke, W. 1997.** *Nutzpflanzenkunde*. George Thieme Verlag, Stuttgart.

Black seeds or *kalonji*, the seeds of *Nigella sativa*

Black seeds on the market

Flowering plant (*Nigella sativa*)

Flower (*Nigella sativa*)

Ocimum basilicum
basil • sweet basil

Ocimum basilicum L. (Lamiaceae); *basiliekruid* (Afrikaans); *besobla* (Amharic); *luo le, yu xiang cai* (Chinese); *basilic* (French); *Basilikum* (German); *babui tulsi* (Hindi); *basilico* (Italian); *bajiru* (Japanese); *kemangi, selaseh* (Malay); *alfavaca* (Portuguese); *albahaca* (Spanish); *horapha* (Thai)

DESCRIPTION Sweet basil leaves are highly aromatic, usually with a rich, spicy aroma but sometimes with lemon, rose, camphor, cinnamon, woody or fruity aromas.[1]

THE PLANT A robust, leafy annual (ca. 0.6 m or 2 ft) with soft-textured, hairless leaves and small white flowers. The numerous hybrids, polyploids and cultivars often differ only in their volatile compounds.[1-3] Holy basil or *tulsi* is *O. tenuiflorum* (*O. sanctum*),[4] a short-lived perennial from India and Malaysia with hairy stems and leaves. Hoary basil or lemon basil is *O. americanum* (*O. canum*),[4] a perennial African species with hairy stems and calyces. Thai basil or *horapha* refers to a form of *O. basilicum* with purple stems and brightly coloured bracts. Thai holy basil or *kraphao* is a cultivar of *O. tenuiflorum* and Thai lemon basil (*O. ×citriodorum*) is a hybrid between *O. basilicum* and *O. americanum*.

ORIGIN Tropical Asia.[4] Cultivated since ancient times in India, the Near East, North Africa and later in Europe.

CULTIVATION Sweet basil is easily grown from seeds but warm conditions are required. Plant out in full sun in rich, well-drained soil at a spacing of 0.3 m (1 ft).

HARVESTING Fresh leaves are harvested regularly to encourage new growth. Drying may result in a loss of flavour – preservation in olive oil is a better option.

CULINARY USES Chopped leaves are used in salads, soups, omelettes, sauces, stuffings, meat dishes and especially pastas. The flavour of sweet basil goes well with tomato salad or any tomato dish. Especially famous is the Italian pesto, a sauce originating in Genoa made from fresh basil pounded with garlic, pine nuts, Parmesan and olive oil. A similar condiment (*pistou*) is made in Provence (but without pine nuts). *Horapa* is the most popular basil in Thailand, where generous amounts of fresh leaves are added to curries, salads and stir-fries. It has an anise or liquorice aroma. *Húng quế*, a similar cultivar but with a cinnamon flavour, is used in Vietnamese cuisine. Thai lemon basil (*maenglak* or *kemangie*) is commonly used in seafood dishes, also in Malaysia and Indonesia. In Indian and Thai cooking, holy basil is mostly added to fish, poultry and meat dishes and is typically cooked to release the flavour.

FLAVOUR COMPOUNDS The traditional sweet basil aroma is due to linalool, methylchavicol and/or 1,8-cineole. The sweetness is ascribed to methylchavicol (= estragole), a phenylpropanoid considered safe only when used in small amounts. The aroma of some cultivars is due to eugenol (clove-scented) or citral (lemon-scented).

NOTES The attractive purple-leaved basil (*Ocimum basilicum* var. *purpurascens*) can be used in the same way as the green variety.

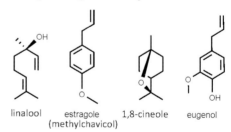

linalool   estragole   1,8-cineole   eugenol
            (methylchavicol)

1. **Simon, J.E., Morales, M.R., Phippen, W.B., Vieira, R.F., Hao, Z. 1999.** Basil: a source of aroma compounds and a popular culinary and ornamental herb. In: Janick, J. (Ed.), *Perspectives on new crops and new uses*. ASHS Press, Alexandria.
2. **Paton, A. 1992.** A synopsis of *Ocimum* L. (Labiatae) in Africa. *Kew Bulletin* 47: 403–435.
3. **Paton, A., Putievsky, E. 1996.** Taxonomic problems and cytotaxonomic relationships between varieties of *Ocimum basilicum* and related species (Labiatae). *Kew Bulletin* 51: 1–16.
4. **Grayer, R.J., Kite, G.C., Goldstone, F.J., Bryan, S.E., Paton, A., Putievsky, E. 1996.** Infraspecific taxonomy and essential oil chemotypes in sweet basil, *Ocimum basilicum*. *Phytochemistry* 43: 1033–1039.
5. **Telci, I., Bayram, E., Yilmaz, G., Avci, B. 2006.** Variability in essential oil composition of Turkish basils (*Ocimum basilicum* L.). *Biochemical Systematics and Ecology* 34: 489–497.

194

Sweet basil, the leaves of *Ocimum basilicum*

Fresh leaves of sweet basil

Dried basil

Holy basil (*Ocimum tenuiflorum*)

Hoary basil (*Ocimum americanum*)

Purple basil (*Ocimum basilicum* var. *purpurascens*)

195

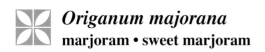

Origanum majorana
marjoram • sweet marjoram

Origanum majorana L. (= *Majorana hortensis* Moench) (Lamiaceae); *marjolein* (Afrikaans); *mah yeuk laahn* (Chinese); *marjolaine* (French); *Majoran* (German); *matzourana* (Greek); *maggiorana* (Italian); *majoramu* (Japanese); *manjerona* (Portuguese); *amàraco* (Spanish)

DESCRIPTION Fresh leaves are small, oval in shape and minutely hairy, with a pleasant sweet and spicy smell and a slightly bitter, aromatic and camphor-like flavour.[1]

THE PLANT A perennial herb (ca. 0.5 m or 1½ ft) with tiny white or pale purple flowers borne in terminal rounded clusters formed by overlapping bracts. Marjoram is closely related to oregano ("wild marjoram") and za'atar ("biblical hyssop"), but it has a milder flavour. Pot marjoram or Turkish oregano (*O. onites*) is sometimes used as a substitute or adulterant – the quality is considered to be inferior. It has cone-like flower spikes arranged in false corymbs and the calyces are one-lipped.[2]

ORIGIN Mediterranean region. Marjoram is one of the most important culinary herbs of Europe, a tradition that dates back to the time of the Egyptians, Greeks and Romans.

CULTIVATION Marjoram is easily propagated from seeds, cuttings or by division and grows well in full sun in almost any soil. It is commonly seen in herb and kitchen gardens.

HARVESTING Leafy stems, cut just before the plants flower, are usually field-dried and cut into fragments (rather than powdered, to retain the flavour). Both fresh and dried marjoram can usually be found in supermarkets.

CULINARY USES Marjoram is especially popular in Britain, Germany, France, Hungary, northern Italy, Spain and the United States. It is used in a wide variety of dishes, including salads, fish dishes, omelettes, fried potatoes, sauces, gravies and stuffings.[3] It goes well with green leafy vegetables and also with beans, cabbage, carrots and cucumbers. The strong, aromatic scent is called for in traditional European meat dishes, including game, meat, liver, poultry, stews and sausages. Together with bay leaves, parsley and thyme it forms the classic *bouquet garni*. The dried herb is much stronger than the fresh herb and is therefore used quite sparingly (less than half a teaspoon is usually sufficient for a meal for six persons).[1] Fresh herb is best added towards the end of the cooking period to avoid losing the flavour.[3] The dried herb or oil has many uses in the food industry to flavour liqueurs and vermouths, as well as charcuterie items (sausages, liverwurst and canned meats).[1]

FLAVOUR COMPOUNDS The characteristic flavour of marjoram oil is ascribed to two main compounds, terpinen-4-ol and (+)-*cis*-sabinene hydrate[4] and structurally related bicyclic sabinyl monoterpenes but it also contains *p*-cymene, γ-terpinene and several other minor constituents.[5]

NOTES The dittany of Crete (*O. dictamnus*) is easily identified by its densely woolly leaves and hops-like flower clusters. This popular garden plant is chemically similar to marjoram and oregano. It has a very long history as culinary and medicinal herb and is mentioned in Greek mythology.

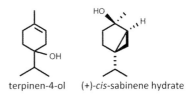

terpinen-4-ol   (+)-*cis*-sabinene hydrate

1. **Farrel, K.T. 1999.** *Spices, condiments and seasonings.* Aspen Publishers, Gaithersburg, USA.
2. **Vokou, D., Kokkini, S., Bessière, J.-M. 1988.** *Origanum onites* (Lamiaceae) in Greece: distribution, volatile oil yield, and composition. *Economic Botany* 42: 407–412.
3. **Larousse. 1999.** *The concise Larousse gastronomique.* Hamlyn, London.
4. **Harborne, J.B., Baxter, H. 2001.** *Chemical dictionary of economic plants.* Wiley, New York.
5. **Vera, R.R., Chane-Ming, J. 1999.** Chemical composition of the essential oil of marjoram (*Origanum majorana* L.) from Reunion Island. *Food Chemistry* 66: 143–145.

Sweet marjoram, the leaves of *Origanum majorana*

Cretan dittany (*Origanum dictamnus*)

Flowers (*Origanum majorana*)

Pot marjoram (*Origanum onites*)

Plant (*Origanum majorana*)

197

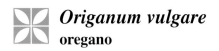

Origanum vulgare
oregano

Origanum vulgare L. (Lamiaceae); *oreganum* (Afrikaans); *satar barri* (Arabic); *wilde marjolein* (Dutch); *origan* (French); *Echter Dost* (German); *origano* (Italian); *orégano* (Spanish); *izmir kekigi* (Turkish)

DESCRIPTION The fresh leaves are oval in shape, sparsely hairy and dark green, with a sweet spicy smell resembling that of thyme, and a bitter aromatic taste.

THE PLANT A perennial herb (up to 0.6 m or 2 ft high, depending on the cultivar) with white to reddish purple flowers borne amongst conspicuous bracts. Oregano closely resembles za'atar, the Middle Eastern "hyssop" mentioned in the Bible (*O. syriacum*).[1] Za'atar (Arabic *za'tar*, also written as *zatar*, *zahtar* or *satar*) can be distinguished by the more prominently veined and more densely hairy leaves.

ORIGIN Oregano is indigenous to Europe and Central Asia. Za'atar (*O. syriacum*) has its natural distribution area in the Near East. Both have been used since ancient times in the Mediterranean region and the Middle East in Roman, Greek and Arabian cuisine.[1] Unlike za'atar, which has remained relatively unknown, oregano has become a major international crop,[2] and is used in Italian pizzas and pastas.

CULTIVATION Oregano is very easy to propagate from seeds, cuttings or by division. It grows best in full sun and slightly alkaline soil. Modern cultivars with colourful leaves are often more decorative than culinary. The white-flowered Greek form (*O. vulgare* subsp. *hirtum*) is considered to be the best for culinary use.

HARVESTING Leafy branches are harvested just before the plants flower and are sold fresh or more often in dried form. To ensure that the flavour is retained, the dried product is not powdered but broken into small fragments.

CULINARY USES Oregano gives the classical pizza aroma, the pungent, thyme-like smell familiar to anyone who regularly eats Italian-style pizzas. Oregano is an important flavour ingredient of southern Italian tomato dishes and pasta sauces and is used in a wide range of vegetable, tomato, egg and meat dishes.[2] Za'atar is very well known in the Middle East and North Africa, where it is the main source of thyme-flavoured herbs, used to prepare an Arabian spice mixture by the same name. The popular and well-known *za'atar* spice mixture is usually made from sesame seeds, dried sumac fruits and za'atar herb but there are many regional variants. Za'atar is used to flavour the Middle Eastern pita bread (almost certainly the original inspiration for the Italian pizza), as well as vegetable and meat dishes.

FLAVOUR COMPOUNDS The main aroma-impact compounds of the essential oil of *O. vulgare* are thymol and carvacrol but several minor compounds in variable proportions are also present.[3] *Origanum syriacum* oil is dominated by carvacrol, with smaller amounts of thymol usually also present.[4]

NOTES The essential oil of *Lippia graveolens* (Mexican oregano) closely resembles oregano oil because it contains thymol and carvacrol as main compounds (see *Lippia adoensis*).

thymol   carvacrol

1. **Fleisher, A., Fleisher, Z. 1988.** Identification of biblical hyssop and the origin of the traditional use of oregano-group herbs in the Mediterranean region. *Economic Botany* 42: 232–241.
2. **Farrel, K.T. 1999.** *Spices, condiments and seasonings.* Aspen Publishers, Gaithersburg, USA.
3. **Raina, A.P., Negi, K.S. 2012.** Essential oil composition of *Origanum majorana* and *Origanum vulgare* subsp. *hirtum* growing in India. *Chemistry of Natural Products* 47: 1015–1017.
4. **Lukas, B., Schmiderer, C., Franz, C., Novak, J. 2009.** Composition of essential oil compounds from different Syrian populations of *Origanum syriacum* L. (Lamiaceae). *Journal of Agricutural and Food Chemistry* 57: 1362–1365.

Oregano, the dried or fresh leaves of *Origanum vulgare*, with za'atar herb, the leaves of *O. syriacum*

Cultivars of oregano (*Origanum vulgare*)

Za'atar herb (*Origanum syriacum*)

Greek oregano (*Origanum vulgare* subsp. *hirtum*)

Za'atar flowers (*Origanum syriacum*)

Pandanus amaryllifolius
pandan • fragrant pandan

Pandanus amaryllifolius Roxb. (Pandanaceae); *skroefpalm* (Afrikaans); *ban lan ye* (Chinese); *pandanus* (French); **Schraubenpalme** (German); *duan pandan* (Indonesian); *pandano* (Italian); *pandan rampeh*, *pandan wangi* (Malay); *pandano* (Portuguese, Spanish); *bai toey hom* (Thai)

DESCRIPTION Wilted pandan leaves are strap-shaped, with minute teeth towards the tip. Fresh leaves become aromatic within a day or two when they wilt.

THE PLANT A palm-like plant of up to 1.5 m (5 ft) in height with erect stems bearing rosettes of sword-shaped leaves and occasional clusters of aerial roots. The plants are almost invariably sterile and hardly ever flower.

ORIGIN Southeast Asia. The exact origin is unknown, as the plant is an ancient cultigen that is now commonly grown in many parts of Asia (from southern India to New Guinea). The rare occurrence of male flowers on the Molucca Islands indicates that the species possibly originated there.[1]

CULTIVATION Seeds are never formed but plants are propagated with ease by simply cutting away partially rooted side branches. High rainfall and tropical conditions are required.

HARVESTING Leafy branches are cut off by hand and tied in large bundles. The slightly wilted leaves are sold individually on fresh produce markets, often tied into a characteristic knot as shown in the photograph.

CULINARY USES Wilted leaves have a delicious aroma similar to that found in scented rice. They are therefore used in Thai, Malay and Indonesian cooking to add flavour to ordinary (bland) rice and to enhance the flavour of rice dishes, including those cooked in coconut milk, such as *nasi lemak* and *nasi kuning*.[2] In Indonesia, rice is steamed in small baskets woven from pandan leaves. The leaves are raked with a fork or lightly crushed to release the flavour.[2] One of the most famous Thai dishes is pandan chicken or *gai ob bai toey* – marinated pieces of chicken wrapped in pandan leaves and deep-fried in a wok[2] (in the same way as lotus leaves are used in China). Pandan is also an important flavour source for sweet snacks, desserts, iced drinks and delicious puddings, made with ice cream and/or coconut milk, glutinous rice and palm sugar as main ingredients.[2] The cooked glutinous rice mixture is poured into baskets made from folded pandan leaves, where it solidifies and becomes jelly-like on cooling. Pandan essence (known as *pandan* or *toey*) is often available in shops and at markets as a substitute for the leaves. Green food dye is added to give a bright green colour. When not available, it is possible to substitute pandan essence with vanilla essence but the flavour is of course quite different.

FLAVOUR COMPOUNDS Pandan is used to flavour rice because it contains the same flavour compound (2-acetyl-1-pyrroline) present in expensive types of scented rice such as jasmine or basmati rice. The aroma of this compound, also known as 2-AP, is described as nutty or popcorn-like. The concentration in pandan leaves was reported to be 10.3 mg/kg.[4] In the same study, the level of 2-AP in a Thai fragrant rice cultivar was determined to be 3.0 mg/kg.[4]

NOTES *Pandanus amaryllifolius* is the species with fragrant leaves, while another species, *P. tectorius*, has fragrant flowers.

2-acetyl-1-pyrroline

1. **Mabberley, D.J. 2008.** *Mabberley's plant-book* (3rd ed.). Cambridge University Press, Cambridge.
2. **Hutton, W. 1997.** *Tropical herbs and spices.* Periplus Editions, Singapore.
3. **Buttery, R.G., Juliano, B.O., Ling, L.C. 1982.** Identification of rice aroma compound 2-acetyl-1-pyrroline in pandan leaves. *Chemistry and Industry (London)* 23: 478.
4. **Wongpornchai, S., Sriseadka, T., Choonvisase, S. 2003.** Identification and quantitation of the rice aroma compound, 2-acetyl-1-pyrroline, in bread flowers (*Vallaris glabra* Ktze). *Journal of Agricultural and Food Chemistry* 51: 457–462.

Pandan, the wilted leaves of *Pandanus amaryllifolius*

Pandan plants (*Pandanus amaryllifolius*)

201

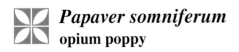

Papaver somniferum
opium poppy

Papaver somniferum L. (Papaveraceae); *papawersaad* (Afrikaans); *maanzaad* (Dutch); *ying su, ya pian* (Chinese); *graines de pavot* (French); *Mohnsamen* (German); *aphim* (Hindi); *seme di papavero* (Italian); *keshi* (Japanese); *kas kas* (Malay); *semilla de amapola* (Spanish)

DESCRIPTION The tiny seeds (less than 1 mm or 0.04 in. in length) are kidney-shaped, white or black (grey or blue) in colour (depending on the cultivar) and have a textured surface. They have a nutty aroma and taste.

THE PLANT An erect and robust annual of up to 1.5 m (ca. 5 ft) in height, with smooth greyish leaves, attractive white, red or purple flowers and characteristic fruit capsules with small openings around the top through which the seeds are dispersed.

ORIGIN Opium poppy is an ancient crop plant and cultigen of unknown origin, thought to be indigenous to southwestern Asia.[1] It has been grown in North Africa, Europe and Asia for many centuries. Opium, obtained from the latex of green fruits, is legally produced in Turkey and India. Turkey, the Czech Republic and India are among the main producers of poppy seeds. Large quantities of morphine, codeine and other alkaloids derived from opium are used each year as commercial painkillers.

CULTIVATION Plants are easily propagated from seeds and are grown as an annual crop. They require warm conditions, full sun and relatively fertile, moist soil.

HARVESTING For seed production, the ripe seed capsules are harvested mechanically. For making poppy seed paste from the tiny seeds, specialized poppy seed grinders are available.

CULINARY USES Poppy seeds are used as spice, condiment or garnish and are best known in Europe and North America for their use in confectionery – inside or more often on top of various types of pastries, breads, hamburger or hot dog buns, bread rolls, poppy seed muffins, rusks, crackers, bagels and sponge cakes. They add a nutty flavour, crunchy texture and a desired colour (white, black or blue). Finely ground seeds are mixed with milk, butter and sugar (also jam, honey or syrup) and used as fillings in pastries and candy bars. Poppy seed paste is commercially available in cans. In India, Pakistan and Southeast Asia, white poppy seeds are often fried to bring out the nutty flavour and are then used in the form of a paste as thickener for Indian and Malay-style curries.[2] Poppy seeds are used as condiment and ingredient in various dishes and desserts, and as flavourant for Chinese rice-flour noodles. Poppy seed oil has a nutty taste and can be used in the same way as olive oil.

FLAVOUR COMPOUNDS Several volatile compounds have been identified in the seed oil,[3] including 2-pentylfuran (considered to be the main aroma compound),[4] 1-hexanal, 1-hexanol, 1-pentanol and caproic acid also known as hexanoic acid. The main seed triglycerides were identified as linoleic, oleic and palmitic acid.[3]

NOTES Poppy seeds contain extremely low levels of opium alkaloids but may nevertheless lead to false positive drug tests and even arrests at airports, so it is safer not to carry poppy seeds or products made from them when travelling.

2-pentylfuran

1-hexanal

1-hexanol

1-pentanol

caproic acid (hexanoic acid)

1. **Mabberley, D.J. 2008.** *Mabberley's plant-book* (3rd ed.). Cambridge University Press, Cambridge.
2. **Hutton, W. 1997.** *Tropical herbs and spices.* Periplus Editions, Singapore.
3. **Krist, S., Stuebiger, G., Unterweger, H., Bandion, F., Buchbauer, G. 2005**. Analysis of volatile compounds and triglycerides of seed oils extracted from different poppy varieties (*Papaver somniferum* L.). *Journal of Agricultural and Food Chemistry* 53: 8310–8316.
4. **Hui, Y.H. 2006.** *Handbook of Food Science, Technology, and Engineering.* CRC Press, Boca Raton.

Poppy seeds, the ripe seeds of *Papaver somniferum* (with poppy fruits)

Poppy seeds (white and grey)

Plant in flower (*Papaver somniferum*)

Flowers (*Papaver somniferum*)

Perilla frutescens
beefsteak plant • shiso • purple mint

Perilla frutescens (L.) Britton (= *Perilla arguta* Benth.) (Lamiaceae); *shiso, perilla* (Afrikaans); *zhou bai su, ri ben bai su* (Chinese); *pérille verte cultivée, pérille rouge cultivée* (French); *Perilla, Chinesische Melisse* (German); *daun shiso* (Indonesian); *shiso* (Japanese); *kkaennip, deulkkae, nag-mon, nga-khi-mon* (Thai)

DESCRIPTION Shiso leaves are green or purple and typically ruffled, with prominent marginal teeth. They have a distinctive odour reminiscent of mint and cinnamon.

THE PLANT An annual herb (up to 1 m or 3 ft 4 in.) with square stems and minute pink flowers borne in terminal spikes. Two basic types are distinguished: var. *frutescens* is used in Korea for seed oil production, while var. *crispa* is used in Japan as a popular garnish and culinary herb. The latter can have green or purple leaves, depending on the cultivar. The green form is known as *ao-shiso* or *aojiso* in Japan and the purple form as *aka-shiso* or *akajiso*.[1]

ORIGIN Himalayas to East Asia.[1,2] It is a traditional Chinese vegetable that has become an important culinary herb in Japan and important oilseed crop (*yegoma*) in Korea.[1,2] Many cultivars have been developed as ornamental plants (purple forms are especially popular) and for the production of essential oil and seed oil (a drying oil with culinary and industrial uses). The plant has become naturalized in the Ukraine[2] and in the eastern parts of the United States.[1]

CULTIVATION Perilla is easily propagated from seeds and does best in a warm position with well-drained, slightly acid soil.

HARVESTING The leaves are picked as required and are available as fresh herbs in Korean and Japanese markets. The seeds are mechanically harvested for the production of seed oil.

CULINARY USES Fresh shiso leaves are important as a garnish, food colourant, culinary herb and vegetable in Japanese and Korean cooking. The main use in Japan is as a garnish for sashimi or tempura (the leaves may also be battered and deep-fried in the same way as the other components of the dish). The purple form is commonly used to add flavour and colour to the traditional pickled and dried sour plums, known as *umeboshi* (*Prunus mume*). The leaves are an ingredient of *beni shoga* (pickled ginger eaten with sushi).

FLAVOUR COMPOUNDS The aroma of shiso is due to the presence of perillaldehyde, the main compound in the essential oil. It co-occurs with smaller amounts of isoegoma ketone. Perillaldehyde is not present in Korean perilla, which has the potentially poisonous perilla ketone as the main aroma-active compound. The artificially produced oxime of perillaldehyde, known as perillartine or perilla sugar, is 2 000 times sweeter than sucrose. It is used to a limited extent as a sweetener in Japan. The main anthocyanin pigment in purple perilla is a malonyl ester of shisonin.

NOTES The seed oil (important in Korean cooking) is rich in linolenic acid.

perillaldehyde   perillartine (perilla sugar)   isoegoma ketone   perilla ketone

1. **Brenner, D.M. 1993.** Perilla: botany, uses and genetic resources. In: Janick, J., Simon, J.E. (Eds), *New crops*. Wiley, New York.
2. **Mabberley, D.J. 2008.** *Mabberley's plant-book* (3rd ed.). Cambridge University Press, Cambridge.
3. **Başer, K.H.C., Demirci, B., Dönmez, A.A. 2003.** Composition of the essential oil of *Perilla frutescens* (L.) Britton from Turkey. *Flavour and Fragrance Journal* 18: 122–123.
4. **Verma, R.S., Padalia, R.C., Chauhan, A. 2013.** Volatile oil composition of Indian Perilla [*Perilla frutescens* (L.) Britton] collected at different phenophases. *Journal of Essential Oil Research* 25: 92–96.

Common perilla, the leaves of *Perilla frutescens*

Leaves and flowers (*Perilla frutescens*)

Leaves and flowers (*Perilla frutescens*)

Leaves (*Perilla frutescens*)

Persicaria odorata
Vietnamese coriander • Vietnamese mint • laksa leaf

Persicaria odorata (Lour.) Soják (= *Polygonum odoratum* Lour.); (Polygonaceae); *Viëtnamese koljander*, *laksaplant* (Afrikaans); *la liao, laksa-yip* (Chinese); *coriandre du Vietnam* (French); *Laksa Blatt, Vietnamesische Mintze* (German); *daun kesum, daun laksa* (Malay); *phak phai* (Thai); *rau ram* (Vietnamese)

DESCRIPTION Fresh leaves are bright green and taper to a narrow tip, each with a distinctive reddish brown marking usually present near the base. They have a warm and fragrant aroma reminiscent of cilantro and a spicy, slightly lemony taste that becomes hot and peppery.

THE PLANT The laksa plant is a spreading perennial herb with soft stems, characteristic sheathing stipules and small pink flowers. It is similar to the semi-aquatic water pepper (*Persicaria hydropiper*) but this species has an erect growth form and leaves are without brown spots. Sikkim knotweed (*P. mollis*) is yet another member of the genus with culinary value. It is used as a potherb.[1]

ORIGIN Southeast Asia.[1] It is nowadays commonly cultivated as a culinary herb in many parts of the world.[2]

CULTIVATION Cuttings are very easily rooted or side shoots with roots may be planted. The plant thrives in tropical conditions but grows best in partial shade. In cold regions, it can be grown in a container and protected or taken indoors in winter. Reduce watering in winter and cut the plants back in summer to stimulate new growth.[2]

HARVESTING Leaves are picked fresh throughout the season when required.

CULINARY USES Leaves are always eaten fresh, never cooked. They are an essential ingredient of *laksa*, the well-known Singaporean and Malaysian seafood noodle soup (sometimes made with chicken). The soup is served with many vegetables, a spicy *bumbu* and a generous helping of freshly chopped laksa leaves as garnish. Fresh leaves are especially popular in Vietnamese and Thai cuisines and are commonly served with noodle soups (into which they are dipped when eaten). The leaves are an ingredient in salads, fresh spring rolls and many other Vietnamese dishes.

FLAVOUR COMPOUNDS Vietnamese coriander is sometimes referred to as a cilantro mimic. The essential oil, known as kesom oil, is variable but dodecanal and decanal are usually the main aroma compounds, with smaller amounts of undecanal.[3] The cilantro-like smell are ascribed mainly to these constituents. Also present are (*Z*)-3-hexenal, (*E*)-2-hexenal and (*Z*)-3-hexen-1-ol, compounds known for their green smell, reminiscent of freshly cut grass.[3]

NOTES Several culinary herbs used in Vietnamese cuisine are rather poorly known to Western cooks or they are not considered to be of culinary value in the rest of the world.[4] Fishwort/fish mint or *giâp cá* (*Houttuynia cordata*) is used as a fresh garnish and leaf vegetable. Knotweed or *rau ngò* (*Polygonum aviculare*) is used in soup and in the popular *hot pot*. Rice paddy herb or *ngò ôm* (*Limnophila aromatica*) has a lemony flavour and is used in *canh chua*, a sweet-and-sour seafood soup.[4] Vietnamese balm or *kinh giói* (*Elsholtzia ciliata*) is used with grilled meats. The leaves and seeds of *Gnetum gnemon* (*rau danh* in Vietnamese, *melinjo* in Indonesian) are ingredients of vegetable dishes and soups, while flour made from ground seeds is used to make deep-fried Indonesian *emping* crackers.

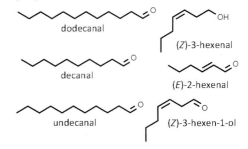

1. **Mabberley, D.J. 2008.** *Mabberley's plant-book* (3rd ed.). Cambridge University Press, Cambridge.
2. **McVicar, J. 2007.** *Jekka's complete herb book*. Kyle Cathie Limited, London.
3. **Quynh, C.T.T., Iijima, Y., Morimitsu, Y., Kubota, K. 2009.** Aliphatic aldehyde reductase activity related to the formation of volatile alcohols in Vietnamese coriander leaves. *Bioscience, Biotechnology, and Biochemistry* 73: 641–647.
4. **Gernot Katzer's Spice Pages** (http://gernot-katzers-spice-pages.com).

Vietnamese coriander, Vietnamese mint or laksa, the leaves of *Persicaria odorata*

Fishwort or *giâp cá* (*Houttuynia cordata*)

Melinjo (*Gnetum gnemon*)

Flowers (*Persicaria odorata*)

207

Petroselinum crispum
parsley

Petroselinum crispum (Mill.) A.W. Hill (Apiaceae); *pieters(i)elie* (Afrikaans); *fan yan sui* (Chinese); *persil* (French); *Petersilie* (German); *prezzemolo* (Italian); *perejil* (Spanish)

DESCRIPTION Parsley leaves are typically bright green, hairless, with curly edges in the common type and flat leaflets in the flat-leaved type. They have a fragrant, fresh and slightly spicy aroma and tangy, sweet, slightly grassy taste.[1] Flat-leaved parsley (var. *neapolitanum*), also called Italian, French, or celery-leaved parsley, has a slightly stronger aroma and is often recommended for culinary use, while the crisped type is the best choice for garnishing. Hamburg or turnip-rooted parsley (var. *tuberosum*)[2] is a root vegetable resembling parsnip.

THE PLANT A biennial herb forming a rosette of leaves in the first year and a stem with small yellow flowers and small dry fruits in the second year.

ORIGIN Mediterranean parts of southern Europe and western Asia.[2] Parsley features in Greek and Roman mythology and early medicinal texts. It is of major commercial importance today and in most countries the supplies are grown domestically.[1]

CULTIVATION Sow seeds in spring and again in summer to ensure a year-round supply. Germination is very slow (up to three weeks or more). Use small pots or plug trays or sow directly, as parsley does not like being transplanted. Plants are spaced 15 cm (6 in.) apart and grow best in rich loam soil with a neutral pH that is kept moist. Regular feeding with liquid fertilizer is recommended.

HARVESTING Fresh leaves can be picked throughout the first growing season. Freezing is the best way to preserve parsley. On a commercial scale, drying or freeze-drying is used, and oil is distilled from both the leaves and the fruits.

CULINARY USES Parsley is the best known and most widely used of all culinary herbs and garnishes. It is the main ingredient of French *fines herbes*, together with chervil, tarragon and chives.[3] Fresh leaves are a popular ingredient of soups, salads, sauces, omelettes, stews and stuffings for meat. Chopped parsley is sprinkled over food and often added to cooked dishes just before serving. It is commonly used to flavour butter, cheese and sauces. The term *persillade* is used in French cooking for a mixture of chopped parsley and garlic that is added to certain dishes at the end of the cooking time (e.g. beef *persillade*).[3] The word *persillé* is used for veined cheeses but also for dishes that are finished with *persillade* or with large amounts of chopped parsley (e.g. ham with parsley or *jambon persillé*).[3] Parsley leaf and fruit oils are commonly used as flavourants in the food industry.[1,4]

FLAVOUR COMPOUNDS The main flavour compound in parsley leaves has been identified as 1,3,8-*p*-menthatriene.[4] Apiole, myristicin and tetramethoxyallylbenzene are also present but in higher concentrations in the fruit oil.[4] Myristicin is the stimulant and hallucinogen principle present in nutmeg. This compound and apiole are toxic at high doses.

NOTES Parsley is a good source of vitamin C (190 mg per 100 g).[4]

apiole

myristicin

1,3,8-*p*-menthatriene

tetramethoxy-allylbenzene

1. **Farrel, K.T. 1999.** *Spices, condiments and seasonings.* Aspen Publishers, Gaithersburg, USA.
2. **Mabberley, D.J. 2008.** *Mabberley's plant-book* (3rd ed.). Cambridge University Press, Cambridge.
3. **Larousse. 1999.** *The concise Larousse gastronomique.* Hamlyn, London.
4. **Harborne, J.B., Baxter, H. 2001.** *Chemical dictionary of economic plants.* Wiley, New York.

Parsley, the leaves of *Petroselinum crispum*

Common parsley (*P. crispum* var. *crispum*)

Italian parsley (*P. crispum* var. *neapolitanum*)

Italian parsley (*P. crispum* var. *neapolitanum*)

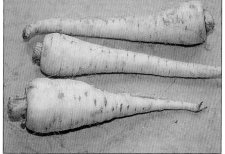

Turnip-rooted parsley (*P. crispum* var. *tuberosum*)

209

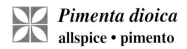

Pimenta dioica
allspice • pimento

Pimenta dioica (L.) Merr. (= *P. officinalis* Lindl.) (Myrtaceae); *pimento, jamaikapeper* (Afrikaans); *piment, poivre de la jamaïque, poivre giroflée* (French); *Piment, Allgewürz, Nelkenpfeffer* (German); *pimento* (Italian); *pimenta-da-jamaica* (Portuguese); *pimienta de Jamaica* (Spanish)

DESCRIPTION The dried, unripe fruits are hard, dark greyish-brown in colour and crowned by a small ring representing the calyx. They are 4–5 mm (ca. ⅕ in.) in diameter in Jamaican allspice or slightly larger (5–6 mm or ca. ¼ in.) in Mexican allspice.[1] The strong fragrance is similar to that of cloves but with elements of cinnamon, nutmeg and black pepper (hence the name allspice).

THE PLANT An evergreen tree of about 12 m (40 ft) high with leathery leaves, small white flowers and fleshy, dark purple fruits. Male and female flowers occur mostly on different plants (hence the Latin name *dioica*).

ORIGIN The tree is indigenous to Central America and the West Indies.[2] Spanish explorers introduced the spice to Europe in the 16th century. Jamaica remains the principal producer of high-quality allspice.[1]

CULTIVATION Plants require humid, tropical conditions and are grow in plantations (called "walks"). Freshly harvested seeds are sown and the seedlings transplanted while still quite small. They require deep, well-drained soil and regular watering and feeding. The trees are killed by frost and drought.

HARVESTING The fully developed but green (unripe) fruits are hand-picked. They are left to "sweat" for a day or two and are then dried in the sun. Essential oil is distilled from the fruits or the leaves.[1]

CULINARY USES Allspice has become popular in some Western and East European countries, initially as a spice to replace cardamom. It is used to flavour a wide range of dishes, including meat stews, sausages, salted beef, pork, meat pastries, pickles, sauces and stuffings. Allspice is an essential ingredient of Caribbean cuisine (e.g. jerk seasoning). It is used in Scandinavian smorgasbord, as well as fish, cheese and vegetable dishes. Mexicans use it in moles. Allspice is popular in Great Britain, where it is used in stews, sauces, confectionery, puddings and the traditional Christmas cake. Jamaican pimento dram and French liqueurs such as Benedictine and chartreuse are flavoured with allspice.[2] The spice, fruit oil or oleoresin extracts are commonly used in food processing, especially to flavour charcuterie items such as sausages, ham, salami and canned meats, as well as curry powders, condiments, relishes, ketchup, pickling spices and gravy mixtures. Leaf oil is used in ice creams, puddings, confectionery and liqueurs.

FLAVOUR COMPOUNDS The flavour is due mainly to eugenol (the dominant component in both the fruit and leaf oil), but minor compounds such as 1,8-cineole, methyleugenol, β-caryophyllene and α-phellandrene add to the complexity.[3] The astringent taste is due to tannins.[4]

NOTES Bay rum oil, distilled from the leaves of *Pimenta racemosa*, is traditionally used in cosmetic products especially for male fragrances (e.g. Old Spice) but this species or its fruits are not used to any extent for culinary purposes.

eugenol   1,8-cineole   β-caryophyllene

1. **Farrel, K.T. 1999.** *Spices, condiments and seasonings.* Aspen Publishers, Gaithersburg, USA.
2. **Mabberley, D.J. 2008.** *Mabberley's plant-book* (3rd ed.). Cambridge University Press, Cambridge.
3. **Harborne, J.B., Baxter, H. 2001.** *Chemical dictionary of economic plants.* Wiley, New York.
4. **Kikuzaki, H., Sato, A., Mayahara, Y., Nakatani, N. 2000.** Galloylglucosides from berries of *Pimenta dioica. Journal of Natural Products* 63: 749–752.

Allspice, the dried fruits of *Pimenta dioica*

Flowers and fruits (*Pimenta dioica*)

Fruits (*Pimenta dioica*)

Bay rum (*Pimenta racemosa*)

Pimpinella anisum
anise • aniseed

Pimpinella anisum L. (Apiaceae); *anys* (Afrikaans); *yan kok, pa chio* (Chinese); *anis vert* (French); *saunf* (Hindi); *Anis* (German); *anice verde* (Italian); *anis, erva-doce* (Portuguese); *anis* (Spanish)

DESCRIPTION The small dry fruits or schizocarps, referred to as "aniseed", are greyish brown in colour, 3–6 mm (⅛–¼ in.) long and have the characteristic sweet and aromatic smell and taste of anise (resembling star anise, fennel and liquorice).

THE PLANT An erect annual herb growing to a height of about 0.6 m (2 ft). It has markedly dimorphic leaves, with the basal ones long-stalked, broad and rounded and the upper ones subsessile and pinnate with narrow segments. The small white flowers are borne in typical umbels.

ORIGIN Eastern Mediterranean (Greece and Egypt) and cultivated since 2000 BC.[1] Anise has a long history as medicinal herb in ancient Greece (e.g. to sweeten the breath) and as a culinary spice used by the Romans, Arabs and British since the Middle Ages.[2]

CULTIVATION Anise is still an important crop, grown commercially in Turkey, Spain, China, Italy, India and other countries.[2] Anise oil has largely been replaced by the less expensive star anise oil from China. Anise requires full sun, good fertile soil and warm growing conditions. Seedlings do not transplant well so it is best to establish the plants by direct sowing.

HARVESTING The above-ground parts are harvested at the end of summer when the fruits start to ripen and are then dried and processed.

CULINARY USES Anise has a strong flavour and is used sparingly as a spice in soufflés, meat dishes (soups, stews, sausages), shellfish, vegetables (cabbage, carrots, turnips), mild cheeses, salad dressings, pickles, fruit dishes, desserts and juices. Chopped fresh leaves can be used in salads, pickled vegetables and fish soups. Star anise is nowadays often used as a substitute but connoisseurs say anise has a more delicate aroma. Anise is well known for its applications in confectionery (breads, biscuits, cakes and sweets) as well as alcoholic and non-alcoholic beverages. Examples of traditional culinary items and sweets are Australian humbugs, Austrian *anisbögen*, British aniseed balls, Dutch *muisjes*, German *Pfeffernüsse* and *Springerle*, Indian candy-coated *saunf* (used as *mukhwas* to freshen the breath after a meal), Italian *pizzelle*, New Mexican *bizcochitos*, New Zealand aniseed wheels, Norwegian *knotts* and Peruvian *picarones*. Well-known anise liqueurs or brandies/liquors (i.e. respectively with or without sugar) include French anisette, pastis and Pernod, Greek ouzo, Middle Eastern arrack, Italian sambuca, Spanish *anís* and Turkish *raki*. These are drunk with a glass of water on the side or more often directly diluted with water, resulting in the familiar ouzo effect (the drink becomes cloudy and milky because the alcohol-soluble anethole is no longer fully soluble and forms an emulsion).

FLAVOUR COMPOUNDS The fruits contain 1–4% essential oil with (*E*)-anethole, also referred to as *trans*-anethole, as the dominant compound (up to 90% or more) and several minor ingredients such as *cis*-γ-himachalene, *trans*-pseudoisoeugenyl 2-methylbutyrate, methylchavicol and *p*-anis-aldehyde.[3] Anethole is a phytoestrogen that also occurs in fennel.

NOTES In Pakistani and Indian cuisine, no distinction is made between anise and fennel – both are called *saunf*. In Southeast Asia, the name for star anise is sometimes shortened to *anis*.

(*E*)-anethole
(*trans*-anethole)

1. **Mabberley, D.J. 2008.** *Mabberley's plant-book* (3rd ed.). Cambridge University Press, Cambridge.
2. **Farrel, K.T. 1999.** *Spices, condiments and seasonings.* Aspen Publishers, Gaithersburg, USA.
3. **Rodrigues, V.M., Rosa, P.T.V., Marques, M.O.M., Petenate, A.J., Meireles, M.A.M. 2003.** Supercritical extraction of essential oil from aniseed (*Pimpinella anisum* L.) using CO_2: Solubility, kinetics, and composition data. *Journal of Agricultural and Food Chemistry* 51: 1518–1523.

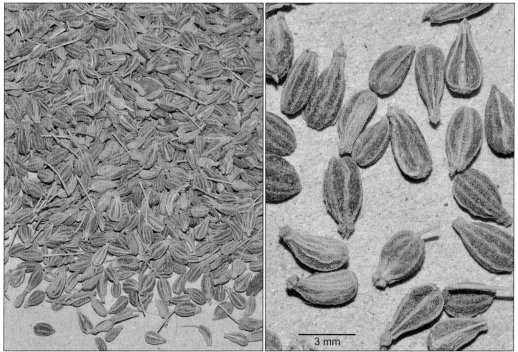

Anise, the small dry fruits of *Pimpinella anisum*

Alcoholic drinks made from anise

Plant (*Pimpinella anisum*)

Flowers (*Pimpinella anisum*)

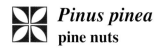

Pinus pinea
pine nuts

Pinus pinea L. (Pinaceae); *dennepitte* (Afrikaans); *cung zi, song zi ren* (Chinese); *pignons, pignoles* (French); *Pinienkernen* (German); *pignolia* (Italian); *matsu no mi* (Japanese); *jat* (Korean); *pinhões* (Portuguese); *piñónes* (Spanish)

DESCRIPTION Ripe seeds (pine nuts) are off-white to pale brown in colour and oblong (European stone pine) or ovoid (Asian and American species) in shape. The seeds and seed oil have a delicious, somewhat resinous, spicy and nutty taste.

THE PLANT The stone pine is a large tree with a typical umbrella-shaped crown. Female cones have overlapping woody scales, each with two seeds on their upper side. About 20 *Pinus* species produce edible nuts. Korean pine (*P. koraiensis*) is the most important in international trade, but stone pine or Italian pine (*P. pinea*), Siberian pine (*P. cembra*), Mexican pine (*P. cembroides*) and the piñon (*P. edulis*) are also of considerable commercial importance.[1]

ORIGIN The use of pine nuts in the Mediterranean region and Asia dates back to ancient times. *Pinus pinea* is the main source of nuts in the Mediterranean, while *P. koraiensis* is the main species for oriental nuts. The latter is widely distributed in northeastern Asia, including China and Japan.

CULTIVATION Pine nuts are mostly obtained from natural stands of pines.

HARVESTING Seeds are gathered by hand and the nuts are mechanically extracted from the hard seed coat. The nuts easily become rancid and are best refrigerated or frozen to retain the flavour.

CULINARY USES Pine nuts are an important component of Mediterranean and Middle Eastern cuisines. They are an essential ingredient of pesto sauce, a popular flavourant and garnish of rice dishes and are used in meat, pasta, fish, poultry and egg dishes, as well as confectionery (bread, croissants, biscuits, tarts and pastries). Roasted nuts are a snack food or used for *piñón* (New Mexican pine nut coffee). Italian American *pignoli* cookies are sprinkled with pine nuts. Pine nuts or pine nut oil adds a delicious mild, nutty flavour to fresh salads (e.g. French *salade landaise*), Levantine kibbeh, Middle Eastern *sambusak* and other types of samosa, as well as Turkish baklava.

FLAVOUR COMPOUNDS The chemical compound(s) responsible for the flavour of pine nuts (and for that matter, most tree nuts except almonds, hazelnuts and Brazil nuts) appear to be as yet unknown.[1] Pinolenic acid (an isomer of γ-linolenic acid or GLA) is a characteristic fatty acid found in the seeds and seed oil of *Pinus* species.[1] The compound has appetite-suppressant effects[2] and may reduce cholesterol. Siberian pine nut oil (rich in GLA) is used in Russia and China to treat peptic ulcers and gastritis.

NOTES Pine nut syndrome (pine mouth) is a unique taste disturbance, resulting in a bitter taste that may last for several weeks, caused by inedible *Pinus armandii* seeds.[3] These seeds have been imported from China to France, Britain and the United States, where many people reported the symptoms. It is speculated that the continuing bitter taste in the mouth is caused by an activation of the bitter taste receptors in the gastrointestinal tract[4] through enterohepatic recirculation of an as yet unidentified chemical compound.

pinolenic acid

γ-linolenic acid

1. **Kamal-Eldin, A., Moreau, R.A. 2009.** Tree nut oils. Chapter 3. In: Moreau, R.A., Kamal-Eldin, A. (Eds), *Gourmet and Health- Promoting Specialty Oils*. AOCS Press, Urbana, Illinois.
2. **Hughes, G.M. et al. 2008.** The effect of Korean pine nut oil (PinnoThin™) on food intake, feeding behaviour and appetite: A double-blind placebo-controlled trial. *Lipids in Health and Disease* 7: 1–10.
3. **Zonneveld, B.J.M. 2011.** Pine nut syndrome: a simple test for genome size of 12 pine nut-producing trees links the bitter aftertaste to nuts of *P. armandii* Zucc. ex Endl. *Plant Systematics and Evolution* 297: 201–206.
4. **Wu, S.V., Rozengurt, N., Yang, M., Young, S.H., Sinnett-Smith, J., Rozengurt, E. 2002.** Expression of bitter taste receptors of the T2R family in the gastrointestinal tract and enteroendocrine STC-1 cells. *Proceedings of the National Academy of Science* 99: 2392–2397.

Pine nuts, the seeds of stone pine (*Pinus pinea*), Siberian pine (*P. cembra*) and Mexican piñon (*P. cembroides*)

Stone pine, Siberian pine and Mexican piñon

Stone pine (*Pinus pinea*)

Siberian pine (*Pinus cembra*)

Mexican piñon (*Pinus cembroides*)

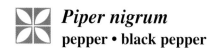

Piper nigrum
pepper • black pepper

Piper nigrum L. (Piperaceae); *peper* (Afrikaans); *hu jiao* (Chinese); *poivre* (French); *Pfeffer* (German); *kaalii mirch* (Hindi); *merica* (Indonesian); *pepe nero* (Italian); *pappaa* (Japanese); *lada hitam* (Malay); *pimenta negra* (Portuguese); *pimienta* (Spanish); *phrik thai* (Thai)

DESCRIPTION The fruits or drupes (peppercorns) are about 5 mm (0.2 in.) in diameter and may be green, black or white. All have a sharp, pungent taste and a delicious, spicy aroma when crushed.

THE PLANT A perennial vine or climber with hanging spikes of small flowers that turn into green and then bright red, fleshy drupes.

ORIGIN Indigenous to South India (Malabar region).[1,2] Pepper featured prominently in the ancient world and was a source of fabulous wealth during the medieval and colonial spice trade. The Dutch and Afrikaans expression "*peperduur*" reflects the high price it once had. Pepper provided the pungency ("pep") of Indian food until it was partially replaced by chilli peppers from the New World. It nevertheless remains the most important and popular of all spices in terms of overall value and trade volume.

CULTIVATION The vines are grown from cuttings, with trees or trellises for support. They thrive only in low-altitude tropical regions. Main suppliers are Vietnam, India, Indonesia, Brazil and China.

HARVESTING Near-ripe (green) or ripe (red) fruits are hand-picked and processed. Black peppercorns are made by briefly cooking the green fruits. The heat disrupts the cells and causes a gradual blackening through enzymatic oxidation of the drupes while they are slowly dried. Green peppercorns are made by pickling or rapidly drying or freeze-drying the green peppercorns to prevent fermentation and to retain the green colour. White pepper is made from fully ripe red fruits by fermentation (retting). The fleshy fruit wall becomes decomposed and easily removed (mechanically or chemically), leaving only the seed which is then bleached and/ or sundried.

CULINARY USES Pepper (usually paired with salt) is a ubiquitous table condiment. A pepper grinder is useful because pre-milled pepper from a shaker can quickly lose much of its flavour. Black pepper is used in practically all savoury dishes and even in sweet ones. Famous examples include pepper steak, *poivrade* sauce, *Pfefferkuchen* and it is even sprinkled on fresh strawberries. Whole peppercorns are added to pickles and marinades, while white pepper typically goes with pale-coloured Chinese dishes and French white sauces. There are countless applications in home cooking as well as commercial food processing.

FLAVOUR COMPOUNDS The pungent principle is piperine (only 1% as hot as capsaicin from chilli peppers). Essential oil in the shrivelled fruit walls adds spicy aromas. White pepper is more pungent and has musty flavours resulting from the fermentation process. The peppery aroma is due to rotundone, a compound also found in Shiraz wines.[3]

NOTES Minor sources of pepper include Ashanti pepper (*P. guineense*), cubebs or tailed pepper (*P. cubeba*) and Indian long pepper (*P. longum*). In ancient times long pepper was the most important item of trade but it was gradually replaced by black pepper. Leaves of *P. sarmentosum* are the *cha plu* of Thai cuisine and those of *P. auritum* the *hoja santa* of Mexican cuisine. See *Schinus molle* for more types of "pepper".

piperine   rotundone

1. **Mabberley, D.J. 2008.** *Mabberley's plant-book* (3rd ed.). Cambridge University Press, Cambridge.
2. **Zeven, A.C. 1995.** Black pepper. In: Smartt, J., Simmonds, N.W. (Eds), *Evolution of crop plants* (2nd ed.), pp. 407–408. Longman, London.
3. **Wood, C., Siebert, T.E., Parker, M., et al. 2008.** From wine to pepper: rotundone, an obscure sesquiterpene, is a potent spicy aroma compound. *Journal of Agricultural and Food Chemistry* 56: 3738–3744.

Green, black and white pepper, the dried fruits of *Piper nigrum*

Leaves and fruits (*Piper nigrum*)

Mexican pepperleaf (*P. auritum*)

Cubebs or tailed pepper (*Piper cubeba*)

Long pepper (*Piper longum*)

Pistacia lentiscus
mastic • masticha • mistki

Pistacia lentiscus L. (Anacardiaceae); *mastik* (Afrikaans); *mastik* (Dutch); *mastic* (French); *Mastix* (German); *mastiha* (Greek); *mastice* (Italian); *goma de masilla* (Spanish)

DESCRIPTION Mastic is solidified drops of tree resin that form small, ivory-coloured, hard, brittle and partly translucent lumps of about 6 mm (¼ in.) in diameter. It has a strong, slightly bitter and resinous taste and a sweet-smelling aroma. The product is quite expensive but will last indefinitely in a sealed container.

THE PLANT An evergreen woody shrub or small tree, barely 1.8 m (6 ft) high, which may reach 6 m (20 ft) under favourable conditions. The leaves are typically compound and lack a terminal leaflet. Male and female flowers are borne on separate plants. The small red fruits first turn red and then black as they ripen.

ORIGIN Mastic is indigenous to the entire Mediterranean region. In Greece and the eastern Mediterranean, the resin has been used since ancient times for chewing gum (*masticha* in Greek, it becomes sticky when chewed), as oral hygiene product (as breath freshener and filler for dental caries), and as ingredient of incense, lotions, perfumes and varnish. Culinary uses apparently developed more recently.

CULTIVATION Trees are capable of surviving in shallow rocky soil under alkaline and saline conditions and can withstand frost and summer drought. Seeds are bird-dispersed. The plants are sometimes cultivated as attractive ornamental street or garden trees. On the southern part of the Greek island of Chios they are grown for mastic production. This form of the species is sometimes referred to as *P. lentiscus* var. *chia*. The mastic-producing villages or cooperatives are known as *Mastichochoria*.

HARVESTING Mastic production on Chios begins in mid-August, when cross-shaped cuts are made in the trunks and larger branches of the trees to start the resin flow. The drops or "tears" take two weeks or more to dry and solidify. They are meticulously cleaned and scratched with a knife, one by one, to remove sand and dirt particles. The next step is to sort and grade them according to size and quality.

CULINARY USES The strong taste and floral fragrance is utilized in Greek and Turkish mastic-flavoured sweets (e.g. halva and *loukoúmia* or Turkish delight), ice cream, confectioneries, sweet yeast breads and biscuits. Small amounts are used in Lebanese and Egyptian soups, stews and desserts, sometimes to mask unwanted smells. It is also a flavourant of some of the many different types of Greek ouzo.

FLAVOUR COMPOUNDS The gum yields small amounts of essential oil with several monoterpenoids. The main volatile compounds in mastic from Chios were determined to be α-pinene, β-myrcene, β-pinene, limonene and β-caryophyllene.[3] It is likely that these compounds in combination give the pine-like or cedar-like fragrance but the main aroma-active compounds appear to be as yet unknown.

NOTES *Pistacia vera* (pistachio nut) is a close relative.

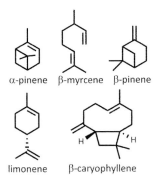

α-pinene β-myrcene β-pinene

limonene β-caryophyllene

1. **Mabberley, D.J. 2008.** *Mabberley's plant-book* (3rd ed.). Cambridge University Press, Cambridge.
2. **Milona, M. (Ed.) 2008.** *Culinaria Greece*. Könemann (Tandem Verlag), Köningswinter.
3. **Koutsoudaki, C., Krsek, M., Rodger, A. 2005.** Chemical composition and antibacterial activity of the essential oil and the gum of *Pistacia lentiscus* var. *chia*. *Journal of Agricultural and Food Chemistry* 53: 7681–7685.

Mastic, the resin drops collected from *Pistacia lentiscus*

Leaves and berries (*Pistacia lentiscus*)

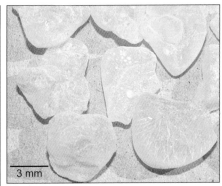

3 mm

Mastic tears

Pistachio nuts (*Pistacia vera*)

Plectranthus amboinicus
Indian borage • Cuban oregano • Spanish sage

Plectranthus amboinicus (Lour.) Sprengel (= *Coleus amboinicus* Lour.) (Lamiaceae); *woud-marog* (Afrikaans); *yin dub bo he*, *pok-hor* (Chinese); *coleus* (French); *Coleus*, *Jamaika-Thymian* (German); *daun kambing*, *daun kuching* (Indonesian); *coleus* (Italian); *daun bangun-bangun* (Malaysian); *oregano* (Tagalog)

DESCRIPTION The leaves are thick and succulent, with prominently toothed margins and a strong spicy smell.

THE PLANT A robust succulent herb of about 0.6 m (ca. 2 ft) high with spreading branches and small off-white to pale violet flowers borne in spikes. It is one of many *Plectranthus* species traditionally used as food and medicine.[1]

ORIGIN The plant is believed to be of southern and eastern African origin, from where it spread to other parts of the world and became known by several somewhat confusing names. These include India ("Indian borage"), Malaysia (*daun bangun-bangun*), the Philippines ("oregano"), and more recently also Australia ("five-in-one"), the United States ("Mexican mint"), the West Indies ("broad-leaf thyme") and South America (e.g. northeastern Brazil where it is known as "orégano" or "*hortelã da folha grossa*").

CULTIVATION New plants are easily established by simply planting the leafy stems. They are often seen in kitchen gardens in rural areas, thriving in partial shade in rich, well-drained soil. When grown in full sun, the leaves tend to become yellow and shrivelled.

HARVESTING Leaves are harvested throughout the year as required and are rarely if ever seen on fresh produce markets.

CULINARY USES The aromatic leaves are used in much the same way as thyme, sage or oregano (note the chemical similarity) and are considered to be useful substitutes (or tropical equivalents) for these temperate herbs. In Malaysia and Indonesia, leaves are used to flavour strong-smelling meat dishes (e.g. goat, fish and shellfish)[2-4] and sometimes also salads.[1] They are used in marinades for beef and chicken.[1] Chopped leaves are used in stuffings and poultry dishes[4] or can be mixed with flour and fried in butter or oil.[1]

FLAVOUR COMPOUNDS The leaves yield small amounts of essential oil, in which carvacrol is the dominant compound, with *p*-cymene, γ-terpinene and β-caryophyllene as minor constituents.[5] Carvacrol is also a main compound in oregano and has a pleasant, spicy aroma and warm, pungent taste on the tongue.

NOTES Several *Plectranthus* species, including *P. amboinicus*, are better known for their medicinal and cosmetic properties than for their culinary uses. Many of them are highly aromatic. An example is *P. neochilus*, an attractive groundcover that is commonly planted on sidewalks, not only as colourful ornamental plant but also to act as a dog repellent.

carvacrol   *p*-cymene   γ-terpinene   β-caryophyllene

1. Lukhoba, C.W., Simmonds, M.S.J., Paton, A.J. 2006. *Plectranthus*: A review of ethnobotanical uses. *Journal of Ethnopharmacology* 103: 1–24.
2. Burkill, I.H. 1966. *A dictionary of the economic products of the Malay Peninsula*, Vol. 1, pp. 634–635. Crown Agents for the Colonies, London.
3. Morton, J.F., 1992. Country borage (*Coleus amboinicus* Lour.): a potent flavoring and medicinal plant. *Journal of Herbs, Spices Medicinal Plants* 1: 77–90.
4. Hutton, W. 1997. *Tropical herbs and spices*. Periplus Editions, Singapore.
5. Mallavarapu, G.R., Rao, L., Ramesh, S. 1999. Essential oil of *Coleus aromaticus* Benth. from India. *Journal of Essential Oil Research* 11: 742–744.

Indian borage, the leaves of *Plectranthus amboinicus*

Flowers (*Plectranthus amboinicus*)

Flowering plant (*Plectranthus amboinicus*)

"Dog gone" (*Plectranthus neochilus*)

Portulaca oleracea
purslane

Portulaca oleracea L. (Portulacaceae); *postelein, porslein* (Afrikaans); *pourpier* (French); *Portulak, Sauburtzel* (German); *verdolaga* (Spanish)

DESCRIPTION Thick fleshy stems, often tinted red, bearing small to medium-sized, succulent leaves with a slightly sour taste.

THE PLANT A weedy annual herb with small yellow flowers opening only for a few hours in the morning. Two basic types are recognized: a variable wild form (subsp. *oleracea*) with ground-hugging stems and small leaves, and a domesticated variant (subsp. *sativa*) with a more upright habit and larger, yellowish green leaves.[1]

ORIGIN Purslane is a cosmopolitan annual and one of the top ten worst weeds in the world.[1] It was apparently present in North America even in pre-Columbian times. The use of purslane dates back to Roman times. It has been used in Europe for pickles since the Middle Ages.[2] The cultivated forms originated in Europe and have become more popular in recent years. One of them, 'Claytone de Cuba', is grown in France and Belgium.[2]

CULTIVATION Purslane is easily grown from seeds. It adapts to any soil type (even poor, compacted soil) and will grow under almost any environmental conditions.

HARVESTING Leafy stems are picked from the garden or wild-harvested as required. They are sometimes available at local farmers' markets. The plant uses the crassulacean acid photosynthetic pathway and therefore forms malic acid during the night. Leaves harvested in the early morning will tend to be more acidic than those picked in the afternoon, when the malic acid has already been metabolised.

CULINARY USES The acidity, spicy flavour and slimy consistency make purslane an old favourite for stews. It is incorporated in fresh salads and may be cooked on its own like spinach or okra, and served with butter, cream or gravy.[2] Leaves can be used as an edible garnish for soups, omelettes and meat dishes in the same way as watercress.[2] French cooks sometimes use fresh leaves to flavour béarnaise (a hot, creamy sauce made from egg yolks, vinegar and butter) or a version thereof known as *paloise*.[2]

FLAVOUR COMPOUNDS The sour and salty taste is linked to oxalic acid (23 mg/g in the fresh raw leaf but much reduced by cooking or pickling)[3] and also to malic acid in daily fluctuations. The herb is exceptionally nutritious, partly due to high levels of omega-3 fatty acids (claimed to be the highest level of all plants)[1] with 1.5–2.5 mg/g of fresh mass in leaves, 0.6–0.9 mg/g in stems and 80–170 mg/g in seeds.[3] The major compound is α-linolenic acid, representing 60% and 40% of total fatty acids in leaves and stems, respectively.[3] Also present is the structurally related but non-essential eicosapentaenoic acid (also known as EPA or timnodonic acid), which has shown potential in the treatment of inflammation and depression.

NOTES Two North American members of the Portulacaceae are occasionally cultivated: winter purslane, also known as miner's lettuce or Cuban spinach (*Montia perfoliata*), and talinum also called flameflower or waterleaf (*Talinum triangulare*).

oxalic acid   malic acid

α-linolenic acid

eicosapentaenoic acid
(EPA, timnodonic acid)

1. **Mabberley, D.J. 2008.** *Mabberley's plant-book* (3rd ed.). Cambridge University Press, Cambridge.

2. **Larousse. 1999.** *The concise Larousse gastronomique*. Hamlyn, London.

3. **Poeydomenge, G.Y., Savage, G.P. 2007.** Oxalate content of raw and cooked purslane. *Food, Agriculture and Environment* 5: 124–128.

4. **Liu, L., Howe, P., Zhou, Y.F., Xu, Z.Q., Hocart, C., Zhan, R. 2000.** Fatty acids and beta-carotene in Australian purslane (*Portulaca oleracea*) varieties. *Journal of Chromatography* 893: 207–213.

Cultivated purslane, the leaves of *Portulaca oleracea*

Wild purslane (*Portulaca oleracea*)

Plant in flower (*Portulaca oleracea*)

Winter purslane (*Montia perfoliata*)

Prunus dulcis
almond

Prunus dulcis (Mill.) D.A. Webb [= *P. amygdalis* (L.) Batsch] (Rosaceae); *amandel* (Afrikaans); *bian tao zi* (Chinese); *amandel* (Dutch); *amande* (French); *Mandel* (German); *badama* (Hindi); *mandorla* (Italian); *amondo* (Japanese); *amêndora* (Portuguese); *almendra* (Spanish)

DESCRIPTION Almonds are seeds (nuts) borne within a bony or papery shell (endocarp) and surrounded by a fleshy outer layer (the mesocarp and exocarp). The seeds become ivory white when the reddish brown seed coat is removed and they are blanched, sliced, chopped, diced or ground to a paste. Sweet almonds (var. *dulcis*) have a delicious, sweet nutty taste and a weak almond aroma, while bitter almonds (var. *amara*) are extremely bitter (and poisonous) but have a strong and sweet-smelling, typical almond or marzipan aroma.[1,2]

THE PLANT A small tree with pale pink or white flowers and velvety fruits that resemble those of peaches (they are closely related)[2] but dry out and split open at maturity.

ORIGIN Western Asia.[1] Almonds have been cultivated for at least 3 000 years in the Near East, North Africa and southern Europe.[1,2]

CULTIVATION Unlike peaches, almonds are not self-fertile, so that two or more cross-pollinating cultivars have to be planted to ensure fruit set. Almonds are cultivated in practically all temperate regions of the world. The United States (California) has become a major producer.[2]

HARVESTING Almond fruits are hand-harvested (or by mechanical tree shakers) when the husks dry out. The seeds are then extracted and further processed.

CULINARY USES Almonds in one form or another are important flavourants, used since the Middle Ages to make almond soup, almond milk, almond butter, almond milk jelly (blancmange) and sweet desserts.[3] They are still important in confectionery (cakes, biscuits, sweetmeats and sweets). Almond paste and marzipan, made from ground almonds mixed with sugar and egg whites are indispensable items, not only for the icing or filling of cakes but also for many types of sweets, sweetmeats (*massepains*), confectionery, breads, cakes, biscuits and pastries.[3] Almonds are used in flavoured butters, sauces, couscous and meat dishes, especially fish and poultry stuffings.[3] Bitter almonds, their essential oil or artificial almond essence may be used sparingly whenever an almond flavour is required. Italian amaretto is an example of an almond-flavoured liqueur.

FLAVOUR COMPOUNDS The typical almond flavour is due to benzaldehyde, a volatile compound that forms rapidly when amygdalin, the main cyanogenic glycoside in almonds, is broken down by enzymatic action. The process only happens when the seed is damaged or eaten, so that the enzymes become mixed with the amygdalin-rich cell contents. The process yields benzaldehyde, hydrogen cyanide (a toxic gas) and two molecules of sugar. Cold-pressed sweet almond oil is valuable as carrier oil in aromatherapy or as ingredient of cosmetics and creams.

NOTES Charoli seeds, obtained from a cultivated shrub (*Buchanaria lanzan*), are used in Indian and Pakistani cuisine to add an almond flavour to milk and milk-based desserts. The small, lentil-sized seeds are often roasted (toasted) to bring out the flavours before use as snack food or as ingredient of sauces and stews.

benzaldehyde

HCN
hydrogen cyanide

amygdalin

1. **Mabberley, D.J. 2008.** *Mabberley's plant-book* (3rd ed.). Cambridge University Press, Cambridge.
2. **Watkins, R. 1995.** Cherry, plum, peach and almond. In: Smartt, J., Simmonds, N.W. (Eds), *Evolution of crop plants* (2nd ed.), pp. 423–428. Longman, London.
3. **Larousse. 1999.** *The concise Larousse gastronomique*. Hamlyn, London.

Almonds, the seeds of *Prunus dulcis*

Almond fruits (*Prunus dulcis*)

Almonds and almond essence

Ripe fruit (*Prunus dulcis*)

Charoli seeds (*Buchanaria lanzan*)

225

Punica granatum
pomegranate • anardana

Punica granatum L. (Lythraceae); *granaat* (Afrikaans); *roman* (Arabic); *shi liu, shi liu pi* (Chinese); *granaatappel* (Dutch); *grenade* (French); *Granatapfel* (German); *anar, anardana* (Hindi); *melograno* (Italian); *delima* (Indonesian); *zakuro* (Japanese); *romã* (Portuguese); *granada* (Spanish); *tap tim* (Thai); *nar, rumman* (Turkish)

DESCRIPTION The spice called *anardana* refers to the seeds of sour cultivars of pomegranate that are dried until they become dark purplish-brown and only slightly fleshy. Ripe seeds with their attractive bright red, fleshy, translucent arils and the red juice or concentrate produced from them have a delicious astringent, sweet and sour taste.

THE PLANT A thorny shrub or small tree. Each of the glossy leaves has a distinctive extrafloral nectar gland at the tip.

ORIGIN An ancient cultigen of Middle Eastern (Persian) origin and perhaps originally indigenous to parts of Turkey and the adjoining Caspian region.[1] From here it spread to all parts of the Middle East, the Mediterranean region, India and the dry parts of Southeast Asia, and in the 17th century also to Central and North America.[2] Pomegranate features prominently in ancient symbols and mythology.[1]

CULTIVATION Seeds are easy to germinate but cuttings are mostly used for commercial plantations.[2] The pomegranate is an old favourite in kitchen gardens in temperate regions because it adapts well to extreme conditions. Cultivars differ mainly in the size and colour of the fruits and the hardness and sweetness of the seeds. Popular commercial cultivars are 'Wonderful' and 'Grenada'.[2]

HARVESTING Fruits are hand-picked when fully ripe. The seeds are extracted and dried in the sun for about two weeks to make *anardana*.

CULINARY USES *Anardana* is an important spice in northwestern Indian cuisine. It is used to acidify curries and sweet-and-sour dishes (in much the same way as is done with tamarind and *kokum*). Pomegranate seeds (fresh or frozen) and juice are used in Mediterranean and Middle Eastern (e.g. Lebanese and Iranian) syrups, sauces, soups, meat dishes, stuffed fish, salads and sweet couscous.[3] Pomegranate has become universally popular in recent years, not only as health drink but also for adding colour and flavour to many dishes. Pomegranate jelly can be made by adding pectin and sugar to the fresh juice.[2] Grenadine is a popular bright red pomegranate-based concentrate used in mixed drinks, aperitifs and cocktails (e.g. tequila sunrise).

FLAVOUR COMPOUNDS The dried fruit aril contains fructose (25%), phytic acid (2.2%), punicalagins (0.57%) and tannins (e.g. ellagic acid and derivatives). The aroma of fresh pomegranate has been described as green, fruity, floral and earthy,[3] and is associated with numerous aroma-active compounds which include hexanal, *cis*-3-hexenal, 1-octen-3-one and β-damascenone. In another study, the aroma-active compounds of the 'Wonderful' cultivar were identified and described as green (hexanal and *cis*-3-hexenal), floral (2-ethyl-hexanol) and fruity (ethyl-2-methylbutanoate, limonene, β-caryophyllene and 2(5*H*)-furanone).[4]

NOTES The juice is exceptionally high in polyphenols with antioxidant activity.[1]

ellagic acid

punicalagin

1. **Mabberley, D.J. 2008.** *Mabberley's plant-book* (3rd ed.). Cambridge University Press, Cambridge.
2. **Morton, J. 1987.** Pomegranate. pp. 352–355. In: *Fruits of warm climates.* Julia F. Morton, Miami, Florida.
3. **Cadwallader, K.R., Tamamoto, L.C., Sajuti, S.C. 2010.** Aroma components of fresh and stored pomegranate (*Punica granatum* L.) juice. *ACS Symposium Series*, No. 136 (Flavors in noncarbonated beverages), American Chemical Society, Washington, pp. 93–101.
4. **Mayuoni-Kirshinbaum, L., Tietel, Z., Porat, R., Ulrich, D. 2012.** Identification of aroma-active compounds in 'Wonderful' pomegranate fruit using solvent-assisted flavour evaporation and headspace solid-phase micro-extraction methods. *European Food Research and Technology* 235: 277–283.

Pomegranate, the fleshy seeds of *Punica granatum*

Pomegranate seeds (*Punica granatum*)

3 mm

Dried pomegranate seeds (*anardana*)

Pomegranate flower (*Punica granatum*)

Pomegranate fruits (*Punica granatum*)

Raphanus sativus
radish

Raphanus sativus L. (Brassicaceae); *radys* (Afrikaans); *luo bo* (Chinese); *radis* (French); **Rettich, Radieschen** (German); *rapani* (Greek); *ravanello* (Italian); *radeisshu* (Japanese); *mu* (Korean); *lobak* (Malay); *rabanete* (Portuguese); *rabanito* (Spanish); **hua phak kat khao** (Thai); *cu cai* (Vietnamese)

DESCRIPTION The fleshy roots are highly variable in shape and colour and have a pungent, peppery taste.

THE PLANT A robust biennial that flowers in the second year. There are four basic types of culinary radishes: small-rooted, short-season European radishes (usually red or red and white) that grow best in temperate regions; large-rooted Asian winter radishes, variable in shape and colour (often very large and white), suitable for more tropical regions; black-rooted European radishes suitable for winter storage and finally, leaf radishes grown in Southeast Asia for their leaves.[1,2]

ORIGIN Europe and Asia.[1] Domestication may have occurred independently in China but direct archaeological evidence is lacking. Early records date back between 5 000 years (Egypt) and 2 000 years (China).[1] Radishes are grown in practically all parts of the world and are staple food items in China (where the maximum genetic diversity is found),[1] Japan, Korea other Asian countries.

CULTIVATION Radishes are one of the easiest crops to grow and is a good choice for young children who want to start vegetable gardening. It is customary to sow seeds every two weeks to ensure a regular supply of fresh radishes throughout the summer. Use a sowing depth of 12 mm (½ in.) for small radishes and up to 40 mm (ca. 1½ in.) for the large daikon type.

HARVESTING Roots of short-season cultivars mature rapidly and are ready to harvest in three or four weeks.

CULINARY USES Radishes are mostly eaten fresh, as a snack, hors d'oeuvre or in green salads. They are appreciated for their crisp texture and sharp, peppery flavour. The roots are eaten whole or sliced, with or without the skin. Radishes have a very diverse application in Asian cooking.[3] Large, oblong radishes, best known as *daikon* (their Japanese name) but called *mooli* in India, are especially popular. In additional to being eaten raw, they are also cooked, steamed or stir-fried. In China, radishes have many different culinary uses and are often pickled.[3] Mashed radishes are mixed with flour to make a dough.[3] Radish leaf (*luo bo ye*) is cooked with meat or fish.[3] Sliced and pickled radishes are a common flavour ingredient of Japanese dishes, and they are often eaten with sushi and sashimi. Radish sprouts (germinated seeds mixed with spices) are popular in Japan (known as *kaiware*).

FLAVOUR COMPOUNDS The sharp taste of radishes is due to mustard oils (glucosinolates). The main flavour compound in daikon radishes has been identified as 4-methylthio-3-butenyl isothiocyanate (MTBITC). The main compound in seeds is raphanin, and therefore also likely to be one of the flavour ingredients of radish sprouts.

NOTES The name radish is derived from *radix*, the Latin word for root.

4-methylthio-3-butenyl isothiocyanate

raphanin

1. **Crisp, P. 1995.** Radish. In: Smartt, J., Simmonds, N.W. (Eds), *Evolution of crop plants* (2nd ed.), pp. 86–89. Longman, London.
2. **Vaughan, J.G., Geissler, C.A. 1997.** *The New Oxford book of food plants.* Oxford University Press, Oxford.
3. **Hu, S.-Y. 2005.** *Food plants of China.* The Chinese University Press, Hong Kong.
4. **Coogan, R.C., Wills, R.B.H. 2008.** Flavour changes in Asian white radish (*Raphanus sativus*) produced by different methods of drying and salting. *International Journal of Food Properties* 11: 253–257.
5. **Ivānovics, G., Horvāth, S. 1947.** Raphanin, an antibacterial principle of the radish (*Raphanus sativus*). *Nature* 160: 297–298.

Radishes, the swollen roots of *Raphanus sativus*

Radishes (*Raphanus sativus*)

Flowers and fruits (*Raphanus sativus*)

Giant daikon radish (*Raphanus sativus*)

Rhamnus prinoides
geisho • gesho

Rhamnus prinoides L.'Herit. (Rhamnaceae); *geisho, blinkblaar* (Afrikaans); *geisho, gesho* (Amharic); *geisho* (French)

DESCRIPTION Geisho is the dried leaves and twigs, often with some flowers or berries included. The leaves are oblong, taper to a narrow tip and have small teeth along the margins. The taste is bitter and the aroma is faintly woody and spicy.

THE PLANT Geisho is a shrub or small tree, up to 6 m (20 ft) high but often not exceeding 1.5 m (5 ft) when cultivated and regularly harvested. The leaves are alternately larger and smaller along the stems. Small yellowish green flowers develop into small fleshy red drupes that become blackish-purple when ripe.[1]

ORIGIN Indigenous to Arabia and sub-Saharan Africa, from southern Africa to Ethiopia.[1] It is an ancient Ethiopian crop that still plays an important role in the rural economy.[2–4]

CULTIVATION Traditionally, seedlings were collected from below wild-growing trees but nowadays selected seeds are used in nurseries to produce plants in containers. They are transplanted and grown in pure stands or may be intercropped with maize, vegetables and other plants.[4] Trees grow well at relatively high altitudes and are resistant to frost but not to drought. Important centres for geisho production include Tigray, North Shoa and the area around Sebeta near Addis Ababa.[4,5]

HARVESTING Plants should be at least three years old before they are first harvested. Geisho is traditionally collected by stripping all the leaves, leaving only those on the branch tips.[4] Cultivated plants can stay productive for up to 50 years or more.[4]

CULINARY USES Geisho is used to flavour traditional Ethiopian alcoholic beverages that are produced and sold everywhere in Ethiopia. The two main products are traditional beer (*tella*) and mead (*tej*) but a traditional home-brewed spirit drink known as *areki* or *katikala* is also made.[4] Geisho is used as bitter flavourant, similar to hops in German-style barley-based beers. The fermentation starter for making traditional beer is called *tinses*. It is made from powdered geisho leaves and stems (with all fruits carefully removed) added to water, together with a small amount of malted grain (usually barley).[4,5] The traditional beer or *tella* (called *sewa* in Tigrinya) is brewed from sorghum, maize or finger millet. *Tej* is also known as *mies*. It is a delicious traditional mead or honey beer/wine, made from local honey, often produced in traditional woven beehives.

FLAVOUR COMPOUNDS The bitter taste of geisho is due to geshoidin (β-sorigenin-8-*O*-β-*D*-glucoside), a napthalenic glycoside.[5] It is possible that various other compounds (e.g. physcion, prinoidin, rhamnocitrin and rhamnazin)[5] also contribute to the flavour and aroma.

NOTES *Tej* is served in a traditional *tej* glass, known as a *berele*, in rural and urban *tej* houses (called *tej betoch*).

geshoidin
(β-sorigenin-8-β-*D*-glucoside)

1. **Vollesen, K. 1989.** Rhamnaceae. In: Hedberg, I., Edwards, S. (Eds), *Flora of Ethiopia and Eritrea*, Vol. 3, pp. 385–398. National Herbarium, Addis Ababa and University of Uppsala, Uppsala.
2. **Jansen, P.C.M. 1981.** Spices, condiments and medicinal plants in Ethiopia, their taxonomy and agricultural significance. *Agricultural Research Reports 906.* Centre for Agricultural Publishing and Documentation, Wageningen.
3. **Pankhurst, R. 1968.** *Economic history of Ethiopia*, p. 194. Haile Selassie I University, Addis Ababa.
4. **Vetter, S. 1997.** *Geisho, its uses, production potential and problems in northern Tigray, Ethiopia.* Institute for Sustainable Development, Addis Ababa.
5. **Abegaz, B.M., Kebede, T. 1995.** Geshoidin: A bitter principle of *Rhamnus prinoides* and other constituents of the leaves. *Bulletin of the Chemical Society of Ethiopia* 9: 107–114.

Geisho, the dried leaves and fruits of *Rhamnus prinoides*

Flowers and fruit (*Rhamnus prinoides*)

Mead (*tej*) flavoured with geisho

Drinking *tej* from a traditional *tej* glass

Field of geisho (*Rhamnus prinoides*) intercropped with rehan (*Artemisia rehan*)

231

Rheum rhabarbarum
rhubarb

Rheum rhabarbarum L. (= *R.* ×*hybridum* Murray) (Polygonaceae); *rabarber* (Afrikaans); *da huang* (Chinese); *rhubarbe* (French); *Rhabarber* (German); *rabarbaro* (Italian); *ruibarbo* (Portuguese, Spanish)

DESCRIPTION Rhubarb leaf stalks (petioles) are uniform in thickness, flattened along the upper surfaces, and green or partially bright red in colour, often mottled with red spots. They have a sour taste and acidic, slightly grassy aroma.

THE PLANT A robust, stemless perennial herb with very large leaves and tall inflorescences growing from a short, thick rhizome below the ground. Many cultivars are available.

ORIGIN *R. rhabarbarum* is of uncertain origin.[1] The common garden rhubarb is sometimes considered to be a hybrid between *R. rhabarbarum* and some other species, and is then called *R.* ×*hybridum* or *R.* ×*cultorum*.[1] The culinary use of rhubarb is believed to have started in England in the 18th century.[2]

CULTIVATION Rhubarb can be grown from seeds but is usually multiplied by division. The plants grow well in temperate regions. They can be brought indoors in winter and can also be forced with special terracotta pots (rhubarb forcers).

HARVESTING Bright pink and succulent stalks can be harvested from early spring onwards (forced rhubarb is available from January to April).[2] They are usually used fresh but can be successfully frozen.[2]

CULINARY USES Rhubarb is popular in European and other Western countries. The stalks, carefully stripped of their outer fibrous layer, are used in much the same way as fruits, for pie fillings, compotes and jams.[2] They are very sour and have to be sweetened with sugar.[2] Minor uses include the making of chutney, wine and aperitifs (e.g. the Italian *Rabarbaro*).[2]

FLAVOUR COMPOUNDS Rhubarb leaves contain 0.5% oxalic acid but the leaf stalks (petioles) have low levels, which are further reduced by cooking and processing.[3,4] The sour, tart taste of cooked rhubarb is due to a wide range of acids, including acetic acid, 2- and 3-hexenoic acid, and 2- and 3-methylbutanoic acid.[4] Uncooked rhubarb displays a range of aroma-active compounds, many with C-6 skeletons ((*E*)-2-hexenal, (*E*)-2-hexanol and (*E*)-2-hexenoic acids).[3] The fruity flavours that develop in cooked rhubarb are reminiscent of caramel and plum jam.[4] They are not caused by a single rhubarb flavour compound but rather by a complex mixture that includes typical Maillard products (e.g. hydroxymethyl-furaldehyde, furfural, 2-formylpyrrole, maltol and derivatives of maltol).[4]

NOTES The rhizomes of rhapontic rhubarb (*R. rhaponticum*) and Chinese rhubarb (*R. palmatum* or *R. officinalis*) are traditional laxative medicines. Excessive ingestion of oxalic acid can lead to poisoning through interference with the absorption of calcium and iron.

oxalic acid · acetic acid · (*E*)-2-hexenal · (*E*)-2-hexenol · (*E*)-2-hexenoic acid · 2-methyl-butanoic acid · 4-methyl-hexanol · 4-methyl-hexanoic acid · hydroxymethyl-furaldehyde · 2-formylpyrrole · maltol

1. **Mabberley, D.J. 2008.** *Mabberley's plant-book* (3rd ed.). Cambridge University Press, Cambridge.
2. **Larousse. 1999.** *The concise Larousse gastronomique.* Hamlyn, London.
3. **Dregus, M., Schmarr, H.G., Takahisa, E., Engel, K.H. 2003.** Enantioselective analysis of methyl-branched alcohols and acids in rhubarb (*Rheum rhabarbarum* L.) stalks. *Journal of Agricultural and Food Chemistry* 51: 7086–7091.
4. **Dregus, M. 2004.** *Untersuchungen flüchtiger Verbindungen in Rhabarber* (Rheum rhabarbarum *L.*). Ph.D. dissertation, Technical University of München, München.

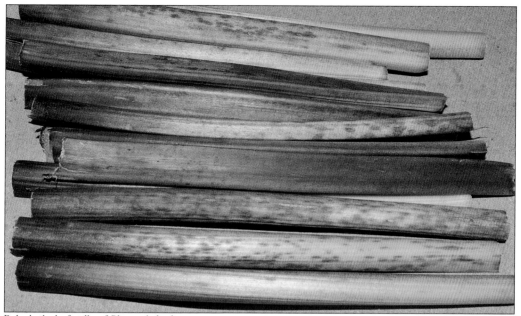
Rubarb, the leaf stalks of *Rheum rhabarbarum*

Flowering plant (*Rheum rhabarbarum*)

Rhubarb for sale

Rhubarb forcers

Rhus coriaria
sumac • sumach • spice sumac

Rhus coriaria L. (Anacardiaceae); *sumak* (Afrikaans); *sumaq* (Arabic); *lan fu mu* (Chinese); *sumak* (Dutch); *corroyère, sumac des épiciers* (French); *Sumach, Gewürzsumach* (German); *soumaki* (Greek); *kankrasing* (Hindi); *sommacco* (Italian); *sumakku* (Japanese); *somagre* (Persian); *sumagre* (Portuguese, Spanish); *sumak, somak* (Turkish)

DESCRIPTION Sumac or sumach is the dried pericarps of the fruits, dark red to purple-red in colour. It has a sour taste and a spicy, woody and slight citrus-like aroma.

THE PLANT A large shrub or small tree bearing large, pinnately compound leaves comprising about six pairs of dentate leaflets. The fruits are globose, velvety and up to 12 mm (ca. ½ in.) in diameter. Additional common names for the species include elm-leaved sumac, tanner's sumac or Sicilian sumac.

ORIGIN Southern Europe.[1] The plant is better known as a traditional source of dye for leather. The dried and powdered leaves were used in the tanning industry for Cordoba and Morocco leather.[1] The dry fruits are an important spice in Middle Eastern cuisine but are relatively poorly known in the rest of the world.[2-4] It is popular in Armenia and Turkey, North Africa (Morocco, Libya, Egypt) and practically the entire Arabian Peninsula.

CULTIVATION Trees are propagated from seeds. They grow in any deep, well-drained soil and do best in a Mediterranean climate of cool, wet winters and hot, dry summers.

HARVESTING The fruits are hand-harvested when fully ripe, dried in the sun and coarsely ground.

CULINARY USES Sumac is an important acidic spice and is usually (but not always) one of the four ingredients of the Middle Eastern spice mixture known as *za'atar* (za'atar herb, sesame seeds, sumac and salt – see notes on *Origanum syriacum* under *O. vulgare*). Sumac is used as a dry seasoning for vegetable dishes, salads, rice dishes, meat kebabs and meatballs. It is sprinkled on mezze dishes, rice and salads in the same way as lemon juice or vinegar, to improve the flavour. Sumac mixed with yogurt is served with kebabs.

FLAVOUR COMPOUNDS The sour taste of sumac is ascribed to malic acid[4] but perhaps also the astringency of gallotannins. Dried pericarps yielded limonene, nonanal and (Z)-2-decenal as main volatile compounds in one study[2] and cembrene, β-caryophyllene, β-caryophyllene alcohol, carvacrol, α-terpineol, α-pinene and farnesyl acetone in another.[3.] Sumac flavour has been determined to be linked mainly to the woody and spicy aromas of cembrene, β-caryophyllene and caryophyllene oxide.[4] Limonene may be partly responsible for the dried lemon balm or citrus-like aroma.[4]

NOTES *Zereshk* or *sereshk* is the dried fruits of barberry (*Berberis vulgaris*), a traditional acidic Persian spice, used in much the same way as sumac. It is cooked with rice to make *zereshk polo*, which is eaten with chicken dishes. The berries are also used for juice, jams and fruit rolls.

malic acid

cembrene

β-caryophyllene

caryophyllene oxide

1. **Mabberley, D.J. 2008.** *Mabberley's plant-book* (3rd ed.). Cambridge University Press, Cambridge.
2. **Kurucu, S., Koyuncu, M., Güvenc, A., Baser K.H.C., Özek, T.** 1993. The essential oils of *Rhus coriaria* L. (sumac). *Journal of Essential Oil Research* 5: 481–486.
3. **Brunke, E.-J., Hammerschmidt, F.-J., Schmaus, G., Akgül, A.** 1993. The essential oil of *Rhus coriaria* L. fruits. *Flavour and Fragrance Journal* 8: 209–214.
4. **Bahar, B., Altug, T. 2009.** Flavour characterization of sumach (*Rhus coriaria* L.) by means of GC/MS and sensory flavour profile analysis techniques. *International Journal of Food Properties* 12: 379–387.

Sumac the dried fruits of *Rhus coriaria* with *za'atar* spice (sumac, za'atar herb and sesame seeds)

Plant with flowers and fruits (*Rhus coriaria*)

Ribes nigrum
blackcurrant • black currant

Ribes nigrum L. (Grossulariaceae); *johannesbessie* (Afrikaans); *hei sui cu li* (Chinese); *cassis* (French); *Schwarze Johannisbeere* (German); *ribes nero* (Italian); *kuro fusa suguri* (Japanese); *casis, grosella negra* (Spanish)

DESCRIPTION Ripe blackcurrants are 20 mm (0.4 in.) in diameter, dark glossy purple and have a characteristic sour and fruity taste and aroma.

THE PLANT A long-lived shrub with five-lobed leaves and pinkish white flowers.

ORIGIN Northern Europe and northern Asia.[1] The fruits of cultivars, developed since the 17th century, do not differ much from those of the wild form of the species.[1] Commercial production is centred in Europe. The crop is making a comeback in the United States after it has been restricted for many years because of it being a carrier of white pine blister rust that threatened the logging industry. New Zealand has also become an important producer, exporting large quantities to Japan.

CULTIVATION At least ten weeks of subzero temperatures during winter are a requirement, so that cultivation is limited to very cold regions. Also important are damp, fertile soil, regular watering, timely pruning and a good choice of cultivar. 'Baldwin' has remained a favourite in Britain for many years,[1] while in France 'Noir de Bourgogne' was considered to have the best flavour.[2] 'Ben Connan', 'Ben Lomond' and 'Ben Sarek' are modern cultivars that have earned the RHS Award of Garden Merit.

HARVESTING The berries should be fully ripe when picked (in Europe, towards the end of June and beginning of July). On a commercial scale, harvesting is mostly done mechanically, using straddle harvesters.

CULINARY USES Blackcurrants are used for making delicious jams, jellies, fruit juices, syrups, iced purées and liqueurs. They have important culinary uses in confectionery and desserts, and combine well with meat dishes, seafood and fresh salads. Blackcurrant leaves are commonly used in Russia as culinary herb to flavour teas, preserves and pickles. The famous *crème de cassis* (blackcurrant liqueur) is used in cooking but is also enjoyed as *Kir, Kir Communard* or *Kir Royale* (a spoonful of blackcurrant liqueur, topped up with dry wine, respectively white, red or champagne).

FLAVOUR COMPOUNDS The most characteristic flavour compound is 4-methoxy-2-methyl mercaptobutane, responsible for the well-known "catty" note[3] (similar sulphur-containing constituents occur in *Agathosma betulina*). Major aroma-impact volatiles that contribute to the complexity and buttery note of the flavour include 2,3-butadiene, methylbutanoate, ethylbutanoate, 1-octen-3-one, terpinen-4-ol and β-damascenone.[4,5] The dark pigments are anthocyanins.

NOTES The jostaberry (*R. ×nidigrolaria*) is a hybrid with gooseberry (*R. uva-crispa*) and Worcesterberry (*R. divaricatum*). Together with redcurrant (*R. rubrum*) and other species, these are also sources of fruity flavours.

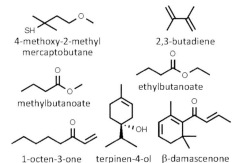

4-methoxy-2-methyl mercaptobutane

2,3-butadiene

methylbutanoate

ethylbutanoate

1-octen-3-one   terpinen-4-ol   β-damascenone

1. **Kepp, E. 1995**. Currants. In: Smartt, J., Simmonds, N.W. (Eds), *Evolution of crop plants* (2nd ed.), pp. 235–239. Longman, London.
2. **Larousse. 1999**. *The concise Larousse gastronomique*. Hamlyn, London.
3. **Le Quere, J.L, Latrasse, A. 1990**. Composition of the essential oils of black currant buds (*Ribes nigrum* L.). *Journal of Agricultural and Food Chemistry* 38: 3–10.
4. **Latrasse A, Riguard J and Sarris J. 1982**. Aroma of the blackcurrant berry (*Ribes nigrum* L). Main odour and secondary notes. *Sciences des Aliments* 2:145–162.
5. **Mikkelsen, B.B., Poll, L. 2002**. Decomposition and transformation of aroma compounds and anthocyanins during black currant (*Ribes nigrum* L.) juice processing. *Journal of Food Science* 67: 3447–3455.

Blackcurrants, the ripe fruits of *Ribes nigrum*

Blackcurrant flowers (*Ribes nigrum*)

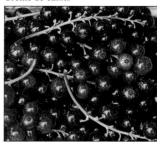

Crème de cassis

Jostaberry (*R. ×nidigrolaria*)

Gooseberry (*R. uva-crispa*)

Redcurrant (*R. rubrum*)

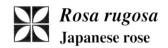# *Rosa rugosa*
Japanese rose

Rosa rugosa Thunb. (Rosaceae); ***Japanese roos*** (Afrikaans); ***mei gui*** (Chinese); ***rosier du Japon*** (French); ***Kartoffel-Rose*** (German); ***hamanasu*** (Japanese); ***haedanghwa*** (Korean)

DESCRIPTION Dried rose petals are usually dark purplish in colour and have a familiar delicate perfume.

THE PLANT A prickly and spiny creeper with attractive pink flowers. It differs from its better known commercial relatives by the coarsely textured leaves and the single flowers.

ORIGIN Roses have been cultivated in China and the Middle East since ancient times, not only for ornamental purposes but also for medicinal and culinary use.[1,2] The Japanese rose is the most important source of culinary rose petals in China, Japan and Korea, while the damask rose (*R. ×damascena*) is an essential element of Middle Eastern cuisine.[2,3] Other major sources of rose petals are the common rose (*R. gallica* and countless hybrids), the tea-scented rose (*R. ×odorata* and its many horticultural proliferations) and the cabbage rose (*R. ×centifolia*).

CULTIVATION Roses have become the most popular and diverse of all ornamental plants and cut flowers. They do best in cool regions and require pruning and regular feeding and watering. Thousands of cultivars are maintained by grafting.

HARVESTING Petals are harvested by hand, an extremely labour-intensive activity. In China, rose farmers gather the petals as soon as the flowers open, in the period between late April and July.[1]

CULINARY USES Rose petals are used in China mainly in the form of rose sugar, made by kneading wilted petals with an equal amount of sugar.[1] In Middle Eastern and Indian cuisine, rose water and rose essence are very important

flavourants of food and confectionery items.[2,3] The petals may be candied or used for rose-petal jam, a traditional delicacy of the Balkan region.[2] They are also used to flavour rose wine, rose honey, rose vinegar, jellies, creams, sorbets, ices, pastries, sweets (such as Turkish delight or *loukoúmia*) and cakes.[2] Dried and powdered rose buds are often one of the components of *ras el hanout* and other spice mixtures used for North African and Middle Eastern sauces, ragouts and rice dishes.

FLAVOUR COMPOUNDS Many different volatile compounds are responsible for the flavour and aroma.[4-6] Major essential oil components such as (−)-citronellol, together with geraniol, nerol, phenethyl alcohol and methyl eugenol, generally have less impact on the odour than some compounds that are present in minute quantities.[5,6] These include the well-known (−)-*cis*-rose oxide but also β-damascenone, β-ionone and several others.[5,6]

NOTES Rose fruits (hips) are widely used to make jams, preserves, jellies and sauces. Hips of the Japanese rose are a traditional food in Japan and are used in China to flavour wine.

(−)-citronellol

phenethyl alcohol

(−)-*cis*-rose oxide

β-damascenone

methyl eugenol

β-ionone

1. **Hu, S.-Y. 2005.** *Food plants of China.* The Chinese University Press, Hong Kong.
2. **Larousse. 1999.** *The concise Larousse gastronomique.* Hamlyn, London.
3. **Widrlechner, M.P. 1981.** History and utilization of *Rosa damascena. Economic Botany* 35: 42–58.
4. **Hashidoko, Y. 1996.** The phytochemistry of *Rosa rugosa. Phytochemistry* 43: 535–549.
5. **Knapp, H., Straubinger, M., Fornari, S., Oka, N., Watanabe, N. 1998.** (S)-3,7-Dimethyl-5-octene-1,7-diol and related oxygenated monoterpenoids from petals of *Rosa damascena* Mill. *Journal of Agricultural and Food Chemistry* 46: 1966–1970.
6. **Jirovetz, L., Buchbauer, G., Stoyanova, A., Balinova, A., Guangjiun, Z., Xihan, M. 2005.** Solid phase microextraction/gas chromatographic and olfactory analysis of the scent and fixative properties of the essential oil of *Rosa damascena* L. from China. *Flavour and Fragrance Journal* 20: 7–12.

Rose petals, rose petal jam and rose water (*Rosa* species)

Japanese rose (*Rosa rugosa*)

Rose petals

Rosa ×centifolia

239

Rosmarinus officinalis
rosemary

Rosmarinus officinalis L. (Lamiaceae); *roosmaryn* (Afrikaans); *romarin* (French); *Rosmarin* (German); *rosmarino* (Italian); *roméro* (Spanish)

DESCRIPTION Rosemary leaves are needle-shaped, dark green above and pale white below, with rolled-in margins. They have a pungent, bitter taste and a strong spicy and slightly camphorous aroma.

THE PLANT An aromatic, evergreen shrub with small, two-lipped flowers that are usually blue or purple but sometimes pink or white.

ORIGIN The plant is indigenous to the Mediterranean region.[1] Rosemary is linked to many myths and legends dating back to ancient Greece and Rome. It features in the mythology of Aphrodite, the Greek goddess of love and beauty, was recommended by Dioscorides as medicinal plant and was often used in wedding and funeral rituals.[1] Rosemary is associated with fidelity, perhaps inspired by the famous quote from Shakespeare's Hamlet (Act IV, scene 5, Ophelia to Laertes): "There's rosemary, that's for remembrance, pray love, remember." Today it is one of the most important of all culinary herbs and of major commercial importance for home cooking[2] as well as industrial food processing.[3,4]

CULTIVATION Rosemary is an old favourite in herb gardens, easily propagated from stem cuttings. It can withstand moderate frost and is exceptionally drought tolerant. Erect and robust cultivars such as 'Tuscan Blue' are particularly productive. It grows well in containers and can be trimmed into various shapes or clipped to form a low hedge.

HARVESTING The leaves are harvested (by hand or mechanically) and are sold fresh or more often in the form of a dried and packaged spice herb. The leaves may be steam-distilled to produce rosemary essential oil.[3,4]

CULINARY USES Leaves are used sparingly to flavour all types of meat dishes (venison, beef, lamb, pork, veal, poultry, grills, sausage meats, ragouts, stews, pâtés and marinades), as well as sauces, soups, salads and salad dressings.[2] It goes well with pasta sauces, baked fish and vegetables, and can be used to flavour herb butter or milk used for a dessert.[2] The bright blue flowers make an attractive garnish for salads or when candied, on cakes and other confectionery. Rosemary is one of the herbs used as ingredient of Benedictine. Rosemary oil is used in the production of alcoholic and non-alcoholic beverages, confectionery, condiments, relishes and processed meats.[3,4]

FLAVOUR COMPOUNDS Steam-distillation yields rosemary oil, with 1,8-cineole, α-pinene, camphor and β-pinene as main compounds.[4] Depending on the chemotype, oil obtained by extraction may yield relatively high levels of minor compounds such as camphene, borneol, verbenone and bornyl acetate, all of which contribute to the complexity of the aroma.[5]

NOTES Rosemary essential oil is used in perfumery, soaps, bath oils, detergents, cosmetics and aromatherapy.[3,4]

α-pinene   1,8-cineole   camphor   camphene

verbenone   borneol   β-pinene   bornyl acetate

1. **Mabberley, D.J. 2008.** *Mabberley's plant-book* (3rd ed.). Cambridge University Press, Cambridge.
2. **Larousse. 1999.** *The concise Larousse gastronomique*. Hamlyn, London.
3. **Farrel, K.T. 1999.** Spices, condiments and seasonings. Aspen Publishers, Gaithersburg, USA.
4. **Harborne, J.B., Baxter, H. 2001.** *Chemical dictionary of economic plants*. Wiley, New York.
5. **Santoyo, S., Cavero, S., Jaime, L., Ibañez, E., Señoráns, F.J., Reglero, G. 2005.** Chemical composition and antimicrobial activity of *Rosmarinus officinalis* L. essential oil obtained via supercritical fluid extraction. *Journal of Food Protection* 68: 790–795.

Rosemary, the leaves of *Rosmarinus officinalis*

Plant with white flowers (*R. officinalis*)

Flowers and leaves (*Rosmarinus officinalis*)

White and pink cultivars (*R. officinalis*)

241

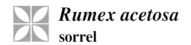

Rumex acetosa
sorrel

Rumex acetosa L. (= *R. rugosus* Campd.) (Polygonaceae); *tuinsuring* (Afrikaans); *zurkel* (Flemish); *oseille* (French); *Sauerampfer* (German); *acetosa* (Italian); *azeda* (Portuguese); *acedera* (Spanish)

DESCRIPTION The leaves of the common garden sorrel are borne on long petioles and are distinctively spear-shaped. They are soft and fleshy in texture and have a bitter and sour taste.

THE PLANT Sorrel is a weedy perennial herb with inconspicuous flowers borne in slender clusters. Several other sorrels are used as culinary herbs.[1,2] French sorrel, also known as buckler-leaved sorrel (*R. scutatus*) can easily be recognized by the rounded leaf shape and prominent ear-like lobes at the base of the leaf lamina. Red-veined sorrel (*R. sanguineus*) has oblong and tapering leaves, beautifully decorated with bright red main and secondary veins. Sheep sorrel (*R. acetosella*) is a common weed with small, spear-shaped leaves that is sometimes wild-harvested.[1,2]

ORIGIN Common sorrel is indigenous to Europe and northern Asia. It has been used as potherb in ancient Egypt, Greece and Rome, and in Europe since medieval times for the traditional early spring sorrel soup. Sorrel and other green herbs were once important as sources of vitamin C to prevent scurvy.

CULTIVATION Sorrel is easily grown from seeds or by division of mature clumps. In very warm regions, it does best in partial shade – the leaves stay green and do not get too bitter.[1] Regular watering and feeding is necessary to maintain healthy plants. Use acidic compost as mulch to ensure that the soil remains acidic, as it will not thrive in alkaline soil.[1] Plants are often grown in kitchen gardens or alongside rural restaurants to ensure a convenient and regular supply of fresh leaves. Sorrel is readily available from supermarkets and fresh produce markets.

HARVESTING Various types of sorrel are wild-harvested in Europe. In the garden, healthy young leaves can be picked throughout spring and summer. Sorrel does not dry well but can be successfully frozen.[1,3]

CULINARY USES Sorrel has many culinary uses in European and especially French cuisine[3] and is traditionally served with fish, veal and egg dishes. Fresh young leaves, usually with the stalks removed, are added to salads, cooked in the same way as spinach or used to prepare *velouté* sauce.[3] It can be puréed to make sorrel purée, often with cream or a white roux added, or used as a filling for omelettes.[3] Sorrel soup is popular in Russia, Poland and most eastern European countries, where it is sometimes referred to as "green borsch". It is served hot or cold, often with a garnish of sour cream (cream changes the acidity of the dish because oxalic acid reacts with calcium). The name sorrel has also been applied to *Oxalis pes-caprae*, a sour-tasting weedy plant in the Cape region of South Africa. It is traditionally used as a substitute for vinegar when making the well-known lamb or mutton stew called *waterblommetjiebredie*. This delicious signature dish of the Cape cuisine is made with the inflorescences of Cape pond weed (*Aponogeton distachyos*).

FLAVOUR COMPOUNDS The sour and bitter taste is due to oxalic acid and perhaps also the astringency caused by gallic acid. Cultivated sorrels tend to have a lower gallic acid content.[4]

NOTES Sorrel leaves or leaf juice is traditionally applied to the skin as a first aid treatment to relieve the pain caused by stinging nettles.

oxalic acid   gallic acid

1. **McVicar, J. 2007.** *Jekka's complete herb book*. Kyle Cathie Limited, London.
2. **Phillips, R., Foy, N. 1992.** *Herbs*. Pan Books, London.
3. **Larousse. 1999.** *The concise Larousse gastronomique*. Hamlyn, London.
4. **Balog, K., Svirčev, E., Lesjak, M., Orcic, D., Beara, I., Francišković, M., Simin, N. 2012.** Phenolic profiling of *Rumex* L. species by means of the LC-MS/MS. *Planta Medica* 78: 12 (Congress abstract).

Sorrel, the leaves of *Rumex acetosa*

Flowering plants (*Rumex acetosa*)

French sorrel (*Rumex scutatus*)

Red-veined sorrel (*Rumex sanguineus*)

243

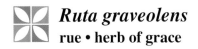

Ruta graveolens
rue • herb of grace

Ruta graveolens L. (Rutaceae); *wynruit* (Afrikaans); *taena adam* (Amharic); *chou cao* (Chinese); *wijnruit* (Dutch); *rue* (French); *Weinraute* (German); *ruta* (Italian); *ru* (Japanese); *duan aroda* (Malay); *arruda* (Portuguese); *ruda* (Spanish)

DESCRIPTION Fresh leaves are bluish green, gland-dotted along the margins, with an intensely bitter taste and an overpowering pungent smell (*graveolens* is Latin for "heavy smell"). The fruits are glandular capsules containing several small black seeds.

THE PLANT A strongly aromatic perennial herb with small yellow flowers. Rue is often confused with the closely similar Aleppo rue (*R. chalepensis*) but this species typically has fringed petals.

ORIGIN Rue is indigenous to southern Europe, while Aleppo rue has a wider distribution around the Mediterranean region and the Arabian Peninsula.[1] No distinction is traditionally made between the two species and both have been used in traditional medicine and to a much lesser extent as potherb and culinary herb. Apart from a biblical reference, it is mentioned by Pliny the Elder (AD 23–79),[2] Shakespeare, Milton (in *Paradise Lost*) and Swift (in *Gulliver's Travels*). Rue features prominently in Lithuanian and Ukrainian folklore and songs. It was used in church ceremonies in England and became known as "herb of grace". It is traditionally associated with regret or remorse.

CULTIVATION Both species are easy to propagate from seeds or cuttings. They are commonly grown in herb gardens, usually for decoration rather than culinary use. Plants grow best in full sun and react well to being cut back each year. It is grown on a small commercial scale as a spice (e.g. in Ethiopia) and also for the production of rue oil.[3]

HARVESTING Leaves or fruits are picked as required. Avoid skin contact because the herb may cause blistering of the skin when it is exposed to sunlight, due to the phototoxicity of furanocoumarins in the plant.

CULINARY USES Leaves are used sparingly as a potherb in Europe, North America, Malaysia and China. In Singapore, Chinese cooks boil it with green beans and in Cantonese cuisine it adds flavour to soups.[4] In eastern Europe, small quantities are used to flavour sausage meat, venison, marinades, cream cheeses and stuffings for meat and poultry. Rue oil is added as flavourant to many food products, including beverages, confectionery, puddings, frozen desserts and gelatins.[3] In Ethiopia, the fruits are used as a spice and fresh leaves are dipped into *buna* (the traditional black coffee), to which it gives, somewhat surprisingly, a delicious flavour. Italian *grappa della ruta* is bottled with a leafy twig of rue inside.

FLAVOUR COMPOUNDS The bitter taste is due to high levels of rutin, a common flavonoid glycoside (the rutinoside of quercetin). The dominant aroma-impact compound of both species is 2-undecanone (so-called rue ketone), which co-occurs with 2-nonanone, 2-dodecanone and several structurally related constituents.[4,5]

NOTES Rue oil is used in detergents, cosmetics, soaps and perfumes.[4]

2-undecanone (rue ketone)

2-nonanone

rutin (R=β-*D*-rutinoside)   2-dodecanone

1. **Mabberley, D.J. 2008.** *Mabberley's plant-book* (3rd ed.). Cambridge University Press, Cambridge.
2. **Farrel, K.T. 1999.** *Spices, condiments and seasonings*. Aspen Publishers, Gaithersburg, USA.
3. **Harborne, J.B., Baxter, H. 2001.** *Chemical dictionary of economic plants*. Wiley, New York.
4. **Hu, S.-Y. 2005.** *Food plants of China*. The Chinese University Press, Hong Kong.
5. **Mejri, J., Abderrabba, M., Mejri, M. 2010.** Chemical composition of the essential oil of *Ruta chalepensis* L.: Influence of drying, hydro-distillation duration and plant parts. *Industrial Crops and Products* 32: 671–673.

Rue, the leaves of *Ruta graveolens*

Coffee flavoured with rue

Fruits (*Ruta chalepensis*)

Fruits and seeds (*Ruta chalepensis*)

Aleppo rue (*Ruta chalepensis*)

Flowers and fruits (*Ruta graveolens*)

245

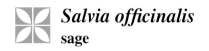

Salvia officinalis
sage

Salvia officinalis L. (Lamiaceae); *salie* (Afrikaans); *sauge* (French); *Echter Salbei*, *Gartensalbei* (German); *faskomilo* (Greek); *salvia* (Italian); *salvia officinal* (Spanish)

DESCRIPTION Sage leaves are typically silver-grey in colour, markedly rugose above and strongly net-veined below, with a bitter taste and a warm, balsamic and spicy aroma. Dried leaf is sold as whole leaf, ground, rubbed (crumbled) or sliced.

THE PLANT A perennial shrublet of up to 0.6 m (2 ft) high with attractive purplish blue flowers. Various colourful leaf forms are shown here, including tricolour sage ('Tricolor'), purple sage ('Purpurascens') and variegated sage ('Icterina').

ORIGIN Sage is indigenous to southern Europe and the eastern Mediterranean region.[1] It has an ongoing reputation as a medicinal plant (*salvia* means to save or heal) that featured in all of the classical Greek, Roman and early European herbals.[1,2]

CULTIVATION The easiest method of propagation is by layering (or by simply removing rooted side shoots). Seeds or cuttings are also an option. Sage dislikes acid soil but can withstand cold temperatures and short periods of drought. Plants can become untidy and need regular pruning or replacement. Commercial cultivation has spread to many parts of the world but the Dalmatian Islands have the reputation of producing the best quality sage and sage oil.[2]

HARVESTING Sage is picked by hand or mechanically, depending on the scale of the operation. Up to three cuts can be made per season but only in the pre-flowering phase.

CULINARY USES Sage is traditionally used to flavour roast meats, game, liver, poultry, stuffings, fish, marinades, omelettes, vegetables, salads, sauces, soups and some cheeses.[3] In Britain and the United States, it is an essential ingredient (with onions) of the stuffing for a roast chicken or turkey (and of roast pork, Derby cheese and Lincolnshire sausages). Italians use it frequently in fatty meat dishes (e.g. *osso buco*, *piccata* and *saltimbocca alla romana*), as well as Tuscan haricot beans (*fagioli*).[3] It is used in German ham and sausages, Greek and Middle Eastern roast mutton and some French sauces, white meats and vegetable soups.[3] Sage oil has many applications in the food industry.[2,4]

FLAVOUR COMPOUNDS The typical aromatic compounds of Dalmation sage and sage oil are α-thujone, β-thujone, camphor, 1,8-cineole, β-caryophyllene and limonene.[2,4] There are strong regional differences in the chemistry and flavour of sage.[2,5]

NOTES Greek sage (*S. fruticosa*) is closely related to common sage but can be distinguished by the lobed leaf bases. It is an adulterant of dried sage and has similar culinary uses. Clary sage (*S. sclarea*) is often seen in herb gardens but is rarely used in the kitchen, its main value being an essential oil that is used to flavour vermouths and liqueurs (apart from uses in perfumery). The colourful pineapple sage (*S. elegans*) with its attractive red flowers has similar culinary limitations despite its delicious fruity and floral odour. Its main use seems to be in garnishing and flavouring cold drinks and cocktails.

α-thujone   β-thujone   1,8-cineole

1. Mabberley, D.J. 2008. *Mabberley's plant-book* (3rd ed.). Cambridge University Press, Cambridge.
2. Farrel, K.T. 1999. *Spices, condiments and seasonings*. Aspen Publishers, Gaithersburg, USA.
3. Larousse. 1999. *The concise Larousse gastronomique*. Hamlyn, London.
4. Harborne, J.B., Baxter, H. 2001. *Chemical dictionary of economic plants*. Wiley, New York.
5. Perry, N.B., Anderson, R.E., Brennan, N.J., Douglas, M.H., Heaney, A.J., McGimpsey, J.A., Smallfield, B.M. 1999. Essential oils from Dalmatian sage (*Salvia officinalis* L.): variations among individuals, plant parts, seasons, and sites. *Journal of Agricultural and Food Chemistry* 47: 2048–2054.

Sage, the leaves of *Salvia officinalis*

Sage flowers (*Salvia officinalis*)

Clary sage (*Salvia sclarea*)

Tricolour sage

Purple sage

Variegated sage

Pineapple sage (*S. elegans*)

Sambucus nigra
elderflowers • elderberries

Sambucus nigra L. (Caprifoliaceae); *vlier* (Afrikaans); *sureau* (French); *Schwarzer Holunder* (German); *sambuco* (Italian); *sabugueiro* (Portuguese); *saúco, sabuco* (Spanish)

DESCRIPTION Elderflowers comprise the small white corollas (five fused petals with five attached stamens) that become dislodged from the flower heads when they are dried. The flowers have an aromatic, musky smell.

THE PLANT Elder is a shrub or small tree (up to 6 m or 20 ft) with compound leaves and small white flowers borne in flat-topped cymes. The fruits (elderberries) are three-seeded drupes (edible only when ripe and preferably cooked). American elder (*S. canadensis*), from the eastern parts of North America, is sometimes considered to be a subspecies of *S. nigra*.[1]

ORIGIN Indigenous to North Africa, Europe and Asia.[1] There are many superstitions and myths associated with elder.[1,2] There is renewed interest in elder as a crop.

CULTIVATION Selected cultivars of both species are grown in plantations, mainly in Europe, Canada and the United States, for the production of flowers and berries. Trees grow well in almost any soil (including chalky soil)[2] and prefer full sun. Propagation is from seeds or semi-hardwood cuttings taken in summer.[2]

HARVESTING Flowers and berries are wild-harvested or collected from commercial plantations. The inflorescences are picked on a dry, sunny day (to preserve the pollen)[2] and are carefully placed upside down to dry so that the flowers can be collected. The fruits are picked when black and fully ripe.

CULINARY USES The aromatic flowers are traditionally soaked in sugar water to make elderflower cordial, fermented to make elderflower wine (which can be still or sparkling) or distilled for alcoholic drinks (such as Italian sambuca and Swedish *fläderblomsnaps*)[2,3]. Flowers are also mixed into a light batter of eggs and flour and fried to make elderflower fritters.[3] Ripe elderberries, with seeds removed, are turned into elderberry wine, vinegar and chutney.[3] They are added to jams, jellies, apple pies (and in Portugal, port wine) to improve the colour and flavour. Pontack sauce[3], made from elderberries, vinegar and spices, is said to be named after a fashionable London tavern called "Pontack's Head", opened in 1666 by François-Auguste de Pontac, an eminent French cook and winemaker.

FLAVOUR COMPOUNDS The aroma of elderflowers is caused by an extremely complex mixture of volatile compounds.[4] The typical musky aroma of the flowers was found to be associated with a few major aroma-impact compounds, which included (–)-*cis*-rose oxide, nerol oxide, hotrienol and nonanal. However, the floral fragrance was found to be caused by linalool, α-terpineol, 4-methyl-3-penten-2-one, and (Z)-β-ocimene, and fruity odours by pentanal, heptanal, and β-damascenone. The presence of hexanal, hexanol, and (Z)-3-hexenol accounted for the fresh and grassy odours also associated with elderflowers and elderflower cordial.[4]

NOTES The pith of old stems are used to hold small botanical specimens for hand sectioning when doing anatomical studies.

(–)-*cis*-rose oxide   nerol oxide   hotrienol   nonanal

1. **Mabberley, D.J. 2008.** *Mabberley's plant-book* (3rd ed.). Cambridge University Press, Cambridge.
2. **McVicar, J. 2007.** *Jekka's complete herb book.* Kyle Cathie Limited, London.
3. **Vaughan, J.G., Geissler, C.A. 1997.** *The New Oxford book of food plants.* Oxford University Press, Oxford.
4. **Jørgensen, U., Hansen, M., Christensen, L.P., Jensen, K., Kaack, K. 2000.** Olfactory and quantitative analysis of aroma compounds in elder flower (*Sambucus nigra* L.) drink processed from five cultivars. *Journal of Agricultural and Food Chemistry* 48: 2376–2383.

Elderflowers and elderberries, from *Sambucus nigra*

Dried flowers (*Sambucus nigra*)

Elderflowers (*Sambucus nigra*)

Leaves and flowers (*Sambucus nigra*)

Elderberries (*Sambucus nigra*)

Sanguisorba minor
salad burnet

Sanguisorba minor Scop. (= *Poterium sanguisorba* L.) (Rosaceae); *pimpernel* (Afrikaans); *pimprenelle* (French); *Bibernelle*, *Pimpernell*, *Kleiner Wiesenknopf* (German); *pimpinella* (Italian)

DESCRIPTION The leaves are pinnately compound and fern-like – bright dark green above, paler below, with several pairs of toothed leaflets. They have a cucumber-like, nutty taste.

THE PLANT A small, perennial, rosette-forming herb with globose heads of tiny wind-pollinated florets. Salad burnet should not be confused with the much taller Eurasian great burnet (*Sanguisorba officinalis*) or the North American bloodroot (*S. canadensis*).

ORIGIN Indigenous to central and southern Europe.[1] Dried rhizomes and roots (and those of *S. officinalis*) have a long history in Europe and China as first aid treatment for wounds and to stop bleeding. The generic name is derived from *sanguis* (blood) and *sorbeo* (absorb). Salad burnet seems to be making a comeback as a popular indigenous culinary herb in Europe.

CULTIVATION In herb gardens, salad burnet is often seen as a decorative edge plant along paths. Seeds are sown in spring or plants can be multiplied by division of mature specimens. Almost any soil conditions are suitable, and the plants adapt to cold, drought, full sun or partial shade.[2] Regular but sparse feeding and watering will ensure that the plants do not become too vigorous and lose their flavour.[2] It is an excellent container plant for those who have no space for a herb garden.

HARVESTING The main advantage of salad burnet is that it is an evergreen perennial that provides a supply of fresh leaves almost throughout the year. Regular harvesting of leaves and cutting back of the flowering stalks as they emerge will ensure that the plants stay in a vegetative growth phase.[2] Leaves do not dry well and are best used fresh.

CULINARY USES Salad burnet can be used in much the same way as parsley – an attractive and edible garnish to decorate salads, soups and casseroles but also ideally suited as the main ingredient, in generous amounts. It can be added to cream cheeses and is used to make herb butter and herb vinegar.[3] Leaves can also be used in cold drinks with a result similar to that of borage.[3] French cooks use salad burnet in the same way as watercress – for salads, soups, cold sauces, marinades and omelettes.[3]

FLAVOUR COMPOUNDS The essential oil is dominated by linalool, accompanied by small amounts of β-caryophyllene.[4] Volatile compounds responsible for the refreshing green flavours include nonanal, (*E,E*)-farnesyl acetate, nonadecane and docosane.[4]

NOTES The styptic and antibiotic activity of *Sanguisorba* species can be ascribed to tannins, visible as red, blood-like exudate when the rhizomes are cut. North American bloodroot (*S. canadensis*) is traditionally used as an antiseptic, spasmolytic and expectorant, and as a war paint and natural dye.[5]

(*E,E*)-farnesyl acetate

linalool   β-caryophyllene   nonanal

nonadecane

docosane

1. **Mabberley, D.J. 2008.** *Mabberley's plant-book* (3rd ed.). Cambridge University Press, Cambridge.
2. **McVicar, J. 2007.** *Jekka's complete herb book*. Kyle Cathie Limited, London.
3. **Larousse. 1999.** *The concise Larousse gastronomique*. Hamlyn, London.
4. **Esmaeili, A., Masoudi, Sh., Masnabadi, N., Rustaiyan, A.H. 2010.** Chemical constituents of the essential oil of *Sanguisorba minor* Scop. leaves, from Iran. *Journal of Medicinal Plants* 35: 67–70.
5. **Harborne, J.B., Baxter, H. 2001.** *Chemical dictionary of economic plants*. Wiley, New York.

Salad burnet, the leaves of *Sanguisorba minor*

Leaf and flowers (*Sanguisorba minor*)

Plant (*Sanguisorba minor*)

Flowering plant (*Sanguisorba minor*)

Satureja hortensis
summer savory

Satureja hortensis L. (Lamiaceae); *bonekruid* (Afrikaans); *chubritsa* (Bulgarian); *bonenkruid* (Dutch); *sarriette* (French); *Bohnenkraut, Pfefferkraut* (German); *santoreggia* (Italian); *seibari* (Japanese); *sabori* (Korean); *segurelha-das-hortas* (Portuguese); *ajedrea, sabroso* (Spanish); *kyndel* (Swedish)

DESCRIPTION Summer savory is the fresh or dried stem tips with leaves. They are soft and succulent with a spicy, peppery taste and an aroma reminiscent of thyme and oregano but with a hint of mint.

THE PLANT An erect, much-branched annual herb with small, pink flowers. The closely related winter savory (*S. montana*) is a perennial shrublet with spreading branches bearing stiff, sharply pointed leaves and white flowers. Thyme-leaved savory (*Satureja thymbra*) is a shrublet with purple flowers arranged in clusters along terminal inflorescences.

ORIGIN Savory is indigenous to southeastern Europe (summer savory) or the Mediterranean region, including southern Europe and North Africa (winter savory).[1] Thyme-leaved savory originates from the southeastern parts of Europe (Balkans, Crete and Greece). Savory has a long history of use as medicine and also for its alleged aphrodisiac properties, as the generic name (derived from "satyr's herb") suggests.[2]

CULTIVATION Plants are easily grown from seeds sown in spring. The perennial *S. montana* and *S. thymbra* are more often propagated from cuttings. Summer savory and the other species will thrive in any fertile, well-drained soil that is regularly fertilized and watered.

HARVESTING The leafy stems tips are picked and used fresh, but dried summer savory is commercially available as a spice herb in some countries. Regular harvesting and pruning is beneficial to prevent flowering and prolong the production of new leaves.

CULINARY USES Fresh summer savory is traditionally used to flavour beans and bean dishes (pulses), especially broad beans.[2,3] It is an important culinary herb in Germany, France, the Balkan region (especially Bulgaria),[3] Georgia and also Canada. The German name *Pfefferkraut* reflects the peppery taste and the historical use as pepper substitute. Summer savory (less often winter savory) goes with salads, mushrooms, ragouts, grilled veal, loin of pork and roast lamb. The fresh or dried herb can be used to flavour pulses (peas, lentils, beans), stews, ragouts, soups, sauces, scrambled egg, poultry, pâtés and stuffings. Dried savory is used in spice mixtures (including commercial versions of *herbes de Provence* and *fines herbes*), The spice or its oil is used in food processing, especially for soup and gravy mixtures but also confectionery, processed vegetables, condiments, meat products and many more. Savory is used to flavour herbal liqueurs, vermouths and various soft cheeses made from sheep and goat's milk.

FLAVOUR COMPOUNDS The flavour is derived from carvacrol, *p*-cymene and γ-terpinene as main essential oil volatiles, but sometimes also with thymol and other minor constituents.[4,5] The presence of linalool and limonene, especially in *S. montana*, accounts for the slight spicy and lemony flavours.

NOTES Essential oil from both species is used in cosmetics, soaps and detergents.

carvacrol   *p*-cymene   γ-terpinene

1. **Mabberley, D.J. 2008.** *Mabberley's plant-book* (3rd ed.). Cambridge University Press, Cambridge.
2. **Larousse. 1999.** *The concise Larousse gastronomique*. Hamlyn, London.
3. **Gernot Katzer's Spice Pages** (http://gernot-katzers-spice-pages.com).
4. **Mihajilov-Krstev, T., Radnović, D., Kitić, D., Zlatković, B., Ristić, M., Branković, S. 2009.** Chemical composition and antimicrobial activity of *Satureja hortensis* L. essential oil. *Central European Journal of Biology* 4: 411–416.
5. **Novak, J., Bahoo, L., Mitteregger, U., Franz, C. 2006.** Composition of individual essential oil glands of savory (*Satureja hortensis* L., Lamiaceae) from Syria. *Flavour and Fragrance Journal* 21: 731–734.

Savory or summer savory, the leaves of *Satureja hortensis*

Flowers of winter savory (*Satureja montana*)

Winter savory plant (*Satureja montana*)

Summer savory (*Satureja hortensis*)

253

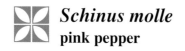 *Schinus molle*
pink pepper

Schinus molle L. (Anacardiaceae); *pienk peper* (Afrikaans); *xiao ru xiang* (Chinese); *roze peper* (Dutch); *poivre rose* (French); *Rosé-Pfeffer, Rosa Pfeffer* (German); *pimenta rosa, aroeira* (Portuguese); *arveira, pimiento de Brasil, terebino* (Spanish)

DESCRIPTION Pink pepper is the ripe fruits of *Schinus molle* or *S. terebinthifolius*, about 6 mm (¼ in.) in diameter, with a small hard seed surrounded by a brittle, bright red or pink outer wall. They have a pungent, resinous and spicy taste.

THE PLANT The pepper tree or Peruvian pepper tree (*Schinus molle*) is a hardy, long-lived tree of up to 15 m (ca. 50 ft) high with characteristic pendulous branches and leaves. The Brazilian pepper tree (*S. terebinthifolius*) is a smaller tree (10 m or ca. 32 ft) with non-drooping branches and broad leaflets. Both species are popular garden trees and have become invasive in warm regions.

ORIGIN *S. molle* is indigenous to the Peruvian Andes in South America, while *S. terebinthifolius* occurs naturally in Brazil.[1] Pink peppercorns are derived from both *S. molle* and *S. terebinthifolius* and their names seem to be used interchangeably, so that it is almost impossible to determine their botanical origin.

CULTIVATION Both species are easily grown from seeds and thrive under almost any conditions.

HARVESTING The fruits occur in large clusters and are hand-picked when fully ripe and dry.

CULINARY USES Pink pepper has become fashionable as a colourful additive to the pepper grinder, often in combination with black and white pepper. It adds to the complexity of the aroma and flavour of the ground pepper but is rarely used on its own. It may not be safe to ingest large amounts. The outer sweet part of the fruits can be fermented to produce a drink or can be boiled to make syrup.

The fruits of *Schinus molle* (*molli* in the local Quechua language) were once very popular in the Central Andean region for making a drink called *chicha*.[2]

FLAVOUR COMPOUNDS The main compound in the essential oil of the fruits of both species is often α-phellandrene, with smaller quantities of β-phellandrene, α-terpineol, α-pinene, β-pinene and *p*-cymene.[3,4] Cardanol, a known skin irritant, occurs in the essential oil of *S. terebinthifolius*.[5] The oil of this species appears to be variable but α- and β-pinene, δ-3-carene, limonene, α- and β-phellandrene, *p*-cymene and terpinolene are often reported as the main compounds.[3]

NOTES Most regions of the world have their own pungent "pepper" plants.[1] These are botanically and chemically mostly unrelated and include African Guinea pepper (*Xylopia aethiopica*), African pepper tree (*Warburgia salutaris*), West African Melegueta pepper (*Aframomum melegueta*), East Asian sansho or Chinese pepper (*Zanthoxylum piperitum*, *Z. simulans* and several others), South American mountain pepper (*Drimys piperita*), Tasmanian mountain pepper (*D. lanceolata*) and, of course, the well-known black pepper and its relatives, as well as the popular chilli peppers.

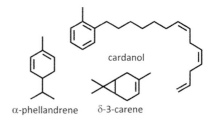

cardanol

α-phellandrene   δ-3-carene

1. **Mabberley, D.J. 2008.** *Mabberley's plant-book* (3rd ed.). Cambridge University Press, Cambridge.
2. **Goldstein, D.J., Coleman, R.C. 2004.** *Schinus molle* L. (Anacardiaceae): *Chicha* production in the Central Andes. *Economic Botany* 58: 523–529.
3. **Bendaoud, H., Romdhane, M., Souchard, J.P., Cazaux, S., Bouajila, J. 2010.** Chemical composition and anticancer and antioxidant activities of *Schinus molle* L. and *Schinus terebinthifolius* Raddi berries essential oils. *Journal of Food Science* 75: 466–472.
4. **Maffei, M., Chialva, F. 1990.** Essential oils from *Schinus molle* L. berries and leaves. *Flavour and Fragrance Journal* 5: 49–52.
5. **Stahl, E., Keller, K., Blinn, C. 1983.** Cardanol, a skin irritant in pink pepper. *Planta Medica* 48: 5–9.

Pink pepper, the fruits of *Schinus terebinthifolius* (left) and *S. molle* (right)

Pepper tree flowers and fruits (*Schinus molle*)

Brazilian pepper (*Schinus terebinthifolius*)

Brazilian pepper tree (*Schinus terebinthifolius*)

Sesamum indicum
sesame

Sesamum indicum L. (Pedaliaceae); *sesam* (Afrikaans); *sim sim* (Arabic); *hu ma* (Chinese); *sesamzaad* (Dutch); *sésame* (French); *gingli* (Hindi); *Sesam* (German); *sesamo* (Italian); *goma, shiro goma* (Japanese); *sésamo* (Portuguese, Spanish); *susam* (Turkish)

DESCRIPTION The ripe seeds are small (about 3 mm or ⅛ in. in length), variable in colour (mostly off-white but often black or various shades of brown) and have a delicious nutty taste and nut-like aroma, especially after being toasted to bring out the flavour. The seeds are sold plain or decorticated and toasted or untoasted.[1]

THE PLANT An erect annual (ca. 1.5 m or 5 ft) with pale pink or purple tubular flowers and oblong seed capsules.

ORIGIN Sesame is an ancient cultigen, thought to be of East African or Indian origin. It is one of the most ancient oil crops in the world, dating back to at least 3000 BC.[2,3] It features prominently in Middle Eastern archaeology[3] and Greek writings.

CULTIVATION Sesame is an important oilseed and food source, with an annual production exceeding 3 million tons. The main production areas are Myanmar, India, China and North African countries.[2] Sesame is exceptionally tolerant of drought (but not frost or waterlogging) and is grown from seeds in tropical, subtropical or temperate regions where no other crop will survive.

HARVESTING Capsules are gathered by hand as they progressively ripen along the stems, but for large-scale production mechanical harvesting is essential.[2]

CULINARY USES Sesame seeds are an important component of the diet in Middle Eastern countries, where they are eaten raw or roasted, but more often converted to tasty sauces and pastes. The best known of these is tahini, a thick paste that is widely used in North Africa and the Near East. It is made from ground sesame seeds, lemon juice, pepper, garlic and various other spices and has many culinary applications. It is mixed with ground chickpeas to make hummus, a popular Middle Eastern condiment served with savoury or sweet dishes, including salads, vegetables, meat and poultry dishes, and beans and peas. Numerous variants of tahini and hummus are used in Indian, African, Japanese, Chinese, Indonesian and Mexican cuisines. Culinary applications include Chinese sesame seed balls, Japanese *gomashio*, Mexican moles, Caribbean sesame seed brittle, Indian *til-patti*, *pinni*, *tilgul* and many more. Sesame seeds are often roasted to create a delicious nutty taste and are commonly sprinkled on bagels, hamburger buns, crackers, bread and cake.[1] Sesame seed oil is popular as cooking oil in Arabian, Chinese and Japanese cooking. Flour made from sesame seed is used for making bread, pancakes, biscuits and other confectionery items. The best-known example is halva, a sweet made from ground sesame seeds, to which ground almonds and sugar are added.

FLAVOUR COMPOUNDS Numerous volatile compounds give roasted sesame its delicious flavour.[4] Examples include 2-furfurylthiol (coffee-like), 4-hydroxy-2,5-dimethyl-3(2H)-furanone (caramel-like) and 2-methoxy-4-vinylphenol (apple-like), as well as sulphur compounds such as 3-methyl-2-buten-1-thiol (sulphurous, meaty flavour) and 4-mercapto-3-hexanone ("catty" flavour, also found in beer).[4]

NOTES Some people are allergic to sesame.

2-furfurylthiol

4-hydroxy-2,5-dimethyl-3(2H)-furanone

2-methoxy-4-vinylphenol

3-methyl-2-buten-1-thiol

4-mercapto-3-hexanone

1. Farrel, K.T. 1999. *Spices, condiments and seasonings*. Aspen Publishers, Gaithersburg, USA.
2. Nayar, N.M. 1995. Sesame. In: Smartt, J., Simmonds, N.W. (Eds), *Evolution of crop plants* (2nd ed.), pp. 404–407. Longman, London.
3. Zohary, D., Hopf, M. 2000. *Domestication of plants in the Old World* (3rd ed.). Clarendon Press, Oxford.
4. Tamura, H., Fujita, A., Steinhaus, M., Takahisa, E., Watanabe, H., Schieberle, P. 2010. Identification of novel aroma-active thiols in pan-roasted white sesame seeds. *Journal of Food and Agricultural Chemistry* 58: 7368–7375.

White and black sesame, the seeds of *Sesamum indicum*

Sesame is an ingredient of dukkah

Sesame seeds (white, brown and black)

Flower (*Sesamum indicum*)

Fruits (*Sesamum indicum*)

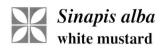 *Sinapis alba*
white mustard

Sinapis alba L. (Brassicaceae); *wit mosterd* (Afrikaans); *bai jie* (Chinese); *gele mosterd* (Dutch); *moutarde blanche* (French); *Weißer Senf* (German); *senape bianca* (Italian); *shiro garashi* (Japanese); *mostarda branca* (Portuguese); *mostaza blanca* (Spanish); *senap* (Swedish)

DESCRIPTION The small, spherical seeds are white or yellowish in some modern cultivars but wild forms originally had black or brown seeds so that seed colour alone is not a reliable way to identify white mustard.[1] Yellow-seeded brown mustards (called "orientals") add to the confusion. Unlike black and brown mustard, white mustard emits no odour when the seeds are crushed but the taste is more pungent and slightly bitter.[1]

THE PLANT A robust and leafy annual, easily identified when in fruit by the hairy capsules which have flat, wing-like tips. In black and brown mustard, the fruits are hairless and the tips are rounded and tapering.

ORIGIN White mustard spread from the eastern Mediterranean region to other parts of Europe during the Middle Ages.[2] Mustard has an interesting history in the Middle East and Europe, as one of the most important condiments of the ancient world for flavouring meat and fish. This tradition has been perpetuated in France and England since the time of the Romans but the style of French and English mustards differ. French mustards are sold in the form of a paste made with verjuice (acidic juice of unripe grapes) and white wine (Dijon mustard), with wine vinegar (Orleans mustard) or with grape must (Bordeaux mustard).[3] Another type is the colourful Meaux mustard, made with vinegar and crushed mustard seeds of different colours.[3] English mustard is a mixture of black and white mustard, either sold as a fine powder, in which turmeric is an ingredient (the mustard is prepared in the kitchen as required), or the powder is mixed with water or vinegar and presented as a ready-to-use paste. Italian Cremona mustard is made from macerated fruits mixed with a sweet and sour mustard sauce.[3]

CULTIVATION White mustard is grown from seeds as an annual crop. Many different cultivars are available, including forms developed for fodder and oil production. The main producers are the United States, Canada, the United Kingdom and parts of Europe.

HARVESTING Seeds are mechanically harvested when fully ripe.

CULINARY USES Mustard is commonly used as a condiment with hot or cold meats, sausages and other charcuterie items and is often served alongside tomato sauce (ketchup) and vinegar in fast food outlets. It is used as a flavour ingredient in many dishes and numerous sauces (e.g. vinaigrette, mayonnaise, Cambridge sauce and barbeque sauce).

FLAVOUR COMPOUNDS The main mustard-oil glycoside in white mustard is not sinigrin but sinalbin (4-hydroxybenzyl-glucosinolate), which is hydrolysed by the enzyme myrosinase to produce the pungent *p*-hydroxybenzyl isothiocyanate.[4] The addition of liquids activates the enzyme. Water and beer give a hot taste, wine gives a pungent, spicy flavour and water or vinegar gives a mild flavour (English mustard).[1]

NOTES Powdered mustard seeds have many uses in the food industry as a flavour enhancer, binder, emulsifier and bulking agent.[1]

p-hydroxybenzyl
isothiocyanate

sinalbin
(4-hydroxbenzyl-glucosinolate)

1. **Farrel, K.T. 1999.** *Spices, condiments and seasonings.* Aspen Publishers, Gaithersburg, USA.
2. **Hemingway, J.S. 1995.** Mustards. In: Smartt, J., Simmonds, N.W. (Eds), *Evolution of crop plants* (2nd ed.), pp. 82–86. Longman, London.
3. **Larousse. 1999.** *The concise Larousse gastronomique.* Hamlyn, London.
4. **Choubdar, N., Li, S., Holley, R.A. 2010.** Supercritical fluid chromatography of myrosinase reaction products in ground yellow mustard seed oil. *Journal of Food Science* 75: 341–345.

White mustard, the seeds of *Sinapis alba*

White, brown and black mustard

Flowering plant (*Sinapis alba*)

Flowers and young fruits (*Sinapis alba*)

Smyrnium olusatrum
alexanders • black lovage

Smyrnium olusatrum L. (Apiaceae); *swart lavas* (Afrikaans); *maceron* (French); *Schwarze Gelbdolde*, *Schwarzer Liebstöckel* (German); *salsa de cavalo, apio dos cavalos* (Portuguese); *apio caballar, olosatro* (Spanish)

DESCRIPTION The leaves are large, pinnately compound and bright green with broad, serrated leaflets. They have a sharp flavour and fragrance reminiscent of myrrh and celery (hence the generic name).

THE PLANT A robust biennial herb of up to 1.5 m (ca. 5 ft) high, growing from a thick taproot and superficially similar to angelica or celery. When it flowers in April to June of the second year, the difference becomes more obvious. The umbels are flattened and bear greenish yellow flowers. The fruits are typical dry schizocarps that split into two mericarps – about 6 mm (¼ in.) long, and black in colour when ripe.

ORIGIN The plant is indigenous to the eastern Mediterranean region, where it grows on cliffs and in marshy areas near the sea.[1,2] The Greeks called it *hipposelinon* which means "horse celery". It became naturalized in Britain, the Netherlands and the Iberian Peninsula after it was introduced by the Romans as a food plant.[3] The name alexanders is said to reflect the popularity of the herb in the time of Alexander the Great (fourth century BC). It was commonly grown in monastery gardens in the Middle Ages and in kitchen gardens (e.g. at Versailles in France). The roots, leaves and stems were eaten as vegetable and culinary herb in much the same way as celery and parsnips, with which it has been replaced around the 18th century.[3,4]

CULTIVATION Plants are easily propagated from seeds, sown in autumn and planted out in early spring.[3] It is a forgotten culinary herb, no longer grown commercially to any extent, but is still commonly seen in herb gardens in Europe.

HARVESTING The stems and leaf stalks are cut in late spring, before the plants flower and the stems become hollow and fibrous. They are peeled in much the same way as rhubarb.[3]

CULINARY USES Young leaves and stems can be added to soups, stews and salads.[3] They can be used on their own, boiled briefly for six to eight minutes and served with butter and black pepper.[3] The flavour is similar to that of celery but stronger and more pungent.[3] The flower buds are also edible.[1,4]

FLAVOUR COMPOUNDS The roots and aboveground parts contain an essential oil rich in furanosesquiterpenoids, of which isofuranodiene is the main constituent (up to 45%).[5] The compound is converted to curzerene when heat is applied during steam distillation.[5] The oil also contains β-myrcene, β-phellandrene and β-caryophyllene.

NOTES Skirret (*Sium sisarum*) and baldmoney (*Meum athamanticum*) are two other umbelliferous herbs that are no longer popular but which were once of considerable importance as root crops in Britain and other parts of Europe, where they were used in much the same way as parsnips. The aboveground parts served as culinary herbs and flavourants for the same purposes as celery does today.

isofuranodiene   curzerene

1. **Mabberley, D.J. 2008.** *Mabberley's plant-book* (3rd ed.). Cambridge University Press, Cambridge.
2. **Randall, R.E. 2003.** *Smyrnium olusatrum* L. *Journal of Ecology* 91: 325–340.
3. **Phillips, R., Foy, N. 1992.** *Herbs*. Pan Books, London.
4. **Kiple, K.F., Ornelas, K.C. (Eds). 2000.** *The Cambridge world history of food*. Cambridge University Press, Cambridge.
5. **Maggi, F., Barboni, L., Papa, F., Caprioli, G., Ricciutelli, M., Sagratini, G., Vittori, S. 2012.** A forgotten vegetable (*Smyrnium olusatrum* L., Apiaceae) as a rich source of isofuranodiene. *Food Chemistry* 135: 2852–2862.

Alexanders or black lovage (*Smyrnium olusatrum*)

Skirret (*Sium sisarum*)

Plant in flower (*Smyrnium olusatrum*)

Baldmoney (*Meum athamanticum*)

Spilanthes acmella
spilanthes • pará cress

Spilanthes acmella (L.) Murr. [= Spilanthes oleracea L., = Acmella oleracea (L.) R.K. Jansen] (Asteraceae); spilantus (Afrikaans); qian ri ju (Chinese); cresson de Para, spilanthe des potagers (French); Parakresse (German); spilante (Italian); agrião do Brasil, agrião do Pará (Portuguese)

DESCRIPTION Fresh leaves are soft, often flushed with purple and have a sweetish taste and a peculiar mouth-tingling and mild anaesthetic effect when eaten.

THE PLANT A weedy annual or short-lived perennial of up about 0.3 m (1 ft) in height, with solitary, yellow and maroon-coloured flower heads. A specific cultivar has been selected or developed in Brazil and is often referred to as Brazilian cress. This plant has been called S. acmella var. oleracea or S. acmella 'Oleracea' (it was formerly regarded as a separate species, S. oleracea).[1,2] An alternative classification is in the genus Acmella, as Acmella oleracea.[3] It is still commonly referred to in the literature as Spilanthes, the name under which is has become well known.[2]

ORIGIN The plant occurs as a cultigen in South America and is believed to be of Peruvian or Brazilian origin, developed from a wild relative, S. alba.[3] It has become a popular culinary herb in almost all tropical parts of the world, including Africa, India, Southeast Asia and China.

CULTIVATION Plants are propagated from seeds sown in spring. It is an easy crop to grow with no particular requirements except that it is not adapted to cold.

HARVESTING Leaves are picked by hand as required. On Madagascar it is sold as a vegetable on fresh produce markets.[2]

CULINARY USES Young leaves are shredded and added to salads, to which they give a tingling and mouthwatering flavour. The plant is used on its own as a potherb and is often added to soups and stews. The name cress can be misleading because pará cress is unrelated to several other culinary herbs from the mustard family (Brassicaceae), such as garden cress (Lepidium sativum), land cress (Barbarea verna) and watercress (Nasturtium officinale). The pungency of these plants comes from mustard oils, which are chemically unrelated to the compounds in spilanthes.

FLAVOUR COMPOUNDS The tingling and anaesthetic sensation caused by spilanthes is due to spilanthol, the most abundant of at least eight N-alkylamides that have been identified in the leaves and especially in the mature flower heads of the plant, where they occur in their biggest concentration.[4,5] Alkylamides are also found in sansho or Chinese pepper (Zanthoxylum species), used in East and Southeast Asia for a similar sensation in the mouth. The flower heads of spilanthes are referred to as "buzz buds" because of their powerful effect on taste buds. It has been shown that spilanthol can permeate the mucosa of the mouth.[5] Experiments have been done to see how the interesting tingling effect, comparable perhaps to that of sherbet but much stronger, can be utilized in soft drinks and sweets.

NOTES Spilanthes is traditionally used as a toothache remedy and to stimulate the flow of saliva. It has interesting antibiotic effects and is poisonous to invertebrates but not to warm-blooded animals.

spilanthol

1. Mabberley, D.J. 2008. Mabberley's plant-book (3rd ed.). Cambridge University Press, Cambridge.
2. Roemantyo, 1993. Spilanthes Jacquin. In: Siemonsma, J.S., Kasem, P. (Eds). Plant Resources of South-East Asia No 8. Vegetables. pp. 264–266. Pudoc Scientific Publishers, Wageningen, Netherlands.
3. Jansen, R.K. 1985. The systematics of Acmella (Asteraceae-Heliantheae). Systematic Botany Monographs 8. 115 pp.
4. Ramsewak, R.S., Erickson, A.J., Nair, M.G., 1999. Bioactive N-isobutylamides from the flower buds of Spilanthes acmella. Phytochemistry 51: 729–732.
5. Boonen, J., Baert, B., Burvenich, C., Blondeel, P., De Saeger, S., De Spiegeleer, B. 2010. LC-MS profiling of N-alkylamides in Spilanthes acmella extract and the transmucosal behaviour of its main bio-active spilanthol. Journal of Pharmaceutical and Biomedical Analysis 53: 243–249.

Spilanthes, the dried inflorescences of *Spilanthes acmella*

Flowering plant (*Spilanthes acmella*)

263

Stevia rebaudiana
stevia • sugar-leaf

Stevia rebaudiana (Bertoni) Bertoni (Asteraceae); *stevia* (Afrikaans); *stevia* (French); *Stevia* (German); *stevia* (Italian); *stevia* (Spanish)

DESCRIPTION Fresh leaves (oblong-elliptic and toothed along the upper margins) or the dried and powdered leaves have an intensely sweet taste with a hint of bittersweet liquorice in the aftertaste.

THE PLANT A perennial herb with erect, somewhat woody branches (up to 0.6 m or 2 ft) bearing small white florets arranged in few-flowered heads at the tips of the stems. The fruits are small, one-seeded, wind-dispersed achenes.

ORIGIN Stevia is indigenous to South America (Paraguay).[1] It is called *caa-ehe* and is traditionally used in Paraguay as a sweetener for local teas and other beverages.[1,2] In the early 1970s, stevia was developed in Japan as a natural non-sugar sweetener because of safety concerns about artificial sweeteners such as cyclamate and saccharin. It is now produced on an industrial scale (mainly in China) and several countries have recently permitted the use of stevia as sweetener, based on the long track record of safety in Japan.

CULTIVATION Seeds do not germinate easily, so that root or stem cuttings (taken in early summer) are usually the best option for propagation.[3] The plants are relatively easy to grow in rich, well-drained soil but do not tolerate shade, drought or frost.

HARVESTING For home use, the leaves can be picked and used fresh or dried. On a commercial scale, the leafy twigs are harvested mechanically, dried and processed. The sweet compounds are extracted with water and purified or crystallized using various solvent systems and nanofiltration.

CULINARY USES The fresh leaves, dried and powdered leaves or standardized leaf extracts are now widely used as natural sweeteners. Fresh leaves are said to be 30 to 45 times sweeter than sucrose, while the pure active compounds are 300 times sweeter than sucrose.[2,3] Branded sweeteners based on stevia are now used on a large scale in carbonated soft drinks.

FLAVOUR COMPOUNDS The sweet taste is due to diterpenoid glycosides, mainly stevioside and rebaudioside A.[4] Stevioside is the main compound but it is slightly less sweet and a little more bitter-tasting than rebaudioside A, so that the latter is preferred for use in commercial sweeteners.[4] Stevia adds little or no calories to the diet, because glucose is released from the steviol glycosides only when they reach the colon, where the sugars are not absorbed but metabolised by bacteria.

NOTES Other natural sweeteners include monatin from the South African *Sclerochiton illicifolius* (Acanthaceae) and thaumatin from the West African *Thaumatococcus daniellii* (Marantaceae).

stevioside

1. **Mabberley, D.J. 2008.** *Mabberley's plant-book* (3rd ed.). Cambridge University Press, Cambridge.
2. **Soejarto, D.D., Compadre, C.M., Medon, P.J., Kamath, S.K., Kinghorn, A.D. 1983.** Potential sweetening agents of plant origin. II. Field search for sweet-tasting *Stevia* species. *Economic Botany* 37: 71–79.
3. **Carneiro, J.W.P., Muniz, A.S., Guedes, T.A. 1997.** Greenhouse bedding plant production of *Stevia rebaudiana* (Bert.) Bertoni. *Canadian Journal of Plant Science* 77: 473–474.
4. **Wölwer-Rieck, U. 2012.** The leaves of *Stevia rebaudiana* (Bertoni), their constituents and the analyses thereof: A review. *Journal of Agricultural and Food Chemistry* 60: 886–895.

Sugar-leaf or stevia, the fresh or dried leaves of *Stevia rebaudiana*

Flowers (*Stevia rebaudiana*)

Flowering plant (*Stevia rebaudiana*)

Leaves and flowers (*Stevia rebaudiana*)

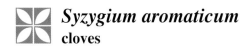

Syzygium aromaticum
cloves

Syzygium aromaticum (L.) Merr. & Perry [= *Eugenia caryophyllus* (C. Spreng.) Bull. & Harr.] (Myrtaceae); *naeltjies* (Afrikaans); *mu ding xiang* (Chinese); *kruidnagel* (Dutch); *clous de girofle* (French); *Gewürznelke* (German); *cingkeh* (Indonesian); *chiodi di garofano* (Italian); *kuroobu, shouji* (Japanese); *bunga cingkeh* (Malay); *clavero, clavo* (Spanish); *garn ploo* (Thai); *hanh con* (Vietnamese)

DESCRIPTION Cloves are unopened, sun-dried flower buds – dark brown, hard in texture and about 12 mm (½ in.) long, with the unopened petals forming a round head. The aroma is described as intensely spicy, woody, musty, fruity and peppery.[1] The flavour is warming, sharp and burning, spicy, fruity, astringent and somewhat bitter with a numbing effect.[1]

THE PLANT An evergreen, medium-sized tree with glossy leaves and small, white flowers bearing numerous stamens.[2]

ORIGIN Indigenous to the North Molucca Islands in Indonesia, the original "Spice Islands" of the colonial era and the associated spice trade.[2] Cultivation started more than 2 000 years ago and the Chinese traded in cloves around 200 BC.[2] Cloves first reached Europe by AD 176 but only became popular during the colonial era. The Portuguese first took control of the clove trade but were displaced by the Dutch in the early 17th century, who established a monopoly by ordering the destruction of clove trees on all but a few islands. The monopoly was broken by the French, who took clove seedlings from the Moluccas to their colonies in Réunion and Mauritius around 1770, from where it reached Madagascar and Zanzibar, and later also the Caribbean. In 1932, the Dutch secretly reintroduced cloves from Zanzibar to Indonesia because smokers of *kretek* cigarettes preferred Zanzibar cloves. These cigarettes are the basis of a very large industry in Southeast Asia. They are made with tobacco and contain 30 to 40% shredded cloves as the second main ingredient. Their popularity has led to Indonesia becoming the main producer of cloves, overtaking Tanzania (Zanzibar) and the Malagasy Republic.[2]

CULTIVATION Trees are grown from seeds or seedlings gathered in existing plantations. Warm and tropical conditions are required.

HARVESTING Trees take up to 10 years to reach reproductive maturity but can be harvested for up to 60 years or more.[2] Commercial harvesting of the clusters of unopened flowers is done by hand. The stalks are removed and the buds are spread out in the sun to dry.

CULINARY USES Cloves are an important spice and are usually included as one of the essential components of spice mixtures.[3] Its main applications in the kitchen are for meat dishes, marinades, gherkins, pickles, pears in red wine, baked apples, mulled wines and some sweet pastries.[3] When seasoning meat, less than 0.6 g (0.02 oz) is used per 45 kg (ca. 100 lbs).[1] Van der Hum, a traditional South African liqueur dating back to the 19th century, is made with tangerine peel and cloves as main flavour ingredients. It is said to be named after a Dutch ship's captain who enjoyed the drink to the point of distraction. Cloves and clove oil have a wide range of applications in the food industry.[1,4]

FLAVOUR COMPOUNDS The main flavour compound in cloves is eugenol, with much smaller quantities of β-caryophyllene, eugenol acetate and other minor volatiles.[1,4] The oil yield of clove buds is about 17% (93% eugenol), that of the stems 6% (83% eugenol) and leaves 2% (80% eugenol).[1] Eugenol also occurs in cinnamon leaves and allspice.

NOTES Clove oil is used in dentistry as an analgesic and disinfectant.

1. **Farrel, K.T. 1999.** *Spices, condiments and seasonings.* Aspen Publishers, Gaithersburg, USA.
2. **Bermawie, N., Pool, P.A. 1995.** Clove. In: Smartt, J., Simmonds, N.W. (Eds), *Evolution of crop plants* (2nd ed.), pp. 375–379. Longman, London.
3. **Larousse. 1999.** *The concise Larousse gastronomique.* Hamlyn, London.
4. **Harborne, J.B., Baxter, H. 2001.** *Chemical dictionary of economic plants.* Wiley, New York.

Cloves, the dried flower buds of *Syzygium aromaticum*

Flower buds of *Syzygium aromaticum*

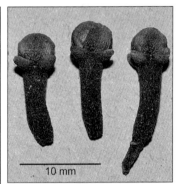

Cloves

Van der Hum liqueur, flavoured with tangerines and cloves

Tamarindus indica
tamarind

Tamarindus indica L. (Fabaceae); *tamarinde, suurdadel* (Afrikaans); *tamar hindi* (Arabic); *suan dou* (Chinese); *tamarin* (French); *Tamarinde, Sauerdattel* (German); *ambli* (Hindi); *tamarindo* (Italian, Japanese); *asam* (Malay); *tamarindeiro* (Portuguese); *tamaríndo* (Spanish); *ma khaam* (Thai); *me* (Vietnam)

DESCRIPTION Tamarind spice is the fleshy, semi-dried, reddish-brown inner wall of the fruits. It is edible and has a delicious sweet and sour taste, with an acidic, fruity fragrance.

THE PLANT A large tree with compound leaves, white and yellow flowers and oblong, pale brown pods.

ORIGIN Indigenous to North Africa, and still grows in parts of Sudan. The tree reached India such a long time ago that it was assumed to be indigenous there. From India it reached the Middle East, hence the Arabic name tamarind, from *tamar* (date) and *hindi* (from India). It also spread eastwards to Malaysia and Indonesia, where it was incorporated in the local cuisines. More recently, the Spanish introduced tamarind to the New World, where it became very popular, especially in Mexico, Puerto Rico and other South American countries. Tamarind is today found in practically all warm regions of the world. The main producers are Thailand, India and Puerto Rico.

CULTIVATION Trees are easily propagated from seeds (that have to be scarified) and thrive in a warm, frost-free climate.

HARVESTING Ripe pods are picked by hand and processed to separate the pulp from the non-fleshy outer skin and to remove the seeds. A large tree can yield about 90 kg of purified pulp.

CULINARY USES The delicious sweet-sour taste of tamarind is used worldwide as an acidifying agent in cooking, often dissolved in water. It is used in sour curries, sweet and sour dishes, lentil soup, chutneys, condiments (e.g. Worcestershire sauce and HP sauce), spice mixtures (e.g. fish curry masala), traditional cold drinks (e.g. Indian tamarind water or *impli panni*), commercial cold drinks, carbonated drinks, jams, sorbets, and fruit and vegetable purées. Commercial tamarind products available in Mexico and Puerto Rico include bottled juice, fruit pulp, jams and sweets.

FLAVOUR COMPOUNDS The sour taste of the fruit pulp is due to tartaric acid, which is also one of the main acids in wine. The fruit pulp contains 10% tartaric acid and 30–40% sugars, hence the distinctive sweet-sour taste. The main volatile compounds in the fruit pulp are 2-phenylacetaldehyde, 2-furfural and hexadecanoic acid (palmitic acid). The aroma of 2-phenylacetaldehyde, also found in flowers and in chocolate, is described as sweet, honey, rose, green and grassy.

NOTES Along the west coast of India, the sun-dried fruit wall of the indigenous kokum tree (*Garcinia indica*) is sometimes the preferred substitute for tamarind. The blackish purple product, known as *kokam* or *aamsul*, is used in sour curries, fish curries, lentil or dhal soups and other dishes. Tamarind pods are superficially similar to the edible and sweet-tasting carob fruits (*Ceratonia siliqua*) which are commonly used in eastern Mediterranean and western Asian cuisines. It is best known in Western countries as a chocolate substitute and main ingredient of carob candy bars, especially popular with people who want to avoid caffeine.

tartaric acid

phenylacetaldehyde

2-furfural

hexadecanoic acid (palmitic acid)

1. **Morton, J. 1987.** Tamarind. pp. 115–121. In: *Fruits of warm climates.* Julia F. Morton, Miami, Florida.
2. **Hutton, W. 1997.** *Tropical herbs and spices.* Periplus Editions, Singapore.
3. **De Caluwé, E., Halamová, K., Van Damme, P., 2010.** *Tamarindus indica* L. – A review of traditional uses, phytochemistry and pharmacology. *Afrika Focus* 23: 53–83.
4. **Pino, J.A., Marbot, R., Vazquez, C. 2004.** Volatile components of tamarind (*Tamarindus indica* L.) grown in Cuba. *Journal of Essential Oil Research* 16: 318–320.

Tamarind, the dried fruit pulp of *Tamarindus indica*

Leaves, flower and fruit (*T. indica*)

Fruit and seeds of tamarind (*T. indica*)

10 mm

Carob fruits (*Ceratonia siliqua*)

Kokum fruit (*Garcinia indica*)

Taraxacum officinale
dandelion

Taraxacum officinale Weber ex Wigg. (Asteraceae); *perdeblom* (Afrikaans); *pu gong ying* (Chinese); *pissenlit* (French); *Löwenzahn* (German); *taraxaco* (Italian); *diente de león* (Spanish)

DESCRIPTION Freshly picked leaves are bright green, oblong in shape, irregularly toothed along the margin and have milky latex exuding from the midrib. They have a bitter, chicory-like flavour.

THE PLANT A common, weedy herb with a rosette of oblong, markedly toothed leaves and solitary, yellow flower heads borne on slender, unbranched, hollow stalks. The small fruits are wind-dispersed. Dandelion is considered to represent several hundreds of cryptic species, but they are usually collectively referred to as *T. officinale* in the broad sense.[1] Several relatives of the Asteraceae, all with milky latex, are used as salad herbs. These include lettuce, endive and sow thistle (see section on SALAD HERBS on page 22).

ORIGIN Originally Europe and Asia, but now a cosmopolitan weed, present in the United States, Australia and all other temperate parts of the world.[1] The roots were once dried and roasted to make a caffeine-free coffee substitute in much the same way as chicory drinks are made from the roots of *Cichorium intybus*.[2,3] The leaves have probably been eaten as a potherb and wild salad for centuries.[3] The popularity of dandelion is reflected in the large number of English common names that have been recorded, often referring to its alleged diuretic activity.

CULTIVATION Dandelion is usually propagated by seeds, sown in pots in early spring, but it can easily be grown from root cuttings as well. Flower heads should be removed to prevent the seeds from spreading all over the garden. Most people would be more interested in learning how to contain or get rid of dandelion than how to grow it! Wild-harvesting is possible in rural areas, but modern cultivars are available to grow as salad herbs.

These have larger leaves with margins variously crisped or curled.[2] The leaves may be blanched by covering the plants with straw or plastic sheeting.[2,3] Dandelion is grown commercially to a limited extent in southern Europe (France and Italy).[3]

HARVESTING Leaves can be picked throughout the season but those harvested before flowering are considered to be the best.[4]

CULINARY USES Young leaves are eaten raw in fresh salads and are appreciated for their refreshing bitter taste (similar to lettuce or endive). In French cuisine, dandelion salad often contains garlic-flavoured croutons, pieces of bacon, walnuts and hard-boiled egg (e.g. *salade de groin d'âne*).[4] Young leaves are stir-fried in China. Dandelion also has other culinary uses. The leaves can be eaten as a green vegetable and cooked like spinach.[4] Open flower heads are traditionally dried and used to make dandelion wine.[2,3]

FLAVOUR COMPOUNDS The bitter taste of dandelion leaves (and roots) is due to the presence of sesquiterpenes (mainly taraxinic acid β-D-glucopyranoside and 11,13-dihydro taraxinic acid β-D-glucopyranoside), as well as *p*-hydroxy phenyl acetic acid and β-sitosterol.[5]

NOTES The typical toothed leaf margins probably gave rise to the English name "dandelion", derived from the French *dent-de-lion* ("lion's tooth").[4]

taraxinic acid β-D-glucopyranoside

11,13-dihydro taraxinic acid β-D-glucopyranoside

1. **Mabberley, D.J. 2008.** *Mabberley's plant-book* (3rd ed.). Cambridge University Press, Cambridge.
2. **Vaughan, J.G., Geissler, C.A. 1997.** *The New Oxford Book of Food Plants.* Oxford University Press, Oxford.
3. **Kiple, K.F., Ornelas, K.C. (Eds). 2000.** *The Cambridge world history of food.* Cambridge University Press, Cambridge.
4. **Larousse. 1999.** *The concise Larousse gastronomique.* Hamlyn, London.
5. **Kuusi, T., Pyysalo, H., Autio, K., 1985.** The bitterness properties of dandelion. II. Chemical investigations. *Lebensmittel-Wissenschaft und Technologie* 18: 347–349.

Dandelion, the fresh leaves of *Taraxacum officinale*

Flowering plant (*Taraxacum officinale*)

Flowering plant (*Taraxacum officinale*)

Leaf and inflorescence (*Taraxacum officinale*)

271

Theobroma cacao
cacao

Theobroma cacao L. (Sterculiaceae); *kakao* (Afrikaans); *cacao* (French); *Kakao* (German); *cacao* (Italian); *cacao real* (Spanish)

DESCRIPTION Ripe seeds (cacao beans) are about 25 mm (or 1 in.) long, with a reddish brown seed coat. They have a bitter taste but the typical chocolate aroma develops only after they are fermented and roasted.

THE PLANT An evergreen tree of about 6 m (ca. 20 ft) high with large, simple leaves and large ovoid fruits formed directly on the trunk. They are green, yellow, red or purplish when ripe and contain 20 to 60 seeds surrounded by a fleshy white pulp.[1]

ORIGIN Central America (upper Amazon basin).[1,2] It has been used by the Maya and Aztec people to make ritual drinks for at least 2 000 years.[1] With sugar, vanilla and milk added, chocolate became fashionable in Europe, long before coffee and tea. Cacao powder and slab chocolate are modern industrial inventions associated with well-known names such as John Cadbury, Henri Nestlé, Rudolphe Lindt, James Baker and Milton Hershey. Chocolate is of major economic importance – the consumption in Switzerland is 9.6 kg (more than 21 lb) per person per year.[2]

CULTIVATION Cacao trees are grown from seeds and require humid tropical conditions.[1] The three main types of cultivars have flattened seeds (common *Forastero* type), rounded seeds (high quality but rare *Criollo* type) or intermediate, being hybrids (*Trinitario* type).[1] About two-thirds of all cacao beans are produced in West Africa (Ivory Coast) and most of the rest in Central America and Southeast Asia.[1,2]

HARVESTING Fruits are hand-picked when ripe and then subjected to a process of fermentation (oxidation), resulting in the development of the important aroma compounds inside the seeds. The processed seeds are washed, dried and packed, ready for the specialists: first the chocolate maker (to create chocolate couverture) and then the chocolatier (to create chocolate bars and chocolate truffles).

CULINARY USES The seeds ("cacao beans") contain about 50% fat (known as cocoa butter) and 50% cocoa solids. The seeds are roasted, pulverized and half of the fat is extracted, while the remaining material is dried and powdered to form cacao powder. Chocolate is made from cacao powder, milk powder and sugar. Cocoa drinks are made from chocolate or cacao powder mixed with water or milk. Chocolate has numerous applications in desserts, confectionery, sweets and truffles. In recent years, bitter chocolate has become fashionable as ingredient of savoury sauces, including chilli-based sauces.

FLAVOUR COMPOUNDS Bitter-tasting alkaloids give the stimulating effect to chocolate, with theobromine as main compound (2–10% or more), accompanied by much lower levels of caffeine (often 0.2%) and phenethylamine. The chocolate flavour is due to at least 24 main aroma-active volatiles, such as 3-methylbutanal (malty) and phenylacetaldehyde (honey-like).

NOTES Chocolate-making is a complicated process and has an interesting history.[4,5]

theobromine   phenethylamine   phenylacetaldehyde   3-methylbutanal

1. Kennedy, A.J. 1995. Cacao. In: Smartt, J., Simmonds, N.W. (Eds), *Evolution of crop plants* (2nd ed.), pp. 472–475. Longman, London.
2. Mabberley, D.J. 2008. *Mabberley's plant-book* (3rd ed.). Cambridge University Press, Cambridge.
3. Schnermann, P., Schieberle, P. 1997. Evaluation of key odorants in milk chocolate and cocoa mass by aroma extract dilution analyses. *Journal of Agricultural and Food Chemistry* 45: 867–872.
4. Bourin, J., Feltwell, J., Bailleux, N., Labanne, P., Perraud, O. 2005. *The book of chocolate*. Flammarion-Pere Castor, Paris.
5. Coe, S.D., Coe, M.D. 1996. *The true history of chocolate*. Thames and Hudson, London.

Cacao beans, the seeds of *Theobroma cacao*

Cacao powder

Fruit and flower (*Theobroma cacao*)

Ripe fruits (*Theobroma cacao*)

Thymus vulgaris
common thyme • garden thyme

Thymus vulgaris L. (Lamiaceae); *tiemie* (Afrikaans); *thym* (French); *Gartenthymian* (German); *timo* (Italian); *tomillo* (Spanish)

DESCRIPTION Thyme leaves are very small (about 6 mm / ¼ in. long), dark green above, pale below and typically with the margins rolled in. They have a sharp, warm and spicy taste and a strong aromatic odour.

THE PLANT A perennial shrublet with slender stems, leaves in opposite pairs and tiny pale mauve flowers (rarely white or purple). Several hundred species and cultivars are grown as culinary herbs and ornamental plants. Amongst the best known are lemon thyme (*T.* ×*citriodora*, a hybrid between *T. pulegioides* and *T. vulgaris*), wild thyme (*T. serpyllum*), caraway thyme (*T. herba-barona*), Pennsylvanian Dutch thyme (*T. pulegioides*) and Spanish thyme (*T. zygis*).[1] Less well known are Ethiopian thyme or *tosign* (*T. schimperi*) and Mongolian thyme or *di jiao ye* (*T. mongolicus*). Lemon thyme is usually a smaller, more spreading plant with rounded leaves (often variegated) and a distinct lemon smell. Wild thyme and Mongolian thyme are the only species used to any extent in East Asia.

ORIGIN Western Mediterranean to southeastern Italy.[1] The use of thyme dates back to ancient Egypt, Greece and Rome. "The smell of thyme" is a Greek compliment, implying courage. Thyme has become one of the most important of all culinary herbs.[2]

CULTIVATION Plants are almost invariably propagated from cuttings or by division, in order to retain the characteristics of the cultivar or clone. Seeds are used in large-scale commercial cultivation. Thyme prefers full sun and should be grown in poor, slightly alkaline soil. Pruning after flowering encourages new growth.

HARVESTING Leafy twigs are picked at any time for use in the kitchen. The commercial dried herb is obtained by cutting and drying the stems until all the leaves have dropped off.

CULINARY USES Thyme is one of the essential culinary herbs of Western and Middle Eastern cuisines. Dried spice herb mixtures sold in supermarkets usually include thyme. It is an important ingredient of all meat dishes, including stuffings, sausages, poultry and fish.[2] Thyme is added to soups, stews, lentils, scrambled eggs, tomato dishes and salads.[2] Together with parsley, bay leaves and rosemary, it is tied into the traditional *bouquet garni* that can easily be removed after the sauce or stock has been flavoured. It is used to flavour Benedictine and other liqueurs. Wild thyme (*serpolet* in French, *farigoule* in Provençal) is typical of mutton, rabbit and trout dishes in the south of France.[2] Lemon thyme is less pungent and adds a citrus flavour to veal, chicken, potatoes and seafood.

FLAVOUR COMPOUNDS The essential oil of common thyme (and several other species) is usually dominated by two phenols, thymol and carvacrol.[3,4] These two compounds also give thyme its characteristic aroma.[5]

NOTES Thymol and carvacrol are antiseptic and have many medicinal uses.

thymol   carvacrol

1. **Mabberley, D.J. 2008.** *Mabberley's plant-book* (3rd ed.). Cambridge University Press, Cambridge.
2. **Larousse. 1999.** *The concise Larousse gastronomique*. Hamlyn, London.
3. **Farrel, K.T. 1999.** *Spices, condiments and seasonings*. Aspen Publishers, Gaithersburg, USA.
4. **Harborne, J.B., Baxter, H. 2001.** *Chemical dictionary of economic plants*. Wiley, New York.
5. **Díaz-Maroto, M.C., Díaz-Maroto Hidalgo, I.J., Sánchez-Palomo, E., Pérez-Coello, M.S. 2005.** Volatile components and key odorants of fennel (*Foeniculum vulgare* Mill.) and thyme (*Thymus vulgaris* L.) oil extracts obtained by simultaneous distillation-extraction and supercritical fluid extraction. *Journal of Agricultural and Food Chemistry* 53: 5385–5389.

Thyme, the fresh or dried twigs and leaves of *Thymus vulgaris* (and related species and cultivars)

Plant (*Thymus vulgaris*)

Dried thyme (*T. vulgaris*)

Dried thyme (*Thymus vulgaris*)

Ethiopian thyme (*T. schimperi*)

Flowers (*Thymus vulgaris*)

Wild thyme (*T. serpyllum*)

Lemon thyme (*T. ×citriodora*)

275

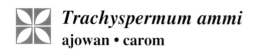

Trachyspermum ammi
ajowan • carom

Trachyspermum ammi L. [= *T. copticum* (L.) Link] (Apiaceae); *ajowan* (Afrikaans); *netch azmud* (Amharic); *ajwan* (Arabic); *xi ye cao guo quin* (Chinese); *ajouan* (French); *Ägyptischer Kümmel, Indischer Kümmel* (German); *ajowan* (Hindi); *ajowan* (Italian); *jintan* (Malay); *ajowan* (Spanish); *phak chi* (Thai)

DESCRIPTION The small fruits ("carom seeds") are up to 2 mm (0.08 in.) long, broadly ovoid and conspicuously warty. They have a bitter, slightly pungent, spicy taste and a strong aroma similar to that of thyme.

THE PLANT An annual with an erect stem, compound, feathery leaves and white florets arranged in flat-topped umbels.[2] A second species, also used as a spice, is ajmud (*T. roxburghianum*). The small fruits of both species resemble those of celery, so that they are often confused or substituted with one another or with celery.

ORIGIN Ajowan is a cultigen of unknown origin, perhaps developed in Egypt or Ethiopia.[1] It is used in a broad region that stretches from Ethiopia and Egypt through the Arabian Peninsula to Afghanistan, Pakistan and India. Recorded names include ajowan, carom, bishop's weed, ajwain, omam, Ethiopian caraway, white cumin and *netch azmud* in Amharic.[3] It should not be confused with true ajmud (*T. roxburghianum*) from South Asia, Southeast Asia and Indonesia.[1] Ajmud is also known as *ajmod* in Hindi and Urdu, *ajamoda* or *ajamodika* in Sanskrit, *radhuni* in Bengali and *kant-balu* in Burmese.

CULTIVATION Ajowan is grown from seeds as a short-lived annual crop. It is adapted to warm and dry conditions and does not survive waterlogging or frost. Ajmud has similar horticultural requirements.

HARVESTING Flowering stalks are collected by hand when the fruits start to mature and are spread out in the sun to dry. The fruits are collected and cleaned, often by sieving and winnowing.

CULINARY USES Ajowan has a powerful aroma and is therefore used sparingly as a spice and condiment. It is appreciated in Indian, Middle Eastern and North African cooking, especially in curry dishes, bean dishes, pickles and chutneys. It is added to or sprinkled on the popular Indian poppadum and other types of breads and biscuits. In Ethiopia and Eritrea, it is an essential ingredient of the hot chilli spice mixture known as *berbere* (in which it is said to reduce and balance the hotness). It is also used to flavour non-pungent sauces, butter and *katikala*, an Ethiopian alcoholic drink.[3] Ajmud is used in spice mixtures and pickles and may replace mustard seeds as one of the five ingredients in a localised version of Bengali *panch phoron*. It is typically tempered in hot oil before use.

FLAVOUR COMPOUNDS The characteristic thyme-like flavour is due to the presence of thymol as the dominant compound in the essential oil.[4] It is accompanied by smaller quantities of *p*-cymene, γ-terpinene and numerous other minor volatile constituents.[4]

NOTES Both ajowan and ajmud have many uses in traditional medicine, mainly for digestive and respiratory ailments.

thymol   *p*-cymene   γ-terpinene

1. **Mabberley, D.J. 2008.** *Mabberley's plant-book* (3rd ed.). Cambridge University Press, Cambridge.
2. **Hedberg, I., Hedberg, O. 2003.** Apiaceae. In: Hedberg, I., Edwards, S., Nemomissa, S. (Eds), *Flora of Ethiopia and Eritrea*, Vol. 4, p. 20. National Herbarium, Addis Ababa and University of Uppsala, Uppsala.
3. **Asfaw, N., Demissew, S. 2009.** *Aromatic plants of Ethiopia*. Shama Books, Addis Ababa.
4. **Masoudi, S., Rustaiyan, A., Ameri, N., Monfared, A., Komeelizadeh, H., Kamalinejad, M., Jami-Roodi, J. 2002.** Volatile oils of *Carum copticum* (L.) C.B. Clarke in Benth. et Hook. and *Semenovia tragioides* (Boiss.) Manden. from Iran. *Journal of Essential Oil Research* 14: 288–289.

Ajowan, the small dry fruits of *Trachyspermum ammi*

Leaves and flowers (*Trachyspermum ammi*)

Flowering plant (*T. ammi*)

Young plants (*T. ammi*)

Trigonella foenum-graecum
fenugreek

Trigonella foenum-graecum L. (Fabaceae); *fenegriek* (Afrikaans); *abish* (Amharic); *hulba, hilbeh* (Arabic); *hu lu ba* (Chinese); *fenegriek* (Dutch); *fénugrec* (French); *Bockshornklee* (German); *trigonella* (Greek); *methi* (Hindi); *fieno greco* (Italian); *koruha, fenu-guriku* (Japanese); *horopa, penigurik* (Korean); *halba* (Malay); *alholva, fenogreco* (Spanish)

DESCRIPTION Fenugreek herb is the fresh or more often dried stems and leaves and have a sweet aroma like hay. Fenugreek spice is the ripe seeds, small and bean-shaped but angular, usually beige or various shades of brown and green. They have a bitter taste and strong smell that becomes more caramel-like when the seeds are roasted and crushed.

THE PLANT A sparse erect annual of about 0.6 m (2 ft) high bearing trifoliate leaves with toothed margins and small white flowers. The fruits are narrow, flat pods with long, slender tips, each producing up to 20 seeds.

ORIGIN Indigenous to the eastern Mediterranean and the Near East.[1] Fenugreek has been grown since ancient times (e.g. by the Assyrians) for food (used like lentils), medicinal purposes (e.g. to stimulate lactation) and as a fodder plant (*foenum-graecum* means "Greek hay"). Charred remains of seeds dating from 4000 BC have been found at Tell Halaf in Iraq and seeds were also found in Tutankhamen's tomb (1325 BC).[1] It is an important spice and pulse in Ethiopia, the Middle East, China and India.[2] The central European blue fenugreek (*T. caerulea*) is important in traditional Georgian cooking.[3]

CULTIVATION Fenugreek is easily grown from seeds in dry temperate and subtropical regions.

HARVESTING Harvesting is done mechanically or by hand. The Hindi names of the products are *methi* (seeds), *sag methi* (fresh leaves) and *kasuri methi* (dried leaves).[2] The last-mentioned has become freely available in supermarkets in cosmopolitan cities with large numbers of people of Indian origin.

CULINARY USES Fenugreek is used mainly as a spice. Roasted and ground seeds are ingredients of pickles, chutneys, spice mixtures and curry powders (e.g. Indian *sambar podi* and Bengali five-spice powder or *panch phoron*, comprising fenugreek, brown mustard or radhuni, cumin, kalonji and fennel) and vegetable stews and sauces (e.g. Ethiopian *shiro* and Yemeni *hilbeh*). The fresh or more often dried leaves are used in North India (e.g. in naan bread and vegetable curries), Turkey (e.g. vegetable and meat stews) and Iran (e.g. in Persian *ghormesabzi* and a mutton *khoresht* made from it).[2] Blue fenugreek is called *utskho suneli* in Georgia.[3] The dried and powdered herb is an important component of the Georgian spice mixture known as *khmeli suneli* but can be substituted with fenugreek.[3]

FLAVOUR COMPOUNDS The strong flavour of roasted fenugreek is due to sotolon (4,5-dimethyl-3-hydroxy-2[5H]-furanone), also known as fenugreek lactone.[4,5] This powerful aroma-active compound is also found in lovage, coffee, artificial maple syrup and molasses and has a detection threshold of 0.001 parts per billion.[5] Also present are volatile compounds such as 2-methyl-2-butenal, δ-elemene and hexanol.[4]

NOTES In Indian cuisine, fenugreek is used as a vegetable in the form of sprouts, microgreens and whole young plants.

sotolon
(fenugreek lactone)

1. Zohary, D., Hopf, M. 2000. *Domestication of plants in the Old World* (3rd ed.). Clarendon Press, Oxford.
2. Gernot Katzer's Spice Pages (http://gernot-katzers-spice-pages.com).
3. Margvelashvili, J. 1991. *The classic cuisine of Soviet Georgia*. Prentice Hall Press, New York.
4. Mazza, G., Di Tommaso, D., Foti, S. 2002. Volatile constituents of Sicilian fenugreek (*Trigonella foenum-graecum* L.) seeds. *Sciences des Aliments* 22: 249-264.
5. Blank, I., Schieberle, P. 1993. Analysis of the seasoning-like flavour substance of a commercial lovage extract (*Levisticum officinale* Koch). *The Flavour and Fragrance Journal* 8: 191–195.

Fenugreek, the seeds or dried leaves of *Trigonella foenum-graecum*

Fenugreek seeds (new and old cultivars)

Flower and fruit (*T. foenum-graecum*)

Plant (*T. foenum-graecum*)

Fenugreek products

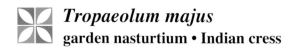

Tropaeolum majus
garden nasturtium • Indian cress

Tropaeolum majus (L.) Kuntze (Tropaeolaceae); *kappertjie* (Afrikaans); *han jin lian* (Chinese); *Oostindische kers* (Dutch); *capucine, cresson d'India* (French); *Kapuzinerkresse, Indische Kresse* (German); *nasturzio d'India, cappucina maggiore* (Italian); *capuchinha-grande, nastúrio* (Portuguese); *capuchina* (Spanish); *Indiankrasse* (Swedish)

DESCRIPTION The leaves, flowers, buds and seeds have a sharp, peppery taste and a sweet, fruity-sulphury aroma.

THE PLANT An annual or weakly perennial creeper with umbrella-shaped leaves and attractive red to yellow, spurred, edible flowers. The common name nasturtium (meaning "nose-twister") is appropriate but potentially confusing (see also watercress, *Nasturtium officinale*). *Tropaeolum* is the only genus of the Tropaeolaceae, a family related to the Brassicaceae.[1]

ORIGIN A cultigen from the Andes region (Peru and Bolivia) and an ancient Inca food plant.[2] Imported from the "Indies" by the Spaniards in the late 17th century, the plant (leaves and flowers) was used in Europe like cress in salads, hence the name "Indian cress" in various European languages.

CULTIVATION The plant is very easily propagated from seeds and often emerges spontaneously in the spring. Seeds can be sown in small pots in the greenhouse in early spring and seedlings planted out as soon as there is no longer any danger of frost. They thrive in full sun or partial shade. Poor soil and infrequent feeding is recommended because if plants become too vigorous they do not produce many flowers.[3] Garden nasturtium is used for companion planting because it attracts aphids. It makes an easy and colourful addition to the herb garden (especially along paths) and is an ideal choice for hanging baskets.[3]

HARVESTING Leaves, flowers, flower buds and green fruits are picked as required (the last-mentioned in as little as four months after sowing).

It is not worthwhile to dry the leaves or flowers but the buds and green fruits can be preserved by pickling in tarragon vinegar.

CULINARY USES The flowers are very popular as a colourful and edible garnish. This is the main culinary use of the plant but the leaves and flowers are also added to fresh salads for their peppery taste. Pickled flower buds or seeds can be used as substitute for capers.

FLAVOUR COMPOUNDS *Tropaeolum* species produce sharp-tasting mustard oils, compounds usually associated with the family Brassicaceae (e.g. mustard, horseradish and wasabi). All parts of the plant but especially the leaves contain high amounts of glucotropaeolin as the characteristic glucosinolate.[4] When cells are damaged, the compound is enzymatically converted (by myrosinases) to the volatile and pungent benzyl isothiocyanate.[4] The fruity aroma, however, is caused by a minor volatile compound called *O,S*-diethyl thiocarbonate, present at a level of only 0.1%.[5]

NOTES The tuber nasturtium (*T. tuberosum*) is an important staple food at high altitudes in the Andes, where potatoes cannot be grown.[1]

O,S-diethyl thiocarbonate

glucotropaeolin   benzyl isothiocyanate

1. **Andersson, L., Andersson, S. 2000.** A molecular phylogeny of Tropaeolaceae and its systematic implications. *Taxon* 49: 721–736.
2. **Mabberley, D.J. 2008.** *Mabberley's plant-book* (3rd ed.). Cambridge University Press, Cambridge.
3. **McVicar, J. 2007.** *Jekka's complete herb book.* Kyle Cathie Limited, London.
4. **Kleinwächter, M., Schnug, E., Selmar, D. 2008.** The glucosinolate-myrosinase system in nasturtium (*Tropaeolum majus* L.): variability of biochemical parameters and screening for clones feasible for pharmaceutical utilization. *Journal of Agricultural and Food Chemistry* 56: 11165–11170.
5. **Breme, K., Guillamon, N., Fernandez, X., et al. 2009.** First identification of *O,S*-diethyl thiocarbonate in Indian cress absolute and odor evaluation of its synthesized homologues by GC-sniffing. *Journal of Agricultural and Food Chemistry* 57: 2503–2507.

Leaves, flowers and seeds of nasturtium (*Tropaeolum majus*)

Ripe and green seeds (*Tropaeolum majus*)

Plant (*Tropaeolum majus*)

Plants (*Tropaeolum majus*)

Vanilla planifolia
vanilla

Vanilla planifolia Andr. (Orchidaceae); *vanielje* (Afrikaans); *vanille* (French); *Echte Vanille* (German); *vaniglia* (Italian); *vainilla* (Spanish)

DESCRIPTION The near-ripe and elaborately cured fruits ("vanilla beans") are black, frosted with white crystals and have the typical flavour and aroma of vanilla.

THE PLANT An epiphytic plant with long climbing stems, fleshy leaves, attractive greenish yellow orchid flowers and long, thin and smooth fruits with thousands of small seeds.

ORIGIN Indigenous to tropical America (mainly Mexico).[1] The Aztecs, but originally Mexicans of Vera Cruz, used vanilla (*tlilxochitl*) to flavour cacao drinks.[1,2] A Mexican monopoly was broken when plantations were established on Réunion and Madagascar by the French and on Java by the Dutch.[1,2] Today, vanilla is also cultivated in other tropical regions, including the West Indies, Central America and Indonesia.

CULTIVATION Plants grow from cuttings and require moist, tropical conditions. The flowers are pollinated by hand when they open in the morning to ensure proper fruit set.[1,2] In nature this task is performed by hummingbirds or bumblebees.

HARVESTING The near-ripe fruits or beans are hand-picked, treated with steam or boiling water to kill the cells, heated in the sun and then "fermented" (enzymatically oxidated) overnight under cover. The heating and fermenting steps are repeated for several weeks to ensure maximum flavour development and quality.[1,3]

CULINARY USES Whole fruits are an expensive but important flavourant for chocolate, ice cream, custards, milkshakes, puddings, various sweet dishes, confectionery, sweets, sugar, preserved fruits, soft drinks and liqueurs.[3] The whole vanilla bean can be added to the dish and removed again for future use. It is often split lengthwise and the soft, aromatic pulp scraped out. The presence of minute black seeds in a homemade ice cream is therefore a sign of quality and authenticity.

FLAVOUR COMPOUNDS The major flavour compound is vanillin (4-hydroxy-3-methoxybenzaldehyde), present in the cured fruits at a level of about 3%.[4,5] The green fruits contain vanillin-glycoside as a non-volatile flavour precursor, which is enzymatically hydrolysed to glucose and vanillin during the curing process, so that the vanillin becomes visible as small white crystals on the surface of cured fruits. There are numerous minor compounds, including vanillic acid, 4-hydroxybenzaldehyde, 4-hydroxybenzoic acid and vanilla vitispirane, all contributing to some extent to the flavour and aroma.[4,5] Synthetic vanillin, produced through semi-synthesis from eugenol (or wood pulp), is a cheap substitute for real vanillin because it lacks the chemical complexity of the natural product.

NOTES Vanillin and related compounds occur in other natural products, such as African white ginger (powdered tubers of *Mondia whitei*) and most famously, oak-aged wine. Seeds of the tonka bean (*Dipteryx odorata*) have a spicy aroma similar to vanilla and almonds and have become an alternative to vanilla for flavouring ice cream, custard and other milk-based desserts.

vanilla | vanillic acid | 4-hydroxy-benzaldehyde | 4-hydroxy-benzoic acid

1. **Mabberley, D.J. 2008.** *Mabberley's plant-book* (3rd ed.). Cambridge University Press, Cambridge.
2. **Kiple, K.F., Ornelas, K.C. (Eds). 2000.** *The Cambridge world history of food*. Cambridge University Press, Cambridge.
3. **Larousse. 1999.** *The concise Larousse gastronomique*. Hamlyn, London.
4. **Pérez-Silva, A., Odoux, E., Brat, P. et al.** 2006. GC–MS and GC–olfactometry analysis of aroma compounds in a representative organic aroma extract from cured vanilla (*Vanilla planifolia* G. Jackson) beans. *Food Chemistry* 99: 728–735.
5. **Schulte-Elte, K.H., Gautschi, F., Renold, W., et al.** 1978. Vitispiranes, important constituents of vanilla aroma. *Helvetica Chimica Acta* 61: 1125–1133.

Vanilla pods, the fruits of *Vanilla planifolia* (with vanilla extracts)

Leaves (*Vanilla planifolia*)

Flowering plant (*Vanilla planifolia*)

White ginger (*Mondia whitei*)

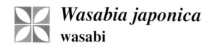

Wasabia japonica
wasabi

Wasabia japonica (Miq.) Matsum. [= *Eutrema wasabi* (Sieb.) Maxim.; = *Wasabia wasabi* (Sieb.) Makino] (Brassicaceae); *wasabi* (Afrikaans); *shan yu cai* (Chinese); *bergstokroos* (Dutch); *raifort du japon, raifort vert* (French); *Wasabi, Japanischer Kren, Japanischer Meerrettich* (German); *wasabi* (Japanese, Korean); *raiz-forte* (Portuguese); *wasábia, rabanete japonês* (Spanish); *Japansk pepparrot* (Swedish)

DESCRIPTION The vertical, underground stems (often referred to as "roots") are thick and fleshy, with scars left where the leaves have been cut off. When chewed, they have an extremely pungent but short-lived flavour and odour.

THE PLANT A stout perennial herb with thick stems, large, pumpkin-like leaves and small white flowers. The species is sometimes placed in the genus *Eutrema* and is then called *E. wasabi*.[1]

ORIGIN Indigenous to Japan[1] and traditionally wild-harvested along mountain streams.[2] In conservative Japanese restaurants, graters made from shark skin are still used when preparing fresh wasabi. It is an integral part of the Japanese culinary tradition and is therefore often referred to as Japanese horseradish (or the equivalent of this term in various languages).

CULTIVATION Propagation is mainly from seeds and this relatively new crop is either grown in normal soil or in running water under 50% shade cloth, using hydroponic methods. Wasabi is very difficult to grow and as a result the products are very expensive. Several cultivars are grown commercially in Japan (including the popular 'Daruma' and 'Mazuma') and in recent years also in New Zealand, Canada and the United States.

HARVESTING Whole plants are harvested by hand and the thick stems (vertical rhizomes) are trimmed to remove the leaves and roots. These fleshy stems are sold fresh or are processed into dried powder or ready-to-use pastes in tubes.[2] Wasabi products sold outside of Japan often do not contain any wasabi but are made from horseradish, mustard, starch and green food colouring.

CULINARY USES Wasabi is extremely pungent, affecting not only the tongue but also the nasal passages and the eyes, causing tears. It can be quite painful if too much is taken at once. There is no long-lasting effect as with chilli peppers because the compounds are volatile, water-soluble and easily diluted or washed away. The pungency of freshly prepared wasabi is short-lived and evaporates after 15 minutes. Wasabi is an important condiment in Japan, alongside soy sauce (or often mixed with it) and *gari* (thin slices of pickled young ginger, usually pinkish in colour). The paste is commonly provided at the table as a dip with sashimi (fresh fish), sushi (the Japanese version of the sandwich but using sweet-and-sour rice instead of bread), tempura (deep-fried vegetables in a light batter) and noodle dishes. Wasabi leaves are also eaten as a vegetable.

FLAVOUR COMPOUNDS Wasabi has the same main isothiocyanate as other members of the family Brassicaceae (e.g. horseradish and black mustard), namely sinigrin, a tasteless compound that is enzymatically converted to the volatile and pungent allyl isothiocyanate when moisture is added). However, what gives the unique flavour to wasabi is the additional presence of methylthioalkyl isothiocyanates, including 6-(methylthio)hexyl-, 7-(methylthio)heptyl- and 8-(methylthio)octyl isothiocyanate.

NOTES Fried or roasted wasabi-coated peas, peanuts and soybeans have become popular as crunchy snack foods.

allyl isothiocyanate 6-(methylthio)hexyl isothiocyanate

1. **Mabberley, D.J. 2008.** *Mabberley's plant-book* (3rd ed.). Cambridge University Press, Cambridge.
2. **Hodge, W.H. 1974.** Wasabi: Native condiment plant of Japan. *Economic Botany* 28: 118–129.
3. **Ina, K., Ina, H., Ueda, M., Yagi, A., Kishima, I., 1989.** ω-methylthioalkylisothiocyanate, in wasabi. *Agricultural Biological Chemistry* 53: 537–538.
4. **Masuda, H., Harada, Y., Tanaka, K., Nakajima, M., Tabeta, H. 1996.** Characteristic odorants of wasabi (*Wasabia japonica* Matum.), Japanese horseradish, in comparison with those of horseradish (*Armoracia rusticana*). ACS Symposium Series, No. 166 (*Biotechnology for Improved Foods and Flavors*), Chapter 6, pp. 67–78.

Stems of wasabi (*Wasabia japonica*)

Flowers (*Wasabia japonica*)

Wasabi products

Wasabi plantation (*Wasabia japonica*)

285

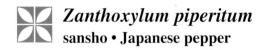

Zanthoxylum piperitum
sansho • Japanese pepper

Zanthoxylum piperitum (L.) DC. (Rutaceae); *Japanese peper, Chinese peper* (Afrikaans); *hua jiao, chuan jiao, chuan chiao* (Chinese); *poivre du Setchuan* (French); *Japanischer Pfeffer, Szechuan-pfeffer, Chinesischer Pfeffer* (German); *sansho* (Japanese); *chopi* (Korean); *ma lar* (Thai)

DESCRIPTION Small, reddish-brown, distinctly warty fruit capsules (pericarps), split open to reveal the smooth whitish interior (the shiny black seeds are tasteless and not considered to be part of the spice). These capsules cause a pungent tingling numbness or "buzz" in the mouth. The leaves and leaf powder have a milder, more citrus-like flavour. The fruits of Chinese pepper or Sichuan pepper (*Z. simulans*) are hard to distinguish from those of Japanese pepper and it seems that these two spices (and even their common names) are used interchangeably.

THE PLANT Small, deciduous trees (ca. 3 m or 10 ft) with thorny stems, pinnately compound leaves and inconspicuous flowers.

ORIGIN *Z. piperitum* is widely distributed in East Asia, while *Z. simulans* is indigenous to China.[1] These are the original pungent spices ("peppers") of Japanese and Chinese cuisine.

CULTIVATION Trees are grown from seeds or cuttings, both in local kitchen gardens and on a commercial scale.[2]

HARVESTING The capsules are picked by hand when ripe and dried in the sun until they split open. Leaves may be dried and powdered.

CULINARY USES Both sansho and Sichuan pepper add a pungent, spicy flavour when ground and sprinkled on soups, noodles, chicken and other Japanese and Chinese dishes.[2,3] They are key ingredients of Chinese five-spices powder and Japanese *shichimi togarashi*.[3] Sansho is one of very few condiments used in Japan[1] because the natural flavours of fresh ingredients are a distinguishing feature of the traditional cuisine. Fresh leaves (*kinome*) or powdered leaves are popular as less pungent and more aromatic flavourants.

FLAVOUR COMPOUNDS The tingling and mouth-numbing pungency of sansho and Sichuan pepper is caused by alkylamides (sanshools), of which hydroxy-α-sanshool is the dominant active ingredient.[4] The psychophysical tingling or buzzing sensation (called *hiri-hiri* in Japanese and *ma* in Chinese; also found in Tasmanian pepper and pará cress)[3] is caused by an activation of somatosensory neurons but the underlying mechanism of action is not yet fully understood.[4] The main aroma-impact compounds of the leaves and leaf oil of sansho are citronellal and citronellol (citrus-like) and (Z)-hexenol (green, grassy).[5]

NOTES There are several *Zanthoxylum* species used as "third pepper" alongside black pepper and chilli pepper. All are thorny shrubs with similar fruits and confusing common names.[2,3] Sansho or Japanese pepper (*Z. piperitum*) is generally called *hua jiao* ("mountain pepper") but this name also applies to what is considered to be the first and original Chinese pepper (*Z. bungeanum*) from northern China.[2] Sichuan pepper or Chinese pepper (*Z. simulans*) is called *chuan jiao* ("Sichuan pepper") and is well known in America and Europe.[2] Koreans use both *Z. piperitum* (called *chopi*) and the less pungent *Z. schinifolium* (called *sancho*). Others include North Indian *Z. rhetsa* (*tippal*), Indonesian *Z. acanthopodium* (*andaliman*), Nepali *Z. alatum* (*timur*), Vietnamese *Z. ailanthoides* (*yue jiao*) and Chinese *Z. armatum* (*zhu ye jiao*).[2,3]

hydroxy-α-sanshool

1. Mabberley, D.J. 2008. *Mabberley's plant-book* (3rd ed.). Cambridge University Press, Cambridge.
2. Hu, S.-Y. 2005. *Food plants of China*. The Chinese University Press, Hong Kong.
3. Gernot Katzer's Spice Pages (http://gernot-katzers-spice-pages.com).
4. Bautista, D.M., Sigal, Y.M., Milstein, A.D., Garrison, J.L., Zorn, J.A., Tsuruda, P.R., Nicoll, R.A., Julius, D. 2008. Pungent agents from Szechuan peppers excite sensory neurons by inhibiting two-pore potassium channels. *Nature Neuroscience* 11: 772–779.
5. Kojima, H., Kato, A., Kubota, K., Kobayashi, A., 1997. Aroma compounds in the leaves of Japanese pepper (*Zanthoxylum piperitum* DC.) and their formation from glycosides. *Bioscience, Biotechnology, and Biochemistry* 61: 491–494.

2 mm

Sansho or Japanese pepper, the ripe fruit capsules of *Zanthoxylum piperitum*

Fruits (*Zanthoxylum piperitum*)

Sichuan or Chinese pepper (*Zanthoxylum simulans*)

Young plants (*Zanthoxylum piperitum*)

Leaves and fruits (*Zanthoxylum piperitum*)

287

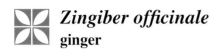

Zingiber officinale
ginger

Zingiber officinale Rosc. (Zingiberaceae); *gemmer* (Afrikaans); *jiang* (Chinese); *gingembre* (French); *Ingwer* (German); *adrak* (Hindi); *zénzero* (Italian); *jinjaa, shouga* (Japanese); *saeng gang* (Korean); *haliya* (Malay); *gengibre, ingever* (Portuguese); *jengibre, anchoas* (Spanish); *khing* (Thai)

DESCRIPTION The rhizomes (either green or mature) are fleshy, pale brown in colour and have a characteristic spicy odour and pungent taste.

THE PLANT A perennial herb (ca. 1 m or 3 ft 4 in. high) with large leaves arising from creeping and branching rhizomes. Yellow flowers are borne in cone-like clusters but plants rarely flower.

ORIGIN Ginger is a cultigen that has not yet been found in the wild.[1,2] It probably originated from northeastern India or southern China and has been grown since ancient times as a spice and medicine (to counteract nausea and improve digestion).

CULTIVATION Humid and tropical or subtropical conditions are required. Pieces of rhizome ("bits") are planted.[2]

HARVESTING Rhizomes are harvested after 5–7 months (for fresh ginger) or 8–9 months (for dried ginger and ginger oil). Dried ginger is produced by scraping the rhizome surfaces before drying or by treating the scraped rhizomes with lime during drying (bleached ginger).[2]

CULINARY USES Ginger is an important spice that is much used in Asian cooking to flavour meat dishes, marinades, fish, curries, soups, sauces, rice dishes and stir-fries. The fresh, pickled or candied root is eaten as a snack or used as a condiment in China and Japan. It is widely used in confectionery such as biscuits, cakes and sweets (e.g. British gingerbread or its French equivalent, *pain d'épice*). Pieces of fresh or preserved ginger are added to preserves and jams. Ginger beer is a fermented, frothy, low-alcohol drink (particularly popular in Britain) prepared from water, sugar, ginger and cream of tartar. Ginger ale (made from ginger essence and colouring added to carbonized water) is used in the same way as club soda in gin- or whisky-based long drinks. Ginger extracts, powdered ginger and the essential oil are used in the food industry and also to flavour ginger teas, wines, brandies and liqueurs.

FLAVOUR COMPOUNDS Ginger and ginger oil contains zingiberene as main flavour compound, as well as several other terpenoids, such as citral (lemon flavour), camphene, β-sesquiphellandrene, β-phellandrene and β-bisabolene.[3] The pungent taste of fresh ginger is due to gingerols, of which [6]-gingerol is the dominant compound.[4] Gingerol is converted to zingerone (less pungent, more sweet and spicy) when ginger is cooked. Dried ginger is more pungent than fresh ginger because the gingerol is converted to the more pungent shogaol during drying.

NOTES Unopened flower buds of the attractive torch ginger (*Etlingera elatior*) are used in Thailand, Singapore and Malaysia as a culinary herb.[5] Known as *kaalaa* (Thai), *bunga kantan* or *ninga kantan* (Malay) or *combrang* (Indonesian), it is used in Malay fish curries, *rujak* salads and Thai *khaao yam* salad.

1. **Burkill, I.H. 1966.** *A dictionary of the economic products of the Malay Peninsula*, Vol. 1, pp. 910–915. Crown Agents for the Colonies, London.
2. **Nayar, N.M., Ravindran, P.N. 1995.** Herb spices. In: Smartt, J., Simmonds, N.W. (Eds), *Evolution of crop plants* (2nd ed.), pp. 491–494. Longman, London.
3. **Harborne, J.B., Baxter, H. 2001.** *Chemical dictionary of economic plants*. Wiley, New York.
4. **Wohlmuth, H., Leach, D.N., Smith, M.K., Myers, S.P. 2005.** Gingerol content of diploid and tetraploid clones of ginger (*Zingiber officinale* Roscoe). *Journal of Agricultural and Food Chemistry* 53: 5772–5778.
5. **Hutton, W. 1997.** *Tropical herbs and spices*. Periplus Editions, Singapore.

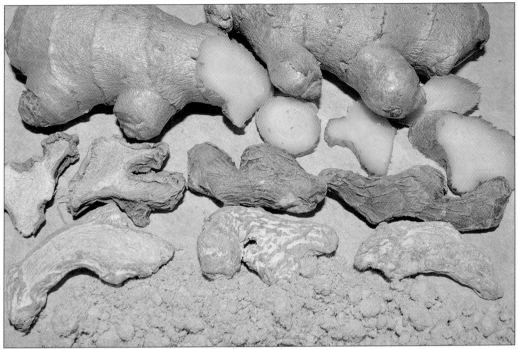
Ginger, the fresh, dried or powdered rhizomes of *Zingiber officinale*

Plants (*Zingiber officinale*)

Green ginger

Ginger beer

Inflorescence (*Z. officinale*)

Torch ginger (*Etlingera elatior*)

289

Quick guide and checklist: culinary herbs and spices of the world

Names of treated and illustrated species are printed in bold; illustrated plant parts are printed in bold.

Abbreviations (region of origin): Afr – Africa; Am – America; As – Asia; Aus – Australia; Bangl – Bangladesh; Braz – Brazil; C – Central; E – East / Eastern; Eur – Europe; Euras – Eurasia; Himal – Himalayas; Ind – India; Indomal – Indonesia-Malaysia; Jap – Japan; Madag – Madagascar; Malay – Malaysia; Med – Mediterranean; Mex – Mexico; N – North / Northern; NE – Northeast; OW – Old World; Pac – Pacific; Philipp – Philippines; S – South / Southern; SE – Southeast; SW – Southwest; Temp – Temperate; Trop – Tropical; Venez – Venezuela; W – West / Western.

Species	Family	Common name	Region of origin	Plant part(s) used
Abelmoschus moschatus	Malvaceae	abelmosk seed; ambrette seed; amber seed; musk mallow; musk okra	E As	seed
Achillea ageratum	Asteraceae	English mace; sweet nancy; sweet yarrow	Eur	leaf; herb
Achillea atrata	Asteraceae	black yarrow	Eur	herb
Achillea clavenae	Asteraceae	bitter yarrow	As; Eur; N Am	leaf
Achillea erba-rotta	Asteraceae	alpine yarrow; musk milfoil; musk yarrow	Eur	leaf; herb
Achillea millefolium	Asteraceae	common yarrow; fragrant yarrow; milfoil	As; Eur; N Am	leaf
Achillea nobilis	Asteraceae	noble milfoil; noble yarrow	As; Eur	leaf
Achyrocline satureioides	Asteraceae	alecrim da parede; macela	Afr; S Am	flower
Acinos arvensis	Lamiaceae	basil thyme; mother of thyme; spring savory	N Am	herb
Acorus calamus	Acoraceae	calamus; sweet flag	As; Eur; N Am	rhizome
Acorus gramineus	Acoraceae	Chinese sweet grass; Japanese sweet flag; grass leaf sweet flag	E As	rhizome
Adansonia digitata	Bombacaceae	baobab	Afr	leaf; seeds
Aegle marmelos	Rutaceae	bael; bel; Bengal quince; Indian bael	As	leaf
Aeollanthus heliotropoides	Lamiaceae	chegadinka; macassa	Afr	leaf
Aframomum alboviolaceum	Zingiberaceae	Cameroon cardamom	Trop Afr	leaf
Aframomum angustifolium	Zingiberaceae	Madagascar cardamom	Madag	seed
Aframomum daniellii	Zingiberaceae	bastard Melegueta; Cameroon cardamom	Trop W Afr	seed
Aframomum exscapum	Zingiberaceae	alligator pepper; grains of paradise	Trop W Afr	seed
Aframomum granum-paradisi	Zingiberaceae	grains of paradise	W Afr	seed
Aframomum hanburyi	Zingiberaceae	Cameroon cardamom	Trop W Afr	seed
Aframomum corrorima	Zingiberaceae	Ethiopian cardamom; korarima	Trop NE Afr	**seed**
Aframomum macrospermum	Zingiberaceae	Guinea cardamom	W Afr	seed
Aframomum melegueta	Zingiberaceae	Melegueta pepper; grains of paradise; alligator pepper	Trop W Afr	**seed**
Aframomum sceptrum	Zingiberaceae	black amomum	W Afr	leaf
Afrostyrax kamerunensis	Styracaceae	Cameroon garlic tree	Trop Afr	bark; seed

Afrostyrax lepidophyllus	Styracaceae	garlic bark (tree)	Trop Afr	bark; seed
Agastache foeniculum	Lamiaceae	anise hyssop	N Am; W As	**leaf**
Agastache mexicana	Lamiaceae	Mexican giant hyssop; lemon hyssop	As	leaf
Agastache rugosa	Lamiaceae	Chinese giant hyssop; Korean mint	N Am; W As	leaf
Agathosma betulina	Rutaceae	buchu; round leaf buchu	S Afr	**leaf**
Agathosma crenulata	Rutaceae	oval leaf buchu	S Afr	**leaf**
Aglaia odoratissima	Meliaceae	mock lime; orchid tree	As; W Pac	flower
Aleurites moluccana	Euphorbiaceae	candle nut	SE As	**seed**
Alliaria petiolata	Brassicaceae	garlic mustard	Eur; Temp As	leaf
Allium altaicum	Alliaceae	Altai onion	C As	bulb
Allium ameloprasum	Alliaceae	leeks	Med; W As	**bulb;** leaf
Allium cepa	Alliaceae	onion	W As; C As	**bulb**
Allium chinense	Alliaceae	Chinese onion; oriental onion; rakkyo	China	**bulb**
Allium fistulosum	Alliaceae	Welsh onion; Japanese bunching onion	China	**bulb; leaf**
Allium giganteum	Alliaceae	giant onion	C As	bulb
Allium kurrat	Alliaceae	kurrat; salad leek; Egypt leek	Med; W As	leaf
Allium moly	Alliaceae	lily leek; moly; yellow onion	Eur	bulb
Allium neopolitanum	Alliaceae	daffodil garlic; Naples garlic; false garlic	Med; W As	bulb; leaf
Allium obliquum	Alliaceae	oblique onion	Eur; As	bulb
Allium oleraceum	Alliaceae	field garlic	Eur	bulb; leaf
Allium oschaninii	Alliaceae	Oschanin onion	C As; W As	bulb
Allium paradoxum	Alliaceae	few-flowered leek	Eur; W As	bulb; leaf
Allium ×proliferum	Alliaceae	Egyptian onion; top onion; tree onion	cultigen	**bulblet;** leaf
Allium pskemense	Alliaceae	Russian onion	As	bulb
Allium ramosum	Alliaceae	Chinese leek	As	bulb; leaf
Allium sativum	Alliaceae	garlic	C As; W As	**bulb**
Allium schoenoprasum	Alliaceae	chive; chives	Med; Eur; As	**leaves**
Allium schorodoprasum	Alliaceae	sand leek; Spanish garlic; rocambole	Eur; As	bulblet; leaf
Allium senescens	Alliaceae	mountain leek	C Eur; As	leaf
Allium sphaerocephalon	Alliaceae	ball leek	Med; Eur	leaf
Allium stipitatum	Alliaceae	drumstick onion	C As	bulb
Allium tricoccum	Alliaceae	ramp; wild leek; wood leek	N Am	bulb; leaf
Allium tuberosum	Alliaceae	garlic chive; Chinese chive	As	**leaf**
Allium ursinum	Alliaceae	bear's garlic; ramsons; wood garlic	Eur; As	**leaf**
Allium vavilovii	Alliaceae	Vavilov's leek	C As	bulb; leaf
Allium victorialis	Alliaceae	alpine leek	Eur; As	leek; bulb
Allium vineale	Alliaceae	crow garlic; false garlic; field garlic	Eur	leaf
Aloysia citrodora (=*Aloysia triphylla*)	Verbenaceae	lemon verbena; vervain; cidron; herb Luisa	S Am	**leaves**
Alpinia calcarata	Zingiberaceae	Indian ginger; snap ginger	As; Pac; E Ind	rhizome
Alpinia conchigera	Zingiberaceae	mussel galangal	SE As	rhizome
Alpinia galanga	Zingiberaceae	galangal; greater galangal	Trop As	**rhizome**
Alpinia nigra	Zingiberaceae	black galangal	Pac; SE As	rhizome
Alpinia officinarum	Zingiberaceae	lesser galangal; small galangal	E As; SE As	**rhizome**
Alpinia purpurata	Zingiberaceae	red ginger	As; Pac; Malay	leaf

Alpinia zerumbet	Zingiberaceae	shell ginger	As; Pac; E As	leaf; rhizome
Alstonia scholaris	Apocynaceae	devil tree; dita bark; palmira alstonia	Indomal; Aus	bark
Alyxia lucida	Apocynaceae	alyxia cinnamon	Indomal	bark
Amelanchier alnifolia	Rosaceae	alderleaf berry; Pacific berry; saskatoon; western serviceberry	N Am	fruit
Ammi majus	Apiaceae	false bishop's weed; greater ammi; lady's lace	Med; NE Afr	fruit
Ammi visnaga	Apiaceae	khella; visnaga; lesser bishop's weed	S Eur; NE Afr	fruit
Ammodaucus leucotrichus	Apiaceae	cafoun	Canary Is; N Afr; W Afr	fruit; seed
Amomum aromaticum	Zingiberaceae	Bengal cardamom; Nepal cardamom; large cardamom	N Ind; Bangl; Nepal	fruit & seed
Amomum compactum	Zingiberaceae	Indonesian cardamom	SE As	fruit & seed
Amomum costatum	Zingiberaceae	Chinese black cardamom	E As	fruit & seed
Amomum globosum	Zingiberaceae	round Chinese cardamom	China	fruit & seed
Amomum gracile	Zingiberaceae	slender cardamom	SE As	leaf
Amomum kepulaga	Zingiberaceae	round cardamom	Trop As	fruit & seed
Amomum krervanh	Zingiberaceae	Cambodian cardamom; krervanh	Trop As	fruit & seed
Amomum maximum	Zingiberaceae	Java cardamom	Trop As	fruit & seed
Amomum ochreum	Zingiberaceae	tepus batu	As	fruit & seed
Amomum subulatum	Zingiberaceae	brown cardamom; greater cardamom; Indian cardamom	As	fruit & seed
Amomum testaceum	Zingiberaceae	ka tepus	As	fruit & seed
Amomum tsao-ko	Zingiberaceae	tsao-ko cardamom; large cardamom	As	fruit & seed
Amomum villosum	Zingiberaceae	Malabar cardamom; Tavoy cardamom; wild Siamese cardamom	As	fruit & seed
Amomum xanthioides	Zingiberaceae	bastard Siamese cardamom; wild Siamese cardamom	As	fruit & seed
Amomum xanthophlebium	Zingiberaceae	elach	As	flower
Anacyclus pyrethrum	Asteraceae	Roman pellitory; Spanish pellitory	Eur; N Afr	root
Anethum graveolens (=*A. sowa*)	Apiaceae	dill; Indian dill	W As; Ind	**leaf; fruit**
Angelica acutiloba	Apiaceae	dong dang gui	Eur; E As	root; leaf
Angelica archangelica	Apiaceae	angelica; garden angelica; archangel	Eur; As	root; **fruit; stem**
Angelica atropurpurea	Apiaceae	American angelica	N Am	root; fruit
Angelica japonica	Apiaceae	Japanese angelica	As	root; fruit
Angelica polymorpha var. *sinensis*	Apiaceae	Chinese angelica; dang gui	As; E As	root
Angelica sylvestris	Apiaceae	wild angelica	Eur; W As	root
Aniba canellila	Lauraceae	Oriniko cinnamon	S Am	bark; leaf
Aniba rosaedora	Lauraceae	rose wood	N Am; S Am	wood
Anthoxanthum odoratum	Poaceae	scented vernal grass; sweet vernal grass	Eur; As	leaf
Anthriscus cerefolium	Apiaceae	chervil; garden chervil; French parsley	As; W As	**leaf**
Apium graveolens	Apiaceae	celery	Eur; As	**leaf; seed**
Aquilaria agallocha	Thymelaeaceae	agar wood; eagle wood	As	wood
Arachis hypogaea	Leguminosae	peanut; ground nut	S Am	seed
Aristolochia serpentaria	Aristolochiaceae	Virginia serpentary; Virginia snakeroot	N Am	rhizome

Armoracia rusticana	Brassicaceae	horseradish	As	**root**
Arnica chamissonis	Asteraceae	leafy leopardsbane	N Am	flower
Arnica montana	Asteraceae	arnica; European arnica; mountain arnica	Eur	flower
Artabotrys hexapetalus	Annonaceae	tail grape	S As	flower
Artemisia abrotanum	Asteraceae	southernwood	Eur?	**leaves**
Artemisia absinthium	Asteraceae	absinthe; common wormwood	Euras; N Afr	**leaves**
Artemisia dracunculoides	Asteraceae	Russian tarragon	Eur; As	**leaves**
Artemisia dracunculus	Asteraceae	tarragon; French tarragon; estragon	Eur; As	**leaves**
Artemisia genipi	Asteraceae	genépi noir	Eur	leaf
Artemisia glacialis	Asteraceae	glacier wormwood	Eur	leaf
Artemisia indica	Asteraceae	Indian mugwort	As	leaf
Artemisia judaica	Asteraceae	graines à vers; zédoire	N Afr	herb
Artemisia ludoviciana	Asteraceae	cudweed; western mugwort; white sage	N Am	herb
Artemisia maritima	Asteraceae	sea wormwood	Eur; As	herb
Artemisia mexicana	Asteraceae	Mexican mugwort	N Am	herb
Artemisia mutellina	Asteraceae	alpine wormwood; white genipi	Eur	herb
Artemisia pallens	Asteraceae	davana	As	herb
Artemisia pontica	Asteraceae	Roman wormwood; small absinth	Eur	**leaf**; herb
Artemisia princeps	Asteraceae	Japanese mugwort	E As	leaf
Artemisia rehan	Asteraceae	rehan; Ethiopian wormwood	Afr	leaf; herb
Artemisia vulgaris	Asteraceae	mugwort; motherwort; sagebrush	Eur; As; N Am	**leaves**
Artocarpus lakoochus	Moraceae	monkey jack	As	male flowers
Averrhoa bilimbi	Oxalidaceae	bilimbi; cucumber tree; tree sorrel	Trop As	fruit
Backhousia citriodora	Myrtaceae	Australian lemon myrtle; citron myrtle	Aus	leaf
Barbarea verna	Brassicaceae	land cress	Med; W As	**leaf**
Berberis vulgaris	Berberidaceae	barberry; zereshk; sereshk	Eur; As	fruit
Bixa orellana	Bixaceae	an(n)atto; achiote; roucou	Trop Am	**seed**
Boesenbergia rotunda (=*Boesenbergia pandurata*)	Zingiberaceae	fingerroot; Chinese keys; krachai	Trop As	**rhizome; root**
Borago officinalis	Boraginaceae	borage	Med; Eur	**leaf; flower**
Brassica juncea	Brassicaceae	brown mustard; Chinese mustard; Indian mustard; leaf mustard; mustard green	C As	**seed**; leaf
Brassica nigra	Brassicaceae	black mustard	W As	**seed**
Buchanaria lanzan	Anacardiaceae	charoli; chironji	As	**seed**
Bunium persicum	Apiaceae	black caraway; black cumin; black zira	Eur; As	**fruit**
Bunium roxburghianum	Apiaceae	ajmud; radhuni	As	fruit
Calamintha cretica	Lamiaceae	dwarf calamint	Med	herb
Calamintha grandiflora	Lamiaceae	showy calamint; showy savory; large flowered calamint	S Eur	herb
Calamintha menthifolia	Lamiaceae	calamint	Eur	**herb**
Calamintha nepeta	Lamiaceae	common calamint; lesser calamint	Eur; As	herb
Calendula officinalis	Asteraceae	pot marigold; marigold flower; calendula	S Eur?	**flower**
Camellia sinensis	Theaceae	tea; chai	S & E As	**leaves**
Canarium pimela	Burseraceae	Chinese black olive; Chinese black canarium	As	**seed**
Canella winterana	Canellaceae	winter cinnamon; white cinnamon	N & C Am	bark

Capparis sicula	Capparaceae	Sicili caper	Med	flower bud; unripe fruit
Capparis spinosa	Capparaceae	capers; caper	Med	**flower bud; unripe fruit**
Capsicum annuum	Solanaceae	paprika; cayenne pepper; chilli (pepper); chili (pepper); green pepper; red pepper	C Am	**fruit**
Capsicum baccatum	Solanaceae	aji; Peruvian pepper	S Am (Bolivia)	**fruit**
Capsicum chinense	Solanaceae	yellow lantern chilli; habanero; Scotch bonnet; Chinese chilli; bonnet pepper	S Am	**fruit**
Capsicum frutescens	Solanaceae	chilli; chili; bird chilli; hot pepper; Tabasco pepper; piri piri	C Am; S Am	**fruit**
Capsicum pubescens	Solanaceae	rocoto; tree chilli	S Am (Andes)	fruit
Carthamus tinctorius	Asteraceae	safflower; false saffron; saffron thistle	N Afr; W As	**flower; fruit**
Carum carvi	Apiaceae	caraway	Med	**fruit**
Centauria benedicta (=Cnicus benedictus)	Asteraceae	blessed thistle; holy thistle	Eur; W As	herb
Centaurium erythraea	Gentianaceae	centaury; feverwort; pink centaury	Eur; As	herb
Ceratonia siliqua	Leguminosae	carob; locust bean; St John's bread	Eur; W As	**fruit**
Chamaemelum nobile	Asteraceae	Roman chamomile	Eur	**flower**
Chenopodium ambrosioides	Chenopodiaceae	blue weed; goose foot; lamb's quarters; pigweed	C Am; S Am	leaf; herb
Chenopodium botrys	Chenopodiaceae	Jerusalem oak; slimy anserine herb	Eur; C As	fresh herb
Chloranthus spicatus	Chloranthaceae	pearl orchid flower; chulan; cha ran	SE As	flower; leaf
Chrysanthemum balsamita	Asteraceae	alecost; costmary	Eur	**herb**
Chrysanthemum coronarium	Asteraceae	chop suey greens; crown daisy; garland chrysanthemum; garland daisy; tangho; Japanese greens	Med	**leaf; seedlings**
Chrysanthemum ×morifolium	Asteraceae	ju hua; florist chrysanthemum	China	**flower heads**
Chrysanthemum vulgare	Asteraceae	tansy	Eur; Med	**leaf**
Chrysopogon zizanioides (=Vetiveria zizanioides)	Poaceae	vetiver (grass); cus cus (grass); khus khus (grass)	Ind	rhizome; root
Cichorium endivia	Asteraceae	endive	Med	**leaf**
Cinchona officinalis	Rubiaceae	quinine; chinabark; Peruvian bark; yellow cinchona; ledger bark; yellow bark	S Am	bark
Cinchona pubescens	Rubiaceae	quinine; red cinchona; red Peruvian; red bark; Jesuit bark	S Am	bark
Cinnamomum aromaticum (=C. cassia)	Lauraceae	cassia; Chinese cinnamon; Chinese cassia	Myanmar	**bark**
Cinnamomum burmanii	Lauraceae	Indonesian cassia; Padang cassia; Batavia cassia; Korintje cassia	SE As	bark; leaf
Cinnamomum camphora	Lauraceae	camphor	As; E As	gum
Cinnamomum loureiroi	Lauraceae	Vietnamese cassia; Saigon cinnamon	SE As	bark; flower
Cinnamomum tamala	Lauraceae	Indian cassia lignea; Indian bark; Malabathri bark	Ind	leaf; bark
Cinnamomum verum (=C. zeylanicum)	Lauraceae	cinnamon; Ceylon cinnamon	Sri Lanka; SW Ind	**bark; leaf**
×*Citrofortunella microcarpa* see *Citrus madurensis*				

Citrus aurantiifolia	Rutaceae	lime; limon; key lime; Mexican lime; sour lime; Persian lime; West Indian lime	Trop As	**fruit; leaves**
Citrus aurantium	Rutaceae	bigarade; bitter orange; Seville orange; orange (American)	Trop As	flower; peel; juice
Citrus bergamia	Rutaceae	bergamot; bergamot orange	Med	peel of ripe fruits
Citrus cavaleriei	Rutaceae	ichang papeda	Trop As	fruit
Citrus deliciosa		Mediterranean mandarin	Trop As	
Citrus hystrix	Rutaceae	lime leaf; makrut lime; papeda	Trop As	**leaf; fruit**
Citrus ichangensis see *Citrus cavaleriei*				
Citrus jambhiri	Rutaceae	rough lemon	Trop As	**fruit**
Citrus ×junos	Rutaceae	yuzu	W China	fruit
Citrus latifolia	Rutaceae	Tahiti lime; Persian lime; seedless lime	Trop As	fruit (juice); peel
Citrus limon	Rutaceae	lemon	Trop As	**peel; fruit**
Citrus madurensis	Rutaceae	calamondin	Trop As	fruit
Citrus medica var. *medica*	Rutaceae	citron; citron peel	Trop As; C As	**fruit**
Citrus medica var. *sarcodactylis*	Rutaceae	Buddha's hand	Trop As	**fruit; peel**
Citrus myrtifolia	Rutaceae	myrtle leaf orange	Trop As	**fruit**; peel
Citrus nobilis	Rutaceae	king mandarin	Trop As	**fruit**
Citrus reticulata	Rutaceae	tangerine; mandarin; clementine; nartjie	E As; China	peeled fruit
Citrus sinensis	Rutaceae	orange; blood orange; navel orange; sweet orange; Valencia orange	Trop As	**fruit**; peel; flower
Citrus trifoliata (=*Poncirus trifoliata*)	Rutaceae	trifoliate orange; citrangequat; Japanese bitter orange	C & N China	fruit
Citrus unshiu	Rutaceae	satsuma mandarin; unshiu mikan	Trop As; China	**fruit**
Clausena anisata	Rutaceae	horsewood; clausena	Afr	leaf
Clausena anisum-olens	Rutaceae	kayumanis; danglais	E As	leaf
Clausena excavata	Rutaceae	pink lime-berry; Hollywood clausena	SE As	leaf
Clausena lansium	Rutaceae	Chinese wampee; wampi	E As	leaf
Cleome gynandra	Capparaceae	African mustard; African spider flower	Afr; As	leaf; fruit; seed
Cnicus benedictus see *Centauria benedicta*				
Cnidium monnieri	Apiaceae	snow parsley	E As	fruit
Cocos nucifera	Arecaceae	coconut	Trop As	**fruit**
Coffea arabica	Rubiaceae	coffee; Arabian coffee	NE Afr (Ethiopia)	**seed**
Coffea canephora	Rubiaceae	robusta coffee; Congo coffee	Afr	seed
Coffea liberica	Rubiaceae	Liberian or Abeokuta coffee	Afr	seed
Cola acuminata	Malvaceae	cola nut; abata cola	W Afr	**seed**
Cola nitida	Malvaceae	gbanja cola	Afr	seed
Coluria geoides	Rosaceae	Siberian avens; clove root; clove oil plant	As	root
Coriandrum sativum	Apiaceae	coriander; cilantro; Chinese parsley	Med; W As; Ind	**fruit; leaf**
Corylus americana	Betulaceae	American hazelnut	N Am	seed
Corylus avellana	Betulaceae	hazelnut	Eur; As	**seed**
Corylus colurna	Betulaceae	Turkish hazelnut	W As	seed

Corylus maxima	Betulaceae	filbert	Eur; C & W As	seed
Cosmos sulphureus	Asteraceae	orange cosmos; yellow cosmos	N Am	herb
Costus afer	Zingiberaceae	ginger lily; spiral ginger	Trop As	rhizome
Crambe maritima	Brassicaceae	seakale; sea kale	Eur; W As	leaf
Crateva religiosa	Capparaceae	garlic pear; sacred barma; temple tree	Madag	fruit
Crateva tapia	Capparaceae	tapia fruit; payagua	Madag	fruit
Crocus sativus	Iridaceae	saffron	Med; W As	stigma
Cryptocarya moschata	Lauraceae	Brazilian nutmeg; (South) American nutmeg	S Am	seed
Cryptotaenia japonica	Apiaceae	mitsuba; Japanese parsley	China	fresh **leaf**
Cuminum cyminum	Apiaceae	cumin	W & C As; Ind	**fruit**
Cunila origanoides	Lamiaceae	American dittany; frost flower; Maryland dittany; mountain dittany; stone mint	N Am	herb
Cunila spicata	Lamiaceae	(American) stone mint	N Am; C Am; S Am	herb
Curcuma amada	Zingiberaceae	mango ginger	S As	**rhizome**
Curcuma aromatica	Zingiberaceae	Bombay or Indian arrowroot; wild turmeric; yellow zedoary	S As	rhizome
Curcuma longa (=C. domestica)	Zingiberaceae	turmeric	Ind	**rhizome**
Curcuma mangga	Zingiberaceae	Indonesian mango ginger	Trop As	**rhizome**
Curcuma zedoaria	Zingiberaceae	zedoary; Japanese turmeric	Ind	rhizome
Cusparia febrifuga see *Galipea officinalis*				
Cymbopogon citratus	Poaceae	lemongrass	cultigen (As?)	**leaf**
Cymbopogon flexuosus	Poaceae	East Indian lemongrass; Malabar lemongrass; Cochin grass	S As	leaf
Cymbopogon iwarancusa	Poaceae	karnkusa grass; khavi grass	As	leaf
Cymbopogon martinii	Poaceae	ginger grass; rosha grass; palmarosa grass	Ind	leaf
Cymbopogon nardus	Poaceae	citronella; Ceylon citronella; nard grass	Trop As	leaf
Cymbopogon schoenanthus	Poaceae	camel grass; geranium grass	N Afr; As	leaf
Cymbopogon winterianus	Poaceae	Java citronella; Java lemongrass	As	leaf
Cynometra cauliflora	Leguminosae	katak puru; nam nam; puki; puru	SE As	cooked fruits
Decalepis hamiltonii	Apocynaceae	mahali kizhangu	S As	root
Deianira nervosa	Gentianaceae	Deianiraktaut	Trop Am	herb
Dianthus caryophyllus	Caryophyllaceae	carnation; clove pink; gilly flower	Med	flower
Dicypellium caryophyllatum	Lauraceae	clove bark; pinkwood bark	S Am	bark
Dillenia indica	Dilleniaceae	chalta tree; elephant apple	Ind; C Malay	fruit pulp
Diplotaxis tenuifolia	Brassicaceae	wild rocket	Eu	**leaf**
Dipteryx odorata	Leguminosae	tonka; tonka bean	Trop Am	seed
Dorystaechus hastata	Lamiaceae	Turkish lavender	SW Turkey	herb
Dracocephalum moldavica	Lamiaceae	Moldavian balm; dragon's head	E Eur; E As	leaf; herb
Dracontomelon dao	Anacardiaceae	argus pheasant	Indomal	fruit; leaf; flower
Drimys lanceolata	Winteraceae	pepper tree; Australian pepper tree; mountain pepper; Tasmania pepper	Tasmania; Aus	fruit
Drimys piperita	Winteraceae	South American mountain pepper	S Am	fruit

Drimys winteri	Winteraceae	drimys bark; winter's bark	S Am	bark
Echinophora tenuifolia	Apiaceae	prickly parsnip	Eur	leaf; stalk
Elettaria cardamomum	Zingiberaceae	cardamom	Ind	**fruit & seed**
Elsholtzia ciliata	Lamiaceae	common elsholtzia; Vietnamese balm	C & E As	herb
Embelia philippinensis	Myrsinaceae	woody vine	Philipp	leaf
Emilia javanica	Asteraceae	tassel flower	Trop Afr; Trop As	leaf; root
Eruca sativa	Brassicaceae	rocket; arugula	Med	**leaf**; seed
Eryngium foetidum	Apiaceae	culantro; eryngo; sawtooth coriander	Trop Am	**leaf**; root
Escobedia scabrifolia	Scrophulariaceae	saffron of Andes; saffron root	Trop Am	root
Etlingera elatior (=Phaeomeria magnifica)	Zingiberaceae	torch ginger; Philippine wax flower	Trop As	**flower** bud
Eupatorium cannabinum	Asteraceae	hemp agrimony	Eur; Med; C As	leaf; herb
Ferula assa-foetida	Apiaceae	asafoetida; devil's dung; hing	W & C As	**gum**
Ferula foetida	Apiaceae	asafoetida; fetida	W As (Iran)	gum
Ferula gummosa	Apiaceae	galbanum	W As (Iran)	gum
Ferula narthex	Apiaceae	asafoetida	Afghanistan	gum
Filipendula ulmaria	Rosaceae	meadow sweet; dropwort	Eur; As	flower
Foeniculum vulgare	Apiaceae	fennel	Eur; Med	**fruit**; **leaf**
Fortunella crassifolia	Rutaceae	large round kumquat; meiwa kumquat	China	fruit
Fortunella hindsii	Rutaceae	Hongkong wild kumquat; Formosan kumquat	E As	fruit
Fortunella japonica	Rutaceae	round kumquat; marumi kumquat	E As (China)	fruit
Fortunella margarita	Rutaceae	oval kumquat; nagami kumquat	E As (China)	fruit
Fortunella polyandra	Rutaceae	Malayan kumquat; hedge lime	E As	fruit
Galipea officinalis	Rutaceae	angostura	Trop Am	bark
Galium odoratum	Rubiaceae	sweet woodruff; woodruff asperule	Eur; As	fresh herb
Garcinia atroviridis	Clusiaceae	asam gelungor	NE Ind	fruit
Garcinia gummi-guta (=G. cambogia)	Clusiaceae	Malabar tamarind; cambodge; goraga	Ind	fruit
Garcinia indica	Clusiaceae	kokam; kokum	Ind	**fruit peel**
Garcinia xanthochymus	Clusiaceae	gamboges; mundu	N Ind	fruit
Gardenia augusta	Rubiaceae	gardenia; Cape jasmine	China	flower; fruit
Gaultheria procumbens	Ericaceae	alpine wintergreen; creeping wintergreen	N Am	essential oil
Gentiana lutea	Gentianaceae	yellow gentian; bitter root	Eur; W As	rhizome
Geum urbanum	Rosaceae	clove root; herb bennet; wood avens	Eur; As; N Am	root
Ginkgo biloba	Ginkgoaceae	gingko; maidenhair tree	E China	seeds
Glechoma hederacea	Lamiaceae	alehoof; gill over the ground; ground ivy	Eur	leaf; herb
Glehnia litoralis	Apiaceae	cork wing	NE As; N Am	leaf
Globba marantina	Zingiberaceae	kapulaga ambon; halia utan; bonelau	E As	rhizome; herb
Glycine max	Leguminosae	soybean; soya bean; soya	As; E As	**seed**
Glycyrrhiza echinata	Leguminosae	Roman liquorice	S Eur	rhizome
Glycyrrhiza glabra	Leguminosae	licorice; liquorice	Med; W As; C As	**rhizome**
Glycyrrhiza lepidota	Leguminosae	American licorice	N Am	rhizome
Glycyrrhiza uralensis	Leguminosae	Chinese licorice; Manchurian liquorice	China; As	root

Gnetum gnemon	Gnetaceae	melinjo; belinjo	Trop As	**leaf**
Guajacum officinale	Zygophyllaceae	lignum vitae	C Am; S Am	wood; resin
Guajacum sanctum	Zygophyllaceae	holywood; lignum vitae	S Am	wood; resin
Hedeoma pulegioides	Lamiaceae	American (false) pennyroyal; mosquito plant; pennyroyal; squaw mint; tickweed	N Am	leaf
Helichrysum italicum	Asteraceae	curry plant; Italian everlasting; immortelle	Med; W As	herb
Hemerocallis aurantiaca	Hemerocallidaceae	orange day lily	E As	flower
Hemerocallis citrina	Hemerocallidaceae	lemon day lily	E As	flower
Hemerocallis fulva	Hemerocallidaceae	fulvous day lily; golden needles; orange needles; tawny day lily	As	flower
Hemerocallis lilio-asphodelus	Hemerocallidaceae	lemon day lily; yellow day lily	E Siberia; Jap	flower
Hemerocallis minor	Hemerocallidaceae	dwarf yellow day lily; grass leaf day lily; little day lily	E As	flower
Hemidesmus indicus	Asclepiadeceae	Indian sarsaparilla; nunnery root	S Ind; SE As; Malay	root
Heracleum persicum	Apiaceae	golpar; Persian cow-parsley	Iran	seed
Hesperethusa crenulata	Rutaceae	beli; tondsha; tor elaga; nayvila; nayibullal	Indomal	fruit
Heterotheca inuloides	Asteraceae	false golden aster; Mexican arnica	S & N Am	flower
Hibiscus sabdariffa	Malvaceae	roselle; hibiscus; karkade; Jamaica sorrel	Trop Afr	calyx
Hierochloe odorata	Poaceae	holy grass; manna grass; seneca grass; sweet grass; vanilla grass	Eur; As; N Am	leaf
Houttuynia cordata	Saururaceae	fishwort; fish mint; saururis; heart leaf	SE As	leaf; **herb**
Humulus lupulus	Cannabaceae	hop; hops	Eur; As; N Am	**female flowers**
Hypericum perforatum	Hypericaceae	St John's wort	Eur; As	herb
Hyptis spicigera	Lamiaceae	black sesame; black beni seed; bush mint	Am; Afr	seed
Hyptis suaveolens	Lamiaceae	Indian horehound; wild spikenard	S Am	leaf
Hyssopus officinalis	Lamiaceae	hyssop	S Eur	**herb**
Illicium verum	Illiciaceae	star anise; Chinese star anise; Chinese anise	China	**fruit**
Inula helenium	Asteraceae	elecampane; scabwort; velvet dock	As; Eur	rhizome
Iris germanica	Iridaceae	flag iris; common iris; German iris; orris	Eur	rhizome
Jasminum officinale	Oleaceae	white jasmine; jessamine	Himal; SW China	flower
Jasminum sambac	Oleaceae	Arabian jasmine; biblical jasmine; samba	As	**flower**
Juglans nigra	Juglandaceae	black walnut	N Am	seed
Juglans regia	Juglandaceae	walnut; English walnut; Persian walnut	C As	**seed**
Juniperus communis	Cupressaceae	juniper	As; Eur; N Am	**fruit**
Juniperus virginiana	Cupressaceae	Virginian cedar; pencil cedar; eastern red cedar	N Am	fruit
Kaempferia galanga	Zingiberaceae	small galangal; kencur; sha jiang	Trop As	**rhizome**; leaf
Krameria lappacea	Krameriaceae	rhatany; Peruvian rhatany	S Am	root
Lactuca sativa	Asteraceae	lettuce	Med	**leaf**
Lantana camara	Verbenaceae	lantana; red sage; sherry pie; shrub verbena	C Am; S Am	leaf
Laser trilobum	Apiaceae	laserwort	Eur; W As	fruit
Laserpitium siler	Apiaceae	laserwort	N Afr; C & S Eu	fruit; root
Laurus nobilis	Lauraceae	laurel; bay leaf; bay laurel; sweet bay	Med	**leaf; fruit**

Lavandula **angustifolia**	Lamiaceae	English lavender; lavender; true lavender; common lavender	Med	**herb**
Lavandula dentata	Lamiaceae	French lavender	Med; W As	**herb**
Lavandula ×intermedia	Lamiaceae	lavandin; Dutch lavender	cultigen	**herb**
Lavandula latifolia	Lamiaceae	broadleaf lavender; spike lavender; spikenard; broad-leaf lavender	Med	herb
Lavandula stoechas	Lamiaceae	Spanish lavender; Arabian lavender; French lavender; Italian lavender	Med	herb
Lepidium draba	Brassicaceae	hoary cress; whitetop	Med; Euras	leaf; seed
Lepidium latifolium	Brassicaceae	dittander; perennial peppergrass	Eur; Med	fresh herb
Lepidium sativum	Brassicaceae	cress; garden cress; pepper grass	N Afr; W & C As	fresh leaf
Leptospermum citratum	Myrtaceae	lemon-scented tea	Aus	leaf
Leucas zeylanica	Lamiaceae	Ceylon leucas; admiration herb	Trop As	leaf
Levisticum officinale	Apiaceae	lovage; garden lovage	Eur; N Am	**herb**; root; **fruit**
Limnophila aromatica	Plantaginaceae	rice-paddy weed	SE As	herb
Limnophila rugosa	Plantaginaceae	hades; selaseh; ayer; selaseh banyu	Trop OW; Pac	leaf; herb
Lippia adoensis	Verbenaceae	koseret	E Afr (Ethiopia)	**herb**
Lippia alba	Verbenaceae	white oregano	N Am	leaf; herb
Lippia citriodora see *Aloysia citrodora*				
Lippia dulcis	Verbenaceae	Mexican lippia; yerba dulce	C Am	leaf; herb
Lippia graveolens	Verbenaceae	Mexican oregano; American oregano; Mexican sage; mintweed	S Am; N Am; C Am	leaf; herb
Lippia javanica	Verbenaceae	fever tea	Afr	leaf; herb
Lippia micromera	Verbenaceae	Spanish thyme; Puerto Rico oregano	N Am; S Am	leaf; herb
Lippia multiflora	Verbenaceae	Gambian tea bush	Afr	leaf
Litsea cubeba	Lauraceae	mountain pepper; pheasant-pepper	As	fruit
Litsea japonica	Lauraceae	Japanese mountain pepper	As	fruit
Litsea pipericarpa	Lauraceae	kulit antarsa; kulit pulaga	As	fruit
Lycopersicon esculentum	Solanaceae	tomato	C Am; S Am	**fruit**
Mammea americana	Guttiferae	mammee; mammee-apple	S Am	flower
Mangifera indica	Anacardiaceae	mango (amchur; amchoor)	Ind; Trop As	**fruit**
Mangifera odorata	Anacardiaceae	kurwini mango; saipan mango	SE As	fruit
Marrubium vulgare	Lamiaceae	horehound; white horehound	Eur	leaf
Matricaria recutita	Asteraceae	German chamomile	Eur	herb
Mediasia macrophylla	Apiaceae	pamir; large-leaved mediasia	As	fruit
Melissa officinalis	Lamiaceae	lemon balm; sweet balm; melissa	Med; As	**leaf**; herb
Mentha alopecuroides	Lamiaceae	apple mint; bowl mint	Eur	leaf; herb
Mentha aquatica	Lamiaceae	water mint	Eur; Afr; W As	**leaf; herb**
Mentha arvensis	Lamiaceae	corn mint; field mint; Japanese mint	Eur; As	**leaf; herb**
Mentha cardiaca see *Mentha ×gracilis*				
Mentha ×cordifolia	Lamiaceae	Kentucky spearmint	N Am	herb
Mentha ×gracilis	Lamiaceae	Scotch mint; Scotch spearmint; ginger mint	Eur	**leaf; herb**
Mentha haplocalyx	Lamiaceae	Japanese peppermint	As	herb
Mentha longifolia	Lamiaceae	horse mint; long-leaf mint	Eur; As; Afr	herb

Mentha ×piperita	Lamiaceae	peppermint	cultigen (Britain)	**leaf**; herb
Mentha ×piperita 'citrata'	Lamiaceae	bergamot mint; eau de Cologne mint; lemon mint; orange mint	Britain	**fresh leaf**
Mentha pulegium	Lamiaceae	pennyroyal; pudding grass	Eur; W As	**leaf; herb**
Mentha requienii	Lamiaceae	Corsican mint; menthella	Eur	**leaf; herb**
Mentha ×rotundifolia	Lamiaceae	round-leaved mint; false apple mint	Eur	**leaf; herb**
Mentha ×smithiana	Lamiaceae	red mint; bergamot mint	Eur	leaf; herb
Mentha spicata	Lamiaceae	spearmint; garden mint; crisp mint; green mint; lamb mint	Eur	**leaf; herb**
Mentha suaveolens	Lamiaceae	apple mint; pineapple mint	Eur	**leaf; herb**
Mentha ×villosa	Lamiaceae	bowl mint; apple mint; woolly mint	Eur	leaf
Menyanthes trifoliata	Menyanthaceae	bogbean; bogmyrtle; marsh-clover; marsh-trefoil; water trefoil	Eur; N Am; As	leaf
Merremia dissecta	Convolvulaceae	alamo vine; dissected merremia	Trop Am	leaf
Meum athamanticum	Apiaceae	baldmoney; bearwort; meu; spignel	N Afr; Eur	root; **leaf**
Michelia champaca	Magnoliaceae	golden champa; yellow champa	C As	flower
Micromeria fruticosa	Lamiaceae	talmud widder hyssop; tea hyssop	E Med; Israel	herb
Minthostachys mollis	Lamiaceae	Ecuadorian mint; tipo leaf	N Andes	leaf; herb
Mirabilis jalapa	Nyctaginaceae	beauty of the night; false jalap; four o'clock plant; marvel of Peru	Mex	seed
Momordica charantia	Cucurbitaceae	balsam pear; bitter cucumber; bitter gourd; bitter melon; carilla plant	Trop OW	pulp
Monarda citriodora	Lamiaceae	lemon bergamot; lemon mint	Mex; USA	leaf
Monarda didyma	Lamiaceae	fragrant balm; Oswego bee balm; Oswego tea	E & N Am	leaf; **herb**
Monarda fistulosa	Lamiaceae	bee balm; horse bergamot; wild bergamot	N Am	leaf
Mondia whitei	Apocynaceae	white ginger	Trop Afr	root
Monodora brevipes	Annonaceae	short-stemmed African nutmeg; yellow flowery nutmeg	Trop Afr	seed
Monodora myristica	Annonaceae	African nutmeg; false nutmeg; calabash nutmeg; Jamaica nutmeg	Trop Afr	seed
Montia perfoliata	Portulacaceae	miner's lettuce; Cuban spinach; winter purslane	N Am	**leaf**
Moringa oleifera	Moringaceae	horseradish tree; drumstick tree	Ind	flower; leaf; fruit
Murraya koenigii	Rutaceae	curry leaf	Ind; Sri Lanka	**leaf**
Murraya paniculata	Rutaceae	jasmine orange; Chinese myrtle; Chinese box-wood; Burmese box-wood; satinwood; Hawaiian orange; orange jasmine	SE As; Aus	flower
Myrica cerifera	Myricaceae	bay berry; candle berry; wax berry; wax myrtle	N Am	leaf
Myrica gale	Myricaceae	bog myrtle; gale meadow fern; sweet gale; sweet myrtle; wax berry	N Am; Euras	herb
Myrica pennsylvanica	Myricaceae	bay berry; candle berry; northern bay-berry; waxy berry	Eastern N Am	herb
Myristica argentea	Myristicaceae	silver nutmeg	New Guinea	seed; aril
Myristica fragrans	Myristicaceae	nutmeg and mace	Moluccas	**seed; aril**
Myristica malabarica	Myristicaceae	Bombay nutmeg; Malabar nutmeg	Indonesia	seed; aril
Myristica speciosa	Myristicaceae	Moluccan nutmeg	Indonesia	seed
Myroxylon balsamum	Leguminosae	balsam of Tolu; balsam of Peru; black balsam; Peruvian balsam	Venez; Peru	balsam

Myrrhis odorata	Apiaceae	cicely; sweet cicely; garden myrrh	Eur	**fresh herb; fruit**
Myrtus communis	Myrtaceae	myrtle; common myrtle	Med	**leaf; fruit**
Nardostachys jatamansi (*N. grandiflora*)	Valerianaceae	jatamansi; Indian nard; spikenard	As	leaf; rhizome
Nasturtium microphyllum	Brassicaceae	wild watercress	Eur	fresh leaf
Nasturtium officinale (=*Rorippa nasturtium-aquaticum*)	Brassicaceae	watercress	Eur	**fresh leaf**
Nelumbo nucifera	Nelumbonaceae	lotus	As; Aus	**herb; leaf**
Nepeta cataria	Lamiaceae	catmint; catnip; lemon catnip	Eur; SW & C As	**herb**
Nepeta ×faassenii	Lamiaceae	blue catmint	Caucasus; Iran	**herb**
Nepeta racemosa	Lamiaceae	mussin catnip	Caucasus; Turkey; N Iran	herb
Nigella arvensis	Ranunculaceae	wild fennel	Euras; Eur; Med	seed; plant
Nigella damascena	Ranunculaceae	love-in-a-mist; wild fennel	Med	seed
Nigella sativa	Ranunculaceae	nigella; kalonji; black seed	Med; As	**seed**
Ocimum americanum (=*O. canum*)	Lamiaceae	hoary basil; lime basil; partminger	Trop OW	**leaf**
Ocimum basilicum	Lamiaceae	sweet basil; basil; common basil lemon basil	Trop As	**leaf**
Ocimum canum see *Ocimum americanum*				
Ocimum ×citriodorum	Lamiaceae	lemon basil; kemanji	cultivated	leaf
Ocimum forskolei	Lamiaceae	mint-leaf basil	cultivated	leaf
Ocimum gratissimum (=*O. suave*)	Lamiaceae	African basil; sweet scented basil clove basil; Russian basil	S As; Trop Afr	leaf; herb
Ocimum kilimandscharicum	Lamiaceae	camphor basil	E Afr	leaf
Ocimum sanctum see *Ocimum tenuiflorum*				
Ocimum tenuiflorum (=*O. sanctum*)	Lamiaceae	holy basil; sacred basil; Thai basil	Trop As	**leaf; herb**
Ocotea cymbarum	Lauraceae	canela	Braz	bark; calyx
Ocotea puchury-major	Lauraceae	louro-puxuri; pichuri; pixurim; puchuiri	S Am	seed
Ocotea quixos	Lauraceae	American cinnamon; ocotea	Ecuador	flower
Ocotea pretiosa	Lauraceae	false sassafras	Braz	bark; leaf
Oenanthe javanica	Apiaceae	water dropwort; water celery; water parsley	Indomal	herb
Origanum dictamnus	Lamiaceae	dittany of Crete	Greece; Crete	**herb**
Origanum majorana	Lamiaceae	marjoram; sweet marjoram	Med; Turkey	**herb**
Origanum onites	Lamiaceae	pot marjoram	Eur; E Med	herb
Origanum syriacum	Lamiaceae	Syrian oregano; za'atar; zatar; white oregano	SE Eur; W As	**herb**
Origanum vulgare	Lamiaceae	oregano; wild marjoram; pizza herb	Eur; C As	**herb**
Osmanthus fragrans	Oleaceae	sweet osmanthus flower; fragrant olive; sweet olive; tea olive	E As (China)	**flower**
Ottelia alismoides	Hydrocharitaceae	water-plantain; ottelia	China; Jap; Aus; NE Afr	leaf
Oxalis acetosella	Oxalidaceae	common wood sorrel; wood sorrel; shamrock	Eur; As	leaf

Oxalis corniculata	Oxalidaceae	creeping wood sorrel; creeping oxalis	Med	leaf
Oxalis pes-caprae	Oxalidaceae	sorrel; yellow sorrel	S Afr	leaf
Pandanus amaryllifolius	Pandanaceae	pandan; fragrant pandan	Trop As (Moluccas)	**wilted leaves**
Pandanus tectorius	Pandanaceae	fragrant screwpine	Trop OW	leaf; male flowers
Papaver somniferum	Papaveraceae	opium poppy; poppy	Med; SW As	**seed**
Parinari curatellifolia	Chrysobalanceae	mobola; mbura	Trop Afr	fruit
Parkia speciosa	Mimosaceae	petai; locust bean	Trop As	seed
Pastinaca sativa	Apiaceae	parsnip	Eur; W As	leaf; root
Peganum harmala	Zygophyllaceae	African rue; Syrian rue; mountain rue; wild rue	Med; As	seed
Pelargonium capitatum	Geraniaceae	rose-scented geranium; rose geranium; rose pelargonium	S Afr	leaf
Pelargonium ×citrosum	Geraniaceae	citrosa geranium; mosquito plant	cultivated	leaf
Pelargonium crispum	Geraniaceae	lemon geranium; curled leaved cranesbill	S Afr	leaf
Pelargonium ×fragrans	Geraniaceae	nutmeg scented geranium	S Afr	leaf
Pelargonium ×graveolens	Geraniaceae	rose geranium; Bourbon geranium; rose pelargonium; sweet-scented geranium	cultivated	leaf
Pelargonium ×limoneum	Geraniaceae	English finger-bowl geranium	cultivated	leaf
Pelargonium odoratissimum	Geraniaceae	apple-scented geranium	S Afr	leaf
Pelargonium radens	Geraniaceae	mint geranium; balsam (scented) geranium	S Afr	leaf
Pelargonium tomentosum	Geraniaceae	peppermint-scented geranium	S Afr	leaf
Peperomia pellucida	Piperaceae	pepper elder; rabbit ear	Trop Am	herb
Perilla frutescens	Lamiaceae	shiso; beefsteak plant; perilla; purple mint	E As	**leaf**; herb
Persicaria hydropiper	Polygonaceae	water pepper	SE As	leaf; herb
Persicaria odorata	Polygonaceae	Vietnamese coriander; Vietnamese mint; laksa leaf	SE As	**leaf**
Petroselinum crispum	Apiaceae	parsley	Eur; W As	**leaf**
Peucedanum ostruthium	Apiaceae	hog fennel	S Eur	rhizome
Peumus boldus	Monimiaceae	boldo	Chile	leaf
Phyllanthus emblica	Euphorbiaceae	emblic; emblic myrobalan; Indian gooseberry	Trop As	fruit
Physalis philadelphica (=P. ixocarpa)	Solanaceae	tomatillo; jamberry	Mex	ripe fruit
Physalis pubescens	Solanaceae	downy ground cherry; ground cherry	N Am	ripe fruit
Picrasma excelsa	Simaroubaceae	bitterwood; Jamaica quassia; quassia wood	West Indies	wood
Picrasma quassioides	Simaroubaceae	bitterwood; quassia wood	E As	wood
Pimenta dioica	Myrtaceae	allspice; pimento; Jamaican pepper	C Am; West Indies	**fruit**
Pimenta racemosa	Myrtaceae	bay rum	Trop Am	fruit
Pimpinella anisum	Apiaceae	anise; aniseed	Med; W As	fruit
Pimpinella major	Apiaceae	greater burnet saxifrage	Eur	root
Pimpinella saxifraga	Apiaceae	burnet saxifrage; garden burnet	Eur	root
Pinus cembra	Pinaceae	Siberian pine nuts	Eur; As	**seed**
Pinus cembroides	Pinaceae	Mexican pine nuts; Mexican piñon	N Am	**seed**

Pinus edulis	Pinaceae	piñon	N Am	seed
Pinus koraiensis	Pinaceae	Korean pine nuts	As	**seed**
Pinus pinea	Pinaceae	pine nuts; pine seeds; pignoles	Med	**seed**
Piper aduncum	Piperaceae	matico; big pepper; Spanish elder; spiked pepper	C Am; S Am	fruit
Piper auritum	Piperaceae	Mexican pepperleaf; alajan pepper	C Am	**leaf**
Piper baccatum	Piperaceae	climbing pepper of Java	SE As	fruit; leaf
Piper borbonense	Piperaceae	Madagascar pepper; poivre sauvage; cubebe du pays; betel marron	Afr (Madag)	fruit
Piper capense	Piperaceae	Cape pepper	E Afr	fruit
Piper clusii	Piperaceae	West African black pepper; African cubebs	W Afr	fruit
Piper cubeba	Piperaceae	cubeb pepper; tailed pepper; Java pepper	Indonesia	**unripe fruit**
Piper guineense	Piperaceae	West African pepper; Ashanti pepper; Benin pepper	Trop Afr	fruit
Piper lolot	Piperaceae	lolot pepper; Vietnamese pepper	SE As	leaf
Piper longum	Piperaceae	Indian long pepper; pippali; pipalli	Trop E Himal	**fruit**
Piper nigrum	Piperaceae	pepper; black pepper	S Ind; Sri Lanka	**fruit; unripe fruit**
Piper retrofractum	Piperaceae	Balinese long pepper; Javanese long pepper	SE As	fruit
Piper saigonense	Piperaceae	Saigon pepper	SE As	fruit
Piper sanctum	Piperaceae	acoyo; acuyo; cordonillo; xihuitl	C Am	herb
Piper sarmentosum	Piperaceae	wild pepper	Indomal	fruit; leaf
Piper umbellatum	Piperaceae	shrubby pepper	Trop Am	fruit; stem
Pistacia lentiscus	Anacardiaceae	mastic; masticha; mistki Chios mastic; lentisc	Med	**resin**
Pistacia terebinthus	Anacardiaceae	terebinth	Med	resin
Pistacia vera	Anacardiaceae	pistachio	W As	seed
Pithecellobium dulce	Leguminosae	Manila tamarind	C Am	fruit; seed
Plectranthus amboinicus	Lamiaceae	Indian borage; Cuban oregano; Spanish sage	E Afr	**fresh leaf**
Pluchea indica	Asteraceae	Indian fleabane; Indian pluchea	Indomal	leaf
Pogostemon cablin	Lamiaceae	patchouli; patchouly	Indomal	leaf
Pogostemon indicus (=*P. heyneanus*)	Lamiaceae	Indian patchouli	Indomal	leaf
Poncirus trifoliata see *Citrus trifoliata*				
Populus balsamifera	Salicaceae	balsam poplar; hackmatack; tacamahaca poplar	N Am; Temp As	leaf bud
Populus nigra	Salicaceae	black poplar; Lombardy poplar	Euras	leaf bud
Porophyllum ruderale	Asteraceae	quillquina; killi	C Am; S Am	fresh leaves
Portulaca oleracea	Portulacaceae	purslane	W As	**leaf**
Portulaca quadrifida	Portulacaceae	single-flowered purslane; wild purslane	S Eur	herb
Potentilla erecta	Rosaceae	tormentill; bloodroot	Eur; As	rhizome
Prostanthera rotundifolia	Lamiaceae	Australian mint bush; round-leaved mint bush	Aus	herb
Prunus dulcis (=*P. amygdalis*)	Rosaceae	almond	W As	**seed**
Prunus laurocerasus	Rosaceae	cherry laurel; laurel	Eur; W As	fresh leaf
Prunus mahaleb	Rosaceae	mahaleb; mahaleb cherry; St Lucie cherry	Eur; As	fruit; **leaf**; seed
Prunus serotina	Rosaceae	American bird cherry; black cherry; rum cherry	E & N Am	leaf

Punica granatum	Punicaceae	pomegranate; anardana	Eur; W As	**seeds**
Pycnanthemum pilosum	Lamiaceae	hairy mountain mint	N Am	leaf; herb
Pycnanthemum virginianum	Lamiaceae	Virginia mountain mint; Virginia thyme	N Am	leaf
Pycnanthus angolensis	Myristicaceae	African nutmeg; Angolan nutmeg; false nutmeg	Trop Afr	seed
Quararibea fieldii	Malvaceae	saha	Trop Am	flower
Quararibea funebris	Malvaceae	chocolate flower	Mex	flower
Quararibea turbinata	Malvaceae	swizzlestick tree	Trop Am	twigs
Quassia amara	Simaroubaceae	bitterwood; Jamaica wood; quassia wood; Surinam quassia	Braz	wood
Quassia excelsa	Simaroubaceae	Jamaica wood	Trop Am	wood
Raphanus sativus	Brassicaceae	radish	cultivated (W As?)	**root**
Ravensara aromatica (*Cryptocarya* sp.)	Lauraceae	clove nutmeg; Madagascar clove; Madagascar nutmeg	Madag	bark; leaf; seed
Renanthera moluccana	Orchidaceae	anggrek merah; bunga karang	Indomal	young leaf
Renealmia alpina	Zingiberaceae	mountain renealmia	Mex	leaf; aril
Reseda odorata	Resedaceae	mignonette; sweet mignonette; sweet reseda	N Afr	flower
Rhamnus prinoides	Rhamnaceae	geisho; gesho	Afr	**herb**
Rhaphidophora lobbii	Araceae	akar asam tebing paya	Pac	leaf
Rheum rhabarbarum (*R. ×hybridum*)	Polygonaceae	rhubarb; garden rhubarb	Eur?	**petiole**
Rheum rhaponticum (*R. ×cultorum*)	Polygonaceae	rhapontic rhubarb	Eur?	**petiole**
Rhododendron tomentosum (=*Ledum palustre*)	Ericaceae	marsh tea; marsh rosemary	Eur; As; N Am	leaf
Rhus aromatica	Anacardiaceae	fragrant sumach; lemon sumach; skunkbush; polecat bush; Sicilian sumac	Eastern N Am	fruit
Rhus coriaria	Anacardiaceae	sumac; sumach; spice sumac	S Eur	**fruit**
Ribes divaricatum		Worcesterberry	N Am	fruit
Ribes ×nidigrolaria		jostaberry	Eur	**fruit**
Ribes nigrum		blackcurrant; black currant	Eur; As	**fruit**
Ribes rubrum		redcurrant	Eur; As	**fruit**
Ribes uva-crispa		gooseberry	Eur	**fruit**
Ricinodendron heudelotii	Euphorbiaceae	African nut	Trop Afr	seed
Rorippa nasturtium-aquaticum see *Nasturtium officinale*				
Rosa ×centifolia	Rosaceae	cabbage rose; Holland rose; Provence rose	Eur	**petals**
Rosa chinensis	Rosaceae	China rose; Bengal rose	China	**petals**
Rosa ×damascena	Rosaceae	damask rose; Portland rose; pink damask rose; Bulgarian rose	SW Eur; W As	**petals**
Rosa gallica	Rosaceae	French rose	Eur; W As	**petals**
Rosa moschata	Rosaceae	musk rose	W Himal	petals
Rosa ×odorata	Rosaceae	tea-scented rose	cultivated	**petals**
Rosa rugosa	Rosaceae	Japanese rose	E As	**petals**
Rosmarinus officinalis	Lamiaceae	rosemary	S & N Eur	**leaf; herb**
Rumex acetosa	Polygonaceae	sorrel; common sorrel; garden sorrel	Eur; As	**fresh leaf**
Rumex acetosella	Polygonaceae	sheep sorrel	N Temp	fresh leaf

Rumex hastatus	Polygonaceae	speared sorrel	E As	leaf
Rumex sanguineus	Polygonaceae	red-veined sorrel	cultivated	**leaf**
Rumex scutatus	Polygonaceae	French sorrel; buckler-leaved sorrel	Eur; As	**leaf**
Rumex vesicarius	Polygonaceae	bladder dock	S Eur; W As; N Afr	leaf
Ruta chalepensis	Rutaceae	fringed rue; Egyptian rue; Syrian rue; Aleppo rue	Med	**herb**
Ruta graveolens	Rutaceae	rue; common rue; herb of grace	Eur	**leaf**; herb
Salvia elegans (=*S. rutilans*)	Lamiaceae	pineapple sage; pineapple-scented sage	C Am	**leaf; herb**
Salvia fruticosa	Lamiaceae	Greek sage; Turkish sage	E Med	leaf
Salvia hispanica	Lamiaceae	chia	C Mex	seed; herb
Salvia lavandulifolia	Lamiaceae	Spanish sage	S Eur	leaf
Salvia officinalis	Lamiaceae	sage; common sage; garden sage	S Eur; Med	**leaf**
Salvia sclarea	Lamiaceae	clary; clary sage	Eur; C As	**leaf**
Sambucus canadensis (=*S. nigra* subsp. *canadensis*)	Caprifoliaceae	American elder	N Am	flower; fruit
Sambucus nigra	Caprifoliaceae	elderflowers; elderberries; black elder; common elder; European elder	Eur; As; Med	**flower; fruit**
Sanguisorba minor	Rosaceae	salad burnet; burnet; small burnet	C & S Eur	**fresh leaf**
Sanguisorba officinalis	Rosaceae	burnet; garden burnet; great burnet	Eur; As	leaf
Santalum album	Santalaceae	white sandalwood; East Indian sandalwood	Ind	wood
Santolina rosmarinifolia	Asteraceae	green santolina	S Eur	fresh herb
Saposhnikova divaricata	Apiaceae	fang-feng	NE As	fresh leaves
Sassafras albidum	Lauraceae	sassafras; filé powder	N Am	**leaf**
Satureja hortensis	Lamiaceae	savory; summer savory; garden savory	Med	**leaf**; herb
Satureja montana	Lamiaceae	winter savory	S Eur	**leaf**; herb
Satureja thymbra	Lamiaceae	thyme-leaved savory; thymbra	SE Eur	leaf
Saussurea costus	Asteraceae	costus	E Himal	root
Schinus molle	Anacardiaceae	pink pepper; Peruvian pepper	Peru	**fruit**
Schinus terebinthifolius	Anacardiaceae	Brazilian pepper; pink pepper; red pepper	Braz	**fruit**
Schisandra chinensis	Schisandraceae	Chinese magnolia vine; five-flavour fruit	China; Jap	fruit; bark
Scyphocephalium mannii	Myristicaceae	West African nutmeg	W Afr	fruit; seed
Scyphocephalium ochocoa	Myristicaceae	ochoco nutmeg	W Afr	seed
Sesamum indicum	Pedaliaceae	sesame; sesam	Afr; Ind	**seed**
Sinapis alba	Brassicaceae	white mustard; yellow mustard; mustard	Med	**seed**
Sinapis arvensis	Brassicaceae	charlock; field mustard; California rape; wild mustard	Eur; Med	seed
Siphonochilus aethiopicus	Zingiberaceae	African ginger	Trop Afr	rhizome; root
Sison amomum	Apiaceae	hedge sison; honewort; stone parsley	Eur; Med	fruit (seed)
Sium bracteatum	Apiaceae	jellico	St Helena	petiole
Sium sisarum	Apiaceae	skirret; crummock	E As	**leaf**
Smilax aristolochiifolia	Smilacaceae	Mexican sarsaparilla; Veracruz sarsaparilla	S & N Am	rhizome; root; bark
Smyrnium olusatrum	Apiaceae	alexanders; black lovage	E Med	**leaf**; fruit
Solanum aethiopicum	Solanaceae	scarlet egg plant; bitter berry	Trop Afr	leaf; fruit

Solanum macrocarpon	Solanaceae	African eggplant; gboma egg plant	W Afr	fresh herb
Sonneratia caseolaris	Lythraceae	red-flowered Pornupan mangrove	Indomal	young fruit
Sorbus aucuparia	Rosaceae	rowan	Eur; SW Asia	fruit
Spilanthes acmella	Asteraceae	spilanthes; pará cress	Braz	**leaf; flower head**
Spondias dulcis	Anacardiaceae	ambarella; Otaheite apple	Pac	fruit
Spondias mombin	Anacardiaceae	caja fruit; yellow mombin; jobo	Trop Am	fruit
Stevia rebaudiana	Asteraceae	stevia; sugar-leaf	S Am	**leaf**
Swertia chirata	Gentianaceae	chirata; chireta	C As	herb
Syzygium aromaticum	Myrtaceae	clove; cloves	Moluccas	**flower buds**
Syzygium polyanthum	Myrtaceae	salam; Indonesian bay-leaf	Indomal	**leaf**
Tagetes erecta (=*T. patula*)	Asteraceae	African marigold; French marigold	Mex; C Am	flower
Tagetes filifolia	Asteraceae	Irish lace marigold	Trop Am	leaf
Tagetes lemmonii	Asteraceae	Mexican marigold	N Am	**petals**
Tagetes lucida	Asteraceae	Mexican marigold mint; Mexican tarragon	Mex; Guatemala	herb
Tagetes maxima	Asteraceae	great marigold	Trop Am	herb; flower
Tagetes minuta	Asteraceae	huacataya; tagette; stinking roger; wild marigold; Mexican marigold	S Am	**flower**; herb
Tagetes tenuifolia	Asteraceae	American saffron; signet marygold; slender leaf marigold; striped Mexican marigold; lemon gem; orange gem	Trop & warm Am	flower; leaf
Talinum triangulare	Portulacaceae	talinum; flameflower; waterleaf	N Am	leaf
Tamarindus indica	Leguminosae	tamarind; Indian tamarind	N Afr	**fruit pulp;** leaf
Tanacetum balsamita	Asteraceae	alecost; costmary	Eur; C As	**herb**
Tanacetum vulgare	Asteraceae	tansy	Euras	**herb**
Taraxacum officinale	Asteraceae	dandelion	Euras	**fresh leaf**
Tetrapleura tetraptera	Leguminosae	aridan; akpa; dawo; essanga; sanga	Trop Afr	fruit
Theobroma cacao	Malvaceae	cacao	S Am (Andes)	**seed**
Thonningia sanguinea	Balanophoraceae	ground pineapple	W Afr	roots
Thymus caespititius	Lamiaceae	Azores thyme; mountain thyme; tiny thyme	Euras	fresh leaves
Thymus capitatus	Lamiaceae	catir; conehead thyme; Cretan thyme; Senegal savory; zatir	Euras	flowering herb
Thymus ×citriodorus	Lamiaceae	lemon thyme	Euras	**fresh leaves; fresh herb**
Thymus herba-barona	Lamiaceae	caraway thyme	Sardinia; Corsica	leaf
Thymus hyemalis	Lamiaceae	lemon thyme; winter thyme	Euras	leaf
Thymus mastichina	Lamiaceae	mastic thyme; Spanish (wild) marjoram; Spanish thyme	Euras	herb
Thymus oenipontanus	Lamiaceae	Austrian thyme	Euras	fresh leaf
Thymus praecox	Lamiaceae	alba thyme; creeping thyme; hairy thyme	Euras	leaf; herb
Thymus pubescens	Lamiaceae	hairy thyme	Euras	leaf; herb
Thymus pulegioides	Lamiaceae	lemon thyme; caraway thyme; wild thyme; herba barona	Eur	herb
Thymus quinquecostatus	Lamiaceae	Japanese thyme; five-ribbed thyme	Euras	herb
Thymus serpyllum	Lamiaceae	wild thyme	Euras	**leaf; herb**

Thymus schimperi	Lamiaceae	Ethiopian thyme	N Afr	**herb**
Thymus vulgaris	Lamiaceae	common thyme; garden thyme	W Med; SE Italy	**leaf; herb**
Thymus zygis	Lamiaceae	Spanish thyme; sauce thyme	W Med	herb
Toddalia asiatica	Rutaceae	Lopez fruit; wild orange tree	Trop OW	fruit
Trachyspermum ammi	Apiaceae	ajowan; bishop's weed; omum; omam; white cumin; Ethiopian caraway	Egypt; Ethiopia	**fruit**
Trachyspermum roxburghianum	Apiaceae	ajmud; radhuni	Indomal	fruit
Treculia africana	Moraceae	African bread fruit; African boxwood	Afr	seed
Trigonella caerulea	Leguminosae	blue trefoil	Eur; W As	seed; herb
Trigonella corniculata	Leguminosae	clustered trefoil	Eur; Med; W As	seed
Trigonella foenum-graecum	Leguminosae	fenugreek	S Eur; W As	**seed; herb**
Triphasia trifolia	Rutaceae	lime berry; trifoliate lime; Chinese lime; myrtle lime	SE As	fruit
Tropaeolum majus	Tropaeolaceae	Indian cress; garden nasturtium	Peru	**bud; leaf; flower**
Turneria diffusa	Turneraceae	damiana	Trop Am	fruit
Urophyllum arboreum	Rubiaceae	ki cengkeh	Trop OW; Jap	bruised leaves
Vaccaria hispanica	Caryophyllaceae	cow cockle; cow herb; dairy pink	Euras; Med	herb
Valerianella locusta	Caprifoliaceae	corn salad; lamb's lettuce; mache	Eur; Med	herb
Vanilla planifolia	Orchidaceae	vanilla; Bourbon vanilla; Mexican vanilla	Trop Am	**fruit**
Verbascum densiflorum	Scrophulariaceae	large flowered mullein; common mullein	C Eur	flower
Verbascum phlomoides	Scrophulariaceae	clasping mullein; orange mullein	S Eur	flower
Verbena officinalis	Verbenaceae	common verbena; European vervain; lemon-scented verbena; vervain	Euras; Afr; Am	leaf; herb
Vernonia amygdalina	Asteraceae	almond veronia; bitter leaf	Trop Afr	leaf
Vetiveria zizanioides see *Chrysopogon zizanioides*				
Viola odorata	Violaceae	common violet; florist's violet; garden violet; sweet-scented violet	Euras; Afr	flower; herb
Vitex agnus-castus	Verbenaceae	chaste pepper	S Eur	fruit
Warburgia salutaris	Canellaceae	pepperbark tree	E Afr	leaf; bark
Wasabia japonica	Brassicaceae	Japanese horseradish; mountain hollyhock; wasabi	Jap	**stem**
Xylopia aethiopica	Annonaceae	African pepper; Ethiopian pepper; Guinea pepper; kimba pepper; negro pepper; grains of Selim; spice tree	Trop W Afr	**fruit**
Xylopia aromatica	Annonaceae	pachinhos; pimenta de macaco; pimenta de negro	E Braz	seed
Xylopia parviflora	Annonaceae	striped African pepper	Liberia	fruit
Xylopia sericea	Annonaceae	hairy peppper	Braz	seed
Zanthoxylum acanthopodium	Rutaceae	adaliman; tomar seed	Indonesia	fruit
Zanthoxylum ailanthoides	Rutaceae	yue jiao	Vietnam	fruit
Zanthoxylum alatum	Rutaceae	timur	Nepal	fruit
Zanthoxylum armatum	Rutaceae	zhu ye jiao; Chinese pepper	China	fruit
Zanthoxylum bungeanum	Rutaceae	Chinese pepper	N China	leaf; fruit

Zanthoxylum piperitum	Rutaceae	sansho; Japanese pepper	E As	**fruit**; leaf
Zanthoxylum rhetsa	Rutaceae	tippal; Indian pepper	N Ind	fruit; seed
Zanthoxylum schinifolium	Rutaceae	sancho	C & E As	fruit; leaf
Zanthoxylum simulans	Rutaceae	Chinese pepper; Sichuan pepper; chuan jiao	China	fruit; leaf
Zanthoxylum tessmannii	Rutaceae	African pepper	Afr	fruit
Zingiber mioga	Zingiberaceae	mioga ginger; Japanese (wild) ginger	Jap	rhizome
Zingiber officinale	Zingiberaceae	ginger; common ginger	Ind	**rhizome**
Zingiber zerumbet	Zingiberaceae	wild ginger; zerumbet ginger	Indomal	rhizome
Ziziphora tenuior	Lamiaceae	ziziphora	C As (Turkey)	leaf

Acknowledgements

The author wishes to thank Briza Publications and the production team, especially Reneé Ferreira for her skill and excellence in managing the editing and publication process. Eben van Wyk is thanked for his support ever since the concept of "Culinary herbs and spices" was first proposed. Thank you also to Christo Reitz and Johan Steenkamp for their help in developing the project.

A special word of thanks goes to the scientific editors at Kew Publishing, for doing a fine job at editing the first draft and for the valuable comments and corrections.

Thank you very much to my wife, Mariana van Wyk, for her help with logistics, especially to obtain samples for photography. Teodor van Wyk deserves a special word of thanks for help with the checklist table, and Margaret Hulley for last-minute help with chemical structures of compounds. The support of Thinus Fourie, Patricia Tilney, Helen Long, Emmy Reinten, Anthony Magee, Sandra Kritzinger, Alvaro Viljoen and Stephan Schneeberger is much appreciated. I also wish to thank the following persons for help in obtaining samples to photograph: Louis van Aswegen (Doonholm Herb Nursery), René Steyn (Moroccan Restaurant), Alvaro Viljoen (Tshwane University of Technology), Ivy Singh-Lim (Bollywood Veggies, Singapore), Heiltje le Roux and Christine Scholtz.

Logistic and financial support from the National Research Foundation of South Africa and the University of Johannesburg are gratefully acknowledged.

Photographic contributions

All photographs were taken by Ben-Erik van Wyk except for those listed below.

Allium chinensis bulbs (Jan-Adriaan Viljoen); *Allium ursinum* flowers (Thomas Brendler); *Boesenbergia rotunda* roots [Marshall Cavendish International (Asia) Pte Ltd, Singapore]; *Cocos nucifera* green fruits (Eben van Wyk); *Eryngium foetidum* leaves (Shutterstock); *Ferula foetida* (Michael Pimenov, Moscow State University); *Origanum syriacum* (Prof K. Hüsnü Can Baser, Anadolu University, Turkey and Mr A. Haluk Turker, Eastern Mediterranean Forestry Institute, Tarsus, Turkey); *Houttuynia cordata* (Wikimedia commons); *Humulus lupulus* female cones and dried cones (Shutterstock); *Pimenta racemosa* (Thomas Brendler); *Piper cubeba* (Shutterstock); *Rhus coriaria* (Mr A. Haluk Turker, Eastern Mediterranean Forestry Institute, Tarsus, Turkey); *Vanilla planifolia* flower (Joreth Duvenhage, Du Roi Nursery); *Wasabia japonica* (Pacific Coast Wasabi Ltd, Vancouver, Canada) *Wasabia japonica* stems (Shutterstock); coffee beans (p. 8), Chinese noodles (p. 13), spices (p. 15), satay (p. 17), *tom yam* soup ingredients (p. 17), Mexican food (p. 19) (all Shutterstock); Spice market (p. 19) (Marten Feiter, Parceval, Wellington).

Index

Plant names and page numbers in **bold** indicate main entries; page numbers in ***bold italics*** indicate illustrated herbs, spices or chemical compounds.

311

313

319